Waves and Stones

Waves and Stones

On the Ultimate Nature of Reality

GRAHAM HARMAN

ALLEN LANE
an imprint of
PENGUIN BOOKS

ALLEN LANE

UK | USA | Canada | Ireland | Australia
India | New Zealand | South Africa

Allen Lane is part of the Penguin Random House group of companies
whose addresses can be found at global.penguinrandomhouse.com.

Penguin Random House UK
One Embassy Gardens, 8 Viaduct Gardens, London sw11 7bw

penguin.co.uk

Penguin
Random House
UK

First published in Great Britain by Allen Lane 2025

001

Set in 12/14.75 pt Dante MT Std
Typeset by Six Red Marbles UK, Thetford, Norfolk
Printed and bound in Great Britain by Clays Ltd, Elcograf S.p.A.

The authorized representative in the EEA is Penguin Random House Ireland,
Morrison Chambers, 32 Nassau Street, Dublin d02 yh68

A CIP catalogue record for this book is available from the British Library

ISBN: 978–0–241–39286–7

Penguin Random House is committed to a sustainable future
for our business, our readers and our planet. This book is made from
Forest Stewardship Council® certified paper.

MIX
Paper | Supporting
responsible forestry
FSC® C018179

Contents

Contents

Contents

'For if it does not belong to the philosopher, then who will be the investigator of whether Socrates and Socrates seated are the same?'

Aristotle

Prologue: The Continuous and the Discrete

Nikos Kazantzakis (1883–1957) is a major figure of modern Greek letters who is most famous for his novels *The Last Temptation of Christ* and *Zorba the Greek*, both of them turned into Hollywood films.[1] Yet he also took the bold step of writing an epic poem, a genre largely absent from recent literature. Entitled *The Odyssey: A Modern Sequel*, it contains a passage where the hero, Odysseus, recalls three moments in his life that towered above all others:

> Sweet, very sweet had been his dread on that first night
> when in the dark he'd laid his hand on a maid's body;
> how like a hawk he'd shrieked, how all the world had sighed
> when in his arms he'd held a son for the first time!
> And then that third dread shriek when on a distant plain
> he'd held on high his foe's slain head for the first time![2]

His first sexual experience, the first time holding a child of his own, and his first time killing someone in battle: it's easy to see why these three moments might stand out as turning points in a human life. Although hoisting the head of one's enemy on a spike is now biographically rare, we can all remember especially dramatic moments in our own lives. Kazantzakis is poetically capturing a vision of the human lifespan as characterized by discrete, monumental instants during which we experience irreversible change.

A different intuition can be found in a passage from Japanese author Yukio Mishima (1925–70), an intense, nationalistic figure who died via ritual suicide at a military base outside Tokyo. His novel *Runaway Horses* offers an account of human life that takes the opposite tack: 'How oddly situated a man is apt to find himself at age thirty-eight! His youth belongs to the distant past. Yet the period of memory beginning with the end of youth and extending to the present has left him not a single vivid impression.'[3] Here Mishima imagines nearly two decades of life as a continuous span during which no particular experience changes us significantly. This uneventful continuum generates the illusion that in our late thirties, and perhaps beyond, we are still close to the now-distant era of youth.

Which author's standpoint seems more accurate? Do our lives consist of a small number of dramatic turning points, or is there nothing but a series of gradual changes from infancy to old age? The same type of question often arises in daily life. In the case of the weather, temperature is the sort of thing we treat as continuous, meaning that changes from day to day are usually experienced as differences of degree rather than radical differences in kind. Yet the freezing and boiling of water happen at specific points on the thermometer: 0 degrees Celsius for freezing and 100 degrees Celsius for boiling, assuming standard atmospheric pressure at sea level.

In American politics there is the constitutional requirement to hold a presidential election in early November every fourth year, and at no other time. This seems to punctuate public life into predictable, discrete segments. But some elections come to be seen as transformational, while others seem in retrospect to have led to nothing but continued business as usual. Within this system elections are usually called transformational if they involve a substantial realignment of previous political coalitions. The Democratic politician Franklin D. Roosevelt was first elected President in 1932, and was eventually elected a total of four times, which is no longer even legal. He led the country out of the Great Depression and through most of World War II with a series of liberal policies that came to typify an entire era. But later, in 1980, a large number of labour union members permanently abandoned their allegiance to the Democratic Party to vote for the Republican candidate Ronald Reagan: 'Reagan Democrats', as they are called. In the wake of the surprising 2024 election, there has been early talk of 'Trump Democrats' as well. In any case, although American politics schedules discrete

four-year administrations in advance, some are largely continu-
ations of the existing spirit of the age while others mark relatively
abrupt shifts in national policy and mood. But are transformational
elections real, or are they merely arbitrary points along a gradually
shifting cultural timeline?

Far from being abstract or arcane, these two ways of looking
at the world – continuous and discrete – have given rise to some
of the most bitter controversies of our time. To list just one: are
there two fully distinct genders existing by nature, or just numerous gradations between the male and female poles?[4] The same
opposition between the continuous and the discrete pulls modern
physics in two directions: Einstein's General relativity treats gravity as something varying continuously, while quantum theory
treats the other three fundamental forces of nature (see p. 242) as
working by way of discontinuous jumps. Incredibly enough, these
two honoured theories remain in contradiction, although both
have been experimentally confirmed to a high degree of accuracy.

This book will show that the issue both predates and exceeds
contemporary physics. Since the days of ancient philosophy, an
equal number of complications have arisen whether we think of
the world as made of tiny discrete units or as a continuum where
everything blends together into a single whole. At the close of his
recent work *Hysteresis*, the Italian philosopher Maurizio Ferraris
(b. 1956) shares an insight drawn from the writings of the German
polymath G. W. Leibniz (1646–1716): 'There are two mazes in
which human thought gets lost. The first is that of predestination,
the second is that of continuity and movement.'[5] Ferraris notes
that both problems boil down to the interplay of the continuous
and the discrete: freedom requires a discontinuous gap between
causation and human decision, while movement is impossible if
space consists of an infinite number of actual discrete points.

But in fact, every area of life is entangled with the paradoxical
relation between the continuous and the discrete. Most other issues
do not cross disciplinary boundaries in this way. While astronomers
and physicists puzzle over the crucial unsolved problems of dark

matter and dark energy, these topics mean very little to researchers in electrical engineering or sociology. The same holds for the question of whether humans first reached the Americas via a land bridge with Asia: an issue pivotal for anthropology and perhaps linguistics, but of little relevance to criminal justice or Shakespeare studies. But in cases where certain topics arise again and again in every place we look, we seem to be dealing with a basic feature of reality. These are the moments when philosophy, sometimes dismissed as idle speculation with little practical value, is called to the scene like Sherlock Holmes to the site of a murder. So it is with the Case of the Continuous and the Discrete: the question of whether reality is made up primarily of sudden jumps or is laid out instead along a gentle gradient, with no clear divisions between the various different things in the world.

But before entering further into this topic, I should introduce myself and the standpoint from which I speak. This will help clarify why the duel of waves and stones – the continuous and the discrete – is so decisive for illuminating the nature of reality. Otherwise we will continue to fall into the same paradoxes as ancient philosophy, which the ancients often grasped more firmly than we do today.

Skipping Stones: The Riddle of Thixis

Object-Oriented Ontology (OOO)

My colleagues and I have been developing Object-Oriented Ontol-ogy (OOO) since the 1990s.[1] This began when I noticed an important glitch in the mainstream interpretation of the influential German philosopher Martin Heidegger (1889–1976), a committed Nazi who nonetheless had some extremely suggestive ideas about reality.[2] Indeed, I am just one of countless commentators who have been fascinated by the analysis of tools found early in his major work *Being and Time*.[3] Although Heidegger was referring to such obvi-ously useful items as hammers, screwdrivers and railway platforms, it seemed to me that his analysis held good for any object at all. His basic insight is that most of the time we do not consciously per-ceive the objects in our midst; instead, we silently rely on them unless something goes wrong, such as a hammer breaking apart in our hands or a train not arriving on time. This was an important insight at a time when phenomenology, the descriptive philosophy of consciousness that Heidegger initially defended, gave excessive priority to our direct mental awareness of things. Whereas phe-nomenology was a rationalism committed to explaining all human experience clearly, Heidegger's mutant version of the doctrine emphasized those aspects of reality that tend to remain hidden: the mysteries of poetic language, the pre-rational impact of human moods, or our dependence on historical structures far predating our birth.

On the whole, I always found Heidegger's tool-analysis convin-cing. What bothered me was an additional claim, to which I was first alerted by my academic advisor Alphonso Lingis (1933–2025). Namely, Heidegger also thought that all tools ultimately blend together in a single system, one that gains its meaning from my human purposes: the house I am trying to build, the money I am trying to earn. But to treat tools as belonging to a single holistic system shaped by each person is to ignore two major complicating factors. First, tools cannot fit snugly into a system, since we know that they sometimes break or otherwise go haywire; this means that the system never fully deploys any tool as a whole. The act of building a house makes use of the hammer, but takes no account of a crack in its handle that eventually causes it to shatter, bringing the construction project to a halt.

Second, and more controversially, it is not just human beings who reduce objects to specific purposes while forgetting their plenitude of hidden qualities. Instead, objects do this to each other as well. Recent philosophy has little or nothing to say about object–object interactions when no human observer is on the scene, and simply assumes that natural science should have a monopoly on this topic. This is one reason why even the early version of OOO faced a good deal of hostility from other philosophers. To this day, in fact, OOO remains more influential in other disciplines, which are professionally less committed to the specific anthropocentric bias of modern philosophy.

The phrase 'object-oriented' was borrowed from computer science, where it refers to a type of programming language. Originally a computer program functioned as a unified whole on which each of its parts was dependent. By contrast, object-oriented programming languages are based on independent modules that perform autonomous tasks and can be moved between different programs with relative ease, given their autonomy from the system as a whole. As an analogy from outside the world of computers, consider a coffee machine. The complex internal functions of the machine are hidden from the user, who deals instead with an interface offering a small number of brewing options. Furthermore, the coffee machine can be moved if necessary from our kitchen to a café or restaurant, assuming that we own one of these businesses.

In other words, the user's coffee-brewing experience is independent of the machine's internal workings (as long as it does not break) and also independent of the context in which it is placed. This is what an object-oriented program is like, and also what an object in OOO is like. As we will see, objects have a degree of independence from both their internal components and their external effects.

As for the 'object' part of OOO, it's important to note that OOO objects are not limited to mid-sized physical things, but refer to anything that can be considered as a unit irreducible to its internal workings or external context. As simplistic as this might sound, we will see that a surprising amount of talk about objects involves reducing them either upward or downward. Although in everyday English 'reduce' refers to making something smaller, in philosophy it indicates that one thing is declared to be merely derivative of another. Note that this is not always a bad thing. If we successfully reduce heat to an effect of atomic motion, or reduce belief in witches to social paranoia and the desire to confiscate the belongings of widows, we have already learned quite a lot. Yet something is always left out when we reduce an object: namely, the flexibility to understand heat or belief in witches on their own terms before explaining them away in terms of something else. In philosophy, for instance, we have seen that Heidegger tends to reduce objects to a wider system that embraces them all. It is a powerful idea, but one that risks losing any robust sense of the independence and individuality of things.

As for the 'ontology' part of OOO, this term stems from ancient Greek, though it was first coined in 1613 by the little-known German philosopher Rudolf Göckel (1547–1628). Ontology can be defined roughly as 'the study of being', and any subtler explanation of the term would only lead to pedantic remarks inconsistent with the unpedantic spirit of this book. Generally speaking, ontology is the most fundamental branch of philosophy. It is concerned with the basic structure of reality as opposed to more specific pursuits such as the philosophies of law, art or language.

For more than a century, professional philosophy has been

polarized between two opposed but partly interlocking traditions. Analytic philosophy, which dominates the elite universities of the Anglo-American world, has a culture that values precisely focused technical research articles in the manner of the natural sciences. Continental philosophy has largely French and German roots, and generally works in a more literary vein. What is most valued here are major books by a relatively small number of superstar thinkers, with the result that others tend mostly to write commentaries on the works of the superstars: Hegel or Heidegger, Hannah Arendt or Walter Benjamin, Judith Butler or Slavoj Žižek. Both traditions have typical strengths and weaknesses. As I see it, one of the strengths of the continental approach is its greater appeal to those who work in disciplines other than philosophy, such as art, architecture, anthropology, psychology or organization studies. This is because continental work, at its best, is less narrowly aimed at a readership of philosophy professors.

But although I come from the continental tradition myself, I am frustrated by some of its most prominent trends. The one most relevant to the present book is that present-day continental thought is overly enamoured of the notion that the world is made up of continuities, turbulent fluxes and flows, and gentle gradations rather than abrupt cut-offs between one thing and the next. As novel and brilliantly counterintuitive as this flux-based model might seem, it is already rather old, and faces defects of the sort that quickly doom any theory. It leads to the idea that reality is what I have called a 'Blend-o-Rama', or what my friend Timothy Morton once colourfully termed an 'everything-is-everything-else Deleuzean Hinduism', referring to the French philosopher Gilles Deleuze (1925–95), a noted advocate of the continuous approach to reality.[4] The price paid by such a theory, and in a different sense by Heidegger's tool-analysis, is an excessive focus on the whole and a weakened ability to account for the status of individual entities.

If we say that so-called objects are just transient patterns in a molten world undergoing constant change, in the manner of a hippy's groovy lava lamp, we still need to explain why it *seems* that

4

there are enduring individual things in the world: hammers, ducks, stars in the sky. Inevitably, the answer from the flux-lovers will be that the biases of human thought and perception deceive us into thinking that objects exist in their own right, when they are really just arbitrary portions of a single vibrating whole. According to this line of thought, the biases of human cognition have the power to misconstrue the nature of reality radically, seeing distinct individuals where there are really just resonant flows. But if that is the case, it means that humans are already being treated as something different enough from the rest of the cosmos that we can make mistakes about it. The contradiction should be obvious: if the whole of reality is in constant flux, then humans too should be part of that flux; hence, we should not be different enough from the rest to be able to misinterpret it in the first place. Some version of this contradiction has haunted every theory of the oneness of reality since the pre-Socratics in ancient Greece.

To summarize, an object-oriented standpoint is important as a counterweight to the view that the world does not contain any genuine discrete entities. If we treat the cosmos as a giant ball of throbbing flux, we strip all agency from individual things. If we quickly assume that everything is in such dynamic pulsation that nothing has enduring identity at all, we fail to grasp the way an object lies concealed behind the many faces it shows us at different times. And if everything is a system, we lose all sense of the resilience, rebelliousness and counterstrike capacity of individual things, including human beings themselves. If the universe is just a raging river, we are not liberated, but left empty-handed.

Two Kinds of Objects

As mentioned, OOO's sense of 'object' is not limited to mid-sized everyday things like horses and trees. Instead, it includes all specific entities of any status or origin: whether they be physical, non-physical, natural, artificial, mathematical, theological, simple, compound, technological, delusional, contradictory or anything else. Readers familiar with philosophy might wonder how this differs from the theory of objects of the Austrian philosopher Alexius Meinong (1853–1920). While there is certainly some overlap, Meinong was interested in objects as possible objects *of thought*, and largely ignored the interaction of objects with each other; in this way his theory still reflected the chief bias of modern philosophy, which places human thought at the centre of everything.[5] In any case, OOO contends that objects can be of any size, but that they come in two and only two basic kinds. These are real objects (which exist apart from their relations) and sensual objects (which exist only for other objects that encounter them).

As for the case of real objects, I will assume that the Convention Centre across the street from my apartment exists when I am not looking at it, even if this cannot be 'proven' by any means other than pointing. It is certainly possible that I am a delusional psychotic who hallucinates buildings that aren't there, but for the most part we can and do assume that human perception has some sort of relation to objects that exist outside it; that topic lies beyond the scope of this book. At any rate, I am convinced that the Convention Centre is a real object even when I'm not looking at it. Others are equally convinced, as seen from the fact that it is scheduled to host the water polo competition during the 2028 Summer Olympics. I find that I have no problem thinking of the building as something that will still exist in that year, even if by then I am dead and buried, or still alive but nowhere in the vicinity, or fallen into a post-accident coma for the duration of the Olympic Games.

But alongside the real Convention Centre there is a sensual version of the building, where 'sensual' refers not to the senses, but to the pleasures of immediate contact: as with a smooth fabric or refreshing liquid. The sensual does not just mean what we know through the five senses: any form of human cognition, including the most abstract forms of logical thought, encounter sensual objects rather than real ones. In the example now at hand, the sensual Convention Centre is entirely dependent on my attention. If I

6

fall asleep or drift into daydream, the sensual building disappears, even though the real one does not. The building might look very different to a cat, mosquito or seagull; it might seem frighteningly large to a young child. To repeat, real objects are what they are no matter what is happening around them. But sensual objects exist only insofar as they are encountered by something else, whether that something be me, my wife, a building inspector, a raindrop, an Olympic athlete or a dog. In principle, all objects can be both real and sensual at the same time, although there are cases (such as hallucination) where sensual objects do not correspond at all to anything beyond our experience. Most importantly, no one's and nothing's encounter with an object (sensual) will ever be inter-changeable with the object in its own right (real). OOO is especially interested in the relations between real and sensual objects.

To repeat, there is no way to gain direct access to real objects. We cannot appeal to some direct intuition of reality, whether through mystical experience, mathematical exactitude, the rigours of logical notation or even a mood such as anxiety (as Heidegger does). Our experience deals solely with sensual objects, and the question of how these correspond to real ones is always a fragile and compli-cated matter, consisting of indirect links between mediated pieces of knowledge. But sensual objects are also not just bundles of qual-ities. The world we encounter is made up of bona fide units, with the same tree or bicycle persisting despite our seeing it from constantly different angles and in different lighting conditions and moods.

Earlier I noted that continental philosophy has become excessively devoted to a model of reality featuring nothing but continuities (waves) while excluding all genuine discrete entities (stones). The cosmos is treated as a continuous field of energy that sometimes gives rise to local illusions of individual things. This tendency increased with the rise of the aforementioned Deleuze, who from the mid-1990s (also the time of his death) began to replace Jacques Derrida as the standard avant-garde continental author. On a per-sonal level I am grateful for the irreverent sense of humour that Deleuze brought to our discipline, which I hope is here to stay.

7

Even so, the almost pornographic level of flux and flow in continental thought today (to which Deleuze is a major contributor) is a grotesque misportrayal of reality. Despite its trendy counterintuitive power, it cannot do justice to both sides of reality simultaneously. The central idea of OOO, instead, is that the world consists of *discrete* objects that also possess *continuous* qualities. And furthermore, these discrete objects also require a continuum where they can interact. Stated differently, not only does reality consist of both waves and stones, but to some extent every individual is half-wave, half-stone, like a hybrid creature from a summer blockbuster film.

Undermining, Overmining and Duomining

Let's turn now to a different aspect of OOO: its strong distrust for the idea that explicit knowledge is the only form of cognition worth having. By explicit knowledge I mean the sort that can be adequately expressed either in clear prose language ('the cat is on the mat') or in equations ($e=mc^2$). After years of considering the matter, I have concluded that there are just two basic forms of knowledge, which I describe with the technical names 'undermining' and 'overmining'. If someone asks you what something is, you can either (a) tell them what it's made of, or (b) tell them what it does. That's it. The point of emphasizing this limitation is to remind the reader that knowledge is just one part of a vast cognitive landscape.

To see how the two types of knowledge work, imagine that a doctor has just prescribed you an unfamiliar medication. Let's invent one called 'Cardiomoxin', which obviously sounds like it has something to do with the heart. Imagine now that a close friend sees the bottle of Cardiomoxin in your cabinet and asks you what it is. In the unlikely event that you and your friend are both chemically literate, you might tell them something like this: 'Cardiomoxin features a complex, multi-ring structure with a trifluoromethyl-substituted phenyl group attached to an imidazole ring.' If this description is over their head (or yours), you can instead give the personal backstory of how the doctor came to prescribe it to you, what medical incident prompted them to do so or something along those lines. In this case you are *undermining* Cardiomoxin (in OOO's technical sense of the term) by reducing it to its physical and auto-biographical underpinnings.[6]

You might also tell your friend the history of how this medicine was discovered, if you happen to know it. This too would be a form of undermining, since it turns our attention away from the object at hand to a discussion of how it was physically and historically produced, or of how its composition became relevant to you personally. Instead of telling your friend directly about the medicine, you are telling them about its subcomponents or about various aspects of its history. In this respect undermining resembles what magicians call 'misdirection', as when they ask you to look closely at the clown gesticulating in the aisle while a rabbit is secretly smuggled into a hat on stage. Undermining is the first of the two forms of knowledge.

The second form of knowledge is called *overmining*. This happens when we reduce an object not downward to its pieces, but upward to its actions, effects, symptoms or visible traits. In the case of our mythical heart medicine, you might tell a knowledgeable friend something like this: 'Cardiomoxin functions as a selective inhibitor of the cardiac sodium channel and exhibits potent antiarrhythmic properties. It stabilizes the inactive state of the sodium channel, reducing the likelihood of aberrant cardiac electrical activity and restoring normal rhythm in cases of ventricular tachycardia and atrial fibrillation.' If this too is over your friend's head (as it would be for most of us) you might simply tell them that Cardiomoxin makes you feel more energetic, or safer, or that it helps you sleep better at night. All of these answers are informative, but all are versions of overmining. If you live in the United States, the land of sky-high drug prices, you might also complain about your $275 out-of-pocket co-payment for Cardiomoxin after insurance. You could add as well that the drug generated $5 billion in sales last year for Pierrot Biotech, the fictitious company that synthesized it, and that this led to the creation of over 700 well-paying laboratory jobs.

These are all examples of overmining. Rather than speaking about Cardiomoxin itself, you have changed the subject to the various things that Cardiomoxin *does*. This is an especially popular way of thinking today. The French philosopher Bruno Latour (1947–2022) treats objects as *actors*, meaning that they are equal to the sum total of everything they do in the world. Among his key inspirations in saying so were the American pragmatist philosophers of the early twentieth century, such as Charles Sanders Peirce (1839–1914), William James (1842–1910) and John Dewey (1859–1952). Peirce (pronounced 'purse', not 'pierce') claimed that the way to make our ideas clear is to rephrase any philosophical dispute as a question about what practical difference it would make if one answer were true rather than another.[7] If there is no such difference, then we have an empty dispute over words rather than a true intellectual difficulty. For instance, a pragmatist might say that it's pointless to argue about whether a world exists outside the mind, since the answer will not affect any of our actions. The more recent Latour made a number of similar statements during his lifetime.

Many people have found the pragmatist principle liberating, since it permits the disposal of seemingly trivial questions inherited from the past. But I happen to think it runs too big a risk. We don't always immediately know the practical consequences of any philosophical debate, and for this reason we should not be too quick to dismiss questions that have no obvious stakes. For example, the seemingly stale old dispute over the existence of a world outside the mind could become crucially important once virtual reality provides a greater portion of human entertainment. Should we be allowed to assault, torture or murder humans (including children) who exist 'only in virtual reality' rather than outside the mind? The legal stakes of this question may prove to be enormous.

Returning to the topic of undermining and overmining, you might now ask what is wrong with undermining an object. Nothing is really wrong with it, since we learn something important when we read about the 'complex, multi-ring structure' of our new medication. Yet when we analyse a thing by breaking it down to its basic elements, we forget that every genuine object is more than the sum of its parts, as the old phrase goes. As a rule, if an object is not just an aggregate of pieces, it will have properties not found in those pieces: an aeroplane can fly, but none of its parts alone is able to fly. To undermine an object is to forget that the object as a whole is just as real as its parts. This idea is so important in philosophy that it has its own technical name. When an object has a surplus of qualities or features exceeding those found in its parts, we call this 'emergence'.[8]

You might also ask: what is wrong with overmining an object? Again, nothing is inherently wrong with talking about an object this way. We learn a lot from overmining, such as the fact that Cardiomoxin 'stabilizes the inactive state of the sodium channel', if we happen to know what that means. But while undermining misses out on emergence, overmining misses out instead on what I call *submergence*. This means that an object always has unexpressed qualities hidden beneath its currently visible effects. If a factory performs 100 tests on a new model of car, we still don't gain a complete picture of the car. It may be capable of other successes and failures that were not ascertained in any of the tests: maybe it handles especially well on a rare type of bridge surface, or perhaps it tends to explode if struck by lightning. The reason that vehicle recalls happen is precisely because no one can foresee all of their possible mechanical troubles.

Overmining also fails in counterfactual historical scenarios, in which we ask how a given human or non-human object might have behaved in a completely different context. Consider the following prompt: 'Imagine that Napoleon had been a French general during World War II. Given his past military strategies, how might he have tried to prevent the Nazi invasion of France in 1940?' While this question could be dismissed as unverifiable speculation, I think that's too harsh a judgement. The reason is that Napoleon isn't just the sum total of things that actually happened in his military career from 1779 to 1815, during which he engaged in sixty battles, winning fifty-three of them while losing only seven. For we can safely assume that the winning battles did not draw on the whole of Napoleon's skill set, and that the losing battles were not sufficiently numerous to exploit all his weaknesses as a commander. Nothing forbids us from speculating about how a World War II version of Napoleon might have fought, despite the later availability of tanks and fighter planes that he never lived to see. In doing so, historians force themselves to reflect more carefully on aspects of Napoleon that were never fully detected amidst the limitations of his era. When I asked ChatGPT to answer this very question, it came up with some feasible answers, such as the likelihood that Napoleon might have persuaded the British military to join him in a pre-emptive campaign in the Rhineland, or that he might have heavily fortified the Ardennes region in northern France.[9]

The point of showing that an object cannot be exhaustively undermined or overmined is to see that it is primarily discrete rather than continuous, in the sense that every object is something distinct from both its components and its actions. This is not to say that there's no relation between the three levels; of course there is. If a molecule of gold contained no gold atoms, it would not be a molecule of gold; if it were gold but somehow did not have the typical colour of that metal, no purchaser would value it as gold. But within certain limits, an object can endure as the same even when its pieces or outward effects change. Looking downward, any chunk of gold will lose atoms from time to time, though we don't stop calling it the same piece of gold for that reason. And looking upward, the gold can be incorporated into different pieces of jewellery without losing its identity. Objects have parts, and objects are usually parts of larger objects in turn. Yet each layer of this process has a certain degree of autonomy, an independence from its smaller components and the larger contexts in which it is placed. This means that while the various objects in the world are somehow capable of connection, they are connected somewhat loosely. Every object has a degree of resistance to what happens to it, a certain deafness to the noise that surrounds it.

The point I am making is that whether we reduce an object downward or upward, similar difficulties arise. Douglas Hofstadter's widely read bestseller *Gödel, Escher, Bach* pits 'reductionism' against what he calls 'holism', the idea that everything is connected.[10] But while he's right to say that reductionism is an extreme way of reducing things downward to their tiniest pieces, I would add that holism is an equally extreme way of reducing things upward to a widest context that governs them all. The problem with holism is that it would be purely arbitrary to say that random objects such as my toothbrush or shoes have systematic relations with the entire cosmos, even though important philosophers such as Alfred North Whitehead (1861–1947) have suggested as much.[11] Essentially, holism is too easy on itself: it simply asserts with cavalier abandon that everything is connected. Yet it fails to explain the mechanics of how limited connections between distinct things are possible without the world being dissolved into a single bubbling pudding.

Returning now to the main point, there are only two basic forms of knowledge: undermining and overmining. I do not believe there are any others. But it's also worth noting that they do not always appear on their own. Instead, it frequently happens that undermining and overmining are performed simultaneously in a gesture we can call 'duomoning'.[12] For instance, we might answer our friend's question about Cardiomoxin by explaining what it's made from *and* what it does. But as comprehensive as this sounds, duomining still does not tell us everything about an object. What makes this combined tour de force of knowledge fall short is that knowledge is not the only form of human cognition. Consider the realm of art, which is not primarily meant to produce knowledge. Imagine that an artist paints a portrait of your bottle of Cardiomoxin: Damien Hirst (b. 1965) comes to mind as a likely suspect. You would never expect this painting to give us knowledge of either the components or the effects of the drug. Yes, Hirst could always undermine the painting by telling us about the specific canvas and pigments he used, or where he got the idea for such a painting, but that would hardly count as an understanding of the artwork itself. He could also try to overmine the painting by going around the room and asking every-one how the painting makes them feel, but this too would fall short of true aesthetic appreciation.

The fact is that the two forms of knowledge (undermining, over-mining) as well as their combination (duomining) give us nothing more than a rough approximation of what an object is. Objects are one thing and knowledge quite another. The painting of Cardio-moxin, like artworks more generally, points to this gap between object and knowledge by alluding to something situated midway between the drug's components and its effects: namely, Cardio-moxin itself. Every object exists at a halfway point between its parts and its effects, irreducible to either of the two forms of knowledge.

I have mentioned art as one form of cognitive activity that helps exhibit a thing in a form that surpasses knowledge. But perhaps an even better example is philosophy itself. In ancient Greek the word is *philosophia*, which means 'love of wisdom', not wisdom pure and simple. As Socrates explains in the Platonic dialogues, humans find themselves midway between the gods and animals, partly but not fully understanding any topic under discussion. Although Socrates is famous for always insisting that people should give definitions of the terms they use, he ought to be more famous for the fact that he is never satisfied with any definition. The reason, as I see it, is that nothing is ever fully describable by a definition, or indeed by any sort of prose statement or mathematical equation. Analytic philosophy has often insisted that philosophy, just like the sciences, should aim at becoming clearer and more exact. I answer that this is a misguided goal, since reality itself is neither clear nor exact; hence knowledge is not always the best way of doing justice to reality. This is a point to which we will return later, when considering the diffi-culty of translating between the continuous and the discrete.

When I walk on the beach in southern California, where I live, I often pick up stones of unusual beauty or intriguing shape. Typi-cally a stone is a durable unit, hardened by millennia of geological pressure. It does not crumble in my hands, and there is never any doubt about its boundaries. This makes stones a perfect metaphor for individual objects. However complex the process that forms an individual stone, to some extent that process is no longer relevant once it is a finished product in my hand; at this late stage it is very clear where the stone ends and the surrounding world begins. By contrast, the waves at the beach alert us to a different face of reality. A wave is less a material object than a pattern of movement that is tied to no particular set of objects; it moves successively through different parts of the ocean before crashing at last into the sand. Although a wave is always finite – it does not directly affect the entire ocean, much less the universe as a whole – we cannot say that it consists of any definite number of parts. Even if we analyse it into smaller and smaller segments, we will never reach a smallest unit of the wave. How different this is from the stone, which cannot be cut at all without destroying it. From this contrast we gain a sense of the two-faced character of reality that is the central topic of this book.

Defining the Terms

Discrete things exist in the world. When we open our eyes in the morning we do not encounter the world as a single lump where everything is melted together. There are hammers and dolphins alongside planets, laws, fictional characters and mathematical objects. It is true that these discrete things are not completely isolated; they interact. But they do not interact with everything, and no interaction is automatic or total. The aforementioned Latour includes an amusing passage to this effect in his 1992 book *Aramis*, which traces the sad demise of an automated Metro system once slated to open in Paris. Here the young engineer narrator is speaking of his mentor, a professor called Norbert who speaks in the final line:

> the violent blow he struck with his fist on the desk had no visible influence on the chapter of Aristotle's *Metaphysics* that was filed

under the letter A at the top of his bookshelf. 'You see: not everything comes together, not everything is connected.'[13]

Even when objects do interact and transform each other, the resulting changes are finite and not always reciprocal. Living in Egypt for
sixteen years changed me in lasting ways, but when I left Cairo in
2016, I was still recognizable to friends and family as the person who
first arrived there in the year 2000. Moreover, whatever reciprocal
influence I may have had on the venerable city of Cairo was surely
minuscule when contrasted with its life-changing effect on me.

Continua (the plural of 'continuum') also exist. A continuum is
something that can be cut up into as many pieces as you please, as
small as you wish, but which is still effectively one thing rather than
a sum total of tiny parts. As an example, consider the number line
you probably remember from your student days; we will return to
it more than once. Please note that most mathematicians would not
agree that the number line is just an idea. For most of them mathematical objects are every bit as real as material physical things, if
not more so. In any case, the number line is usually drawn with zero
in the middle, negative numbers extending to the left, and positive
numbers to the right. The reason the number line is continuous is
because there are no holes in it. Imagine that someone wrongly
claimed to notice a gap on the line between the numbers 2 and 3.
This person could easily be refuted by adding 2.5 between them. If
that person complained further that there are still holes on the line,
we could add as many new numbers as needed to keep filling up
the supposed gaps. This process would never end, since even if we
have a rather tiny gap between numbers (say, between 2.49978361
and 2.49978362) we can still add infinitely many more numbers into
any gap we find. That is what makes the number line a continuum.

Space and time are also continua as far as we know, though in
advanced physics the jury is still out on this question. When I teach a
philosophy seminar at the Southern California Institute of Architecture (SCI-Arc), it lasts in principle for two hours and fifty minutes.
Just as with the number line, we can mentally split the length of my

seminar into three parts, seven parts, 4,000 parts, a billion parts, or as many as we like, to the verge of infinity. That's because time is a continuum and consists of no definite number of smallest instants. But in another sense, the full length of the seminar is just a single unified stretch of time. The reason is that if the class were truly made up of infinitely many moments, I could never pass through them all and reach the end of the lecture. That is what it means to say that even though a continuum can be carved up indefinitely, it always remains one thing.

The same holds for space, another important continuum. When I fly from Los Angeles to Istanbul during summer vacation, the air route can be divided into four, twenty-five, or 7,986 parts at my whim, and I can always increase the number of divisions as far as I want. But if there were really infinitely many points in space, as with the infinitely many numbers on the number line, it would be impossible for me to fly from California to anywhere else. More than that, it would be impossible even to take off from the runway at LAX Airport, since there would be infinitely many points to cross before the plane could even begin to fly. This type of thinking will soon lead us to consider Zeno's Paradoxes, which ranked among the most prominent intellectual puzzles in ancient Greece and are still discussed today.

It's also important to distinguish between a continuum and the related but looser notion of a spectrum. A good example of a spectrum is the range of political positions, running from Left to Right, represented by the parties of democratic nations. Take present-day Germany, for instance. We can put the Green Party towards the left of the spectrum, with the Social Democrats closer to the centre. At right of centre you'll find the classically liberal Free Democrats, followed by the Christian Democrats further to the right, and finally the radical anti-immigrant Alternative für Deutschland party. Why do we call this a spectrum and not a continuum? Because the German political system consists of a finite number of voters and parties, so that we cannot always find another party between any two given parties. Yes, someone could always try to launch a new party midway between the Greens and the Social Democrats, but there is not endless room to create an infinite number of subtly different parties in that section of the spectrum. Even if we imagine the German population growing at a completely out-of-control rate, eventually reaching billions of residents, there will always be a finite number of parties and hence there will always be gaps between them. This makes the German political spectrum (or any other) very different from the continuum between the numbers 4 and 5, which truly contains an infinite number of numbers. As an analogy, while our continuum can be understood as a wave, a spectrum is more like a finite series of stones ranging from largest to smallest, or brightest to darkest in colour.

Radical Stances

This book claims that the most common mistake about the continuous and the discrete is the assumption that one of them must be *more fundamental* than the other. In other words, when it comes to waves and stones, there is disagreement between one set of theories that treats reality as a continuous, pulsating flux, and another that asserts the absolute individuality and mutual separation of specific things. We will soon meet with examples of both extreme positions.

At present, there is no question that the model of unremitting change has the sexier reputation of the two, whether in philosophy departments, popular science books or creative fields such as literature and the visual arts. The widespread desire to embrace ceaseless turbulence is captured in the following passage from the eminent microbiologist Carl R. Woese (1928–2012):

> Imagine a child playing in a woodland stream, poking a stick into an eddy in the flowing current, thereby disrupting it. But the eddy quickly reforms. The child disperses it again. Again it reforms, and the fascinating game goes on. There you have it! Organisms are resilient patterns in a turbulent flow– patterns in an energy flow.[14]

The prolific American philosopher Thomas Nail (b. 1979) writes in a similar spirit, presenting this river-like model of reality as the irrevocable verdict of science and history alike. In fact, he links his grand theory of motion both to state-of-the-art physics and to the increasing concern with migration and refugee issues in contemporary politics.[15] As Nail puts it:

> the flux, turbulence, and movement of energy are more primary than the relative or metastable fixity of classical bodies . . . The old paradigm of a static cosmos, linear causality, fundamental particles, and classical space-time no longer fits the twenty-first-century reality of cosmic acceleration, turbulence, and continuously vibrating fields.[16]

Rein Raud (b. 1961), one of Estonia's most prominent intellectuals, has made similar arguments in his fine book *Being in Flux*.[17]

It's easy to see why the idea of replacing stagnant things with dynamic processes is thrilling to so many people. If the everyday world of solid physical objects, monotonous daily errands and unjust political and economic conditions seems dull and familiar, how revolutionary it feels to think that everything is in constant motion! From this daring standpoint, all apparently stable forms or patterns are nothing but derivative products of an energetic turbulence in the heart of things. But how can this type of theory account for the fact that individual things at least *seem* to exist? As we will see, Parmenides and his disciple Zeno said that our senses are simply deluded when they display many different things; the world is really an unarticulated primal whole. Other philosophers have conceded that individual things do exist, but only because human thought arbitrarily cuts a single world-lump into pieces that meet its own practical needs.

If we insist that reality is just a continuum in constant motion, we avoid addressing the most important problem of all: how to account for the fact that the continuous and the discrete interact. Individual things (which are discrete) engage in relations in space and time (which are continuous). An object (which is discrete) possesses qualities (which can be continuously varied, say, from brighter to darker red). A photon (a discrete quantum of light) moves through space warped by gravity (a continuous effect). To view the world as nothing but a constantly raging firehose of change is to pretend that there are no discrete entities at all, when in fact there are.[18]

In the history of philosophy there have also been extreme positions on the other side of the debate: theories claiming that everything is so discrete that individuals are completely cut off from mutual contact. Such positions are rare today, since they were linked with a style of theologically based philosophy that has long been out of fashion. The name for this way of thinking is occasionalism; it flourished in early medieval Islam and again in early modern Europe. The occasionalist idea of how radically discrete beings influence each other is to say that God is not just the only creator, but the only causal agent at all. In this way all causal power is removed from individual things. When objects collide, they only seem to collide; God alone makes contact with the various things in the world. The problem with this extreme form of discreteness is not that it draws on religion, which has been widely disdained by secular intellectuals since the Enlightenment. Instead, the true problem is that if we say God alone is capable of causal influence when nothing else in the universe is capable, this does nothing to clarify the dynamics of interrelation between things, even if such a theory quickly satisfies whatever degree of piety we might have. For instance, what features of God can we deduce from his purported ability to touch an individual thing when nothing else is able to do so? What enables a homely terrestrial object to make contact with the divine? Such questions are usually ignored in favour of vague appeals to God's power and goodness. But as missionaries know very well, if philosophical arguments are to be effective they must also address the unconverted.

In the modern era a more widely accepted view is that causation is not monopolized by God, but by the human mind. This idea was made popular by the eighteenth-century philosophers David Hume and Immanuel Kant. Both argued in different ways that we can only grasp causation as a feature of human experience and cannot assume it exists in the world itself. While this position has a better chance of convincing secular-minded intellectuals, it is not much of an improvement on theological occasionalism. The same difficulty remains: the causal monopoly granted by occasionalism to God is transferred to the human mind, in a manner similar to sweeping dust from one part of the floor to another. We still have no idea how discrete entities affect each other in continuous time and space. This is no small problem, since hasty appeals to God or the structure of human experience prevent us from considering the interaction of objects among themselves.

But there has long existed an alternative to the radical stances of pure continuity and pure discreteness, in the writings of philosophers who recognize that we must account for both. One example is the ancient master Aristotle, who devoted one of his great books (the *Physics*) to the continuous character of nature, and another (the *Metaphysics*) to the discrete character of individual things. Another example is the more recent French thinker Henri Bergson, who, despite being known as a remorseless thinker of flux and flow, was aware that we must also address the apparent endurance of individual things. Since neither the continuous nor the discontinuous can single-handedly account for the whole of reality, we need a model of the cosmos that recognizes both while clarifying how they coexist and interact. To put it plainly, both the discrete and the continuous exist: often simultaneously, and even within the same object, insofar as an individual thing is not identical with the continuous qualities it also possesses. A continuous number line would mean nothing without the existence of specific numbers, and a discrete individual horse cannot run except through continuous space and time. Likewise, a stone could not disrupt a wave if it simply melted into a wave-like continuous universe. To understand the paradoxical interaction between the continuous and the discrete requires an exploration of the very fabric of reality, and could change our deepest sense of how the world works.

This book will pose the following basic question: what sort of relation do the continuous and the discrete have to each other? In one sense the two have highly distinct properties, which means they are both discrete; in another sense they do interact, which means they are also continuous. That is our ultimate problem: how can the continuous and the discrete be both continuous and discrete with respect to each other without leading to a mess of contradictions? Let's break it down. When it comes to the continuous and the discrete, three important questions appear at centre stage:

1. Given that entities are discrete, how can they touch at all?
2. Given that each continuum is one, how can it contain infinitely many parts?
3. Given that the continuous and the discrete are not the same thing, they are discrete with respect to each other. That being the case, how can they also become continuous with each other? If this were not possible, they would not be able to interact.

It's easy to imagine someone making the following complaint about our first question: 'Objects can obviously touch and interact. We see this happen constantly, so why waste time asking about it?' I have heard this objection frequently. But we can see how misguided it is when we apply it to other cases. For example, imagine someone saying this: 'Obviously children inherit features from both of their parents – so why waste time finding genetic explanations?' Or this: 'Obviously human personality is shaped by both heredity and environment, so why waste time figuring out the exact mechanisms of each?' Or this: 'We already know that earthquakes happen. So why waste time figuring out how?' The fact that something is known to happen is not an argument against studying how it occurs. Quite the contrary: the existence of a thing or phenomenon is an argument *in favour* of such study; to acknowledge that something exists is an opening for conversation rather than an end. Stated differently, our goal is not to 'prove' that objects affect each other (no such proof is needed) but to determine the mechanisms that bring it about. If we proceed in the opposite direction and assume that everything is fundamentally related to everything else – remember our holists – then we lose sight of the fact that while some objects have tremendous mutual influence, others sit side by side for years without affecting each other at all. For instance, the genetic writings of Gregor Mendel lay mostly unread for decades before Darwinians turned Mendel into a powerful ally in the combined doctrine known as neo-Darwinism. Another example: quite often a book sits on my shelf for years or even decades before I rediscover it exactly when needed.

Our second question is focused on the continuum. As we will see when discussing Zeno's Paradoxes in the coming chapter, a continuum cannot actually consist of infinitely many points, since this would lead to impossible results. For one thing, we would not be able to travel from one place to the next, given that any change of place would need to pass through infinitely many points to do so. For another, we would not be able to experience an hour or even a second of time: after all, we would first need to pass through infinitely many moments before completing the full hour or second. Nonetheless, it's impossible for a continuum *only* to be a unified whole, since we do not move through any continuum instantly, but only cover part of it during any interval of time. Although a walking trip from central Manhattan to central Brooklyn can be treated as a single journey, the route this entails is not a simple whole, since throughout the journey we are always at some specific place along the route.

The third question is the one that really gets to the heart of the matter. How can something continuous interact with something discrete, given that the two play by such different rules? What we'll see is that two continua can only make contact through something discrete, and that two discrete objects can only make contact in a continuum. Everyday life teaches that the discrete and the continuous do interact, since discrete objects exist in continuous space and time; conversely, continua can be cut up into discrete slices. As stated earlier, what we need is less a proof that interactions between things actually happen (few would deny it) and more an explanation of how these interactions occur, or how and why discrete objects and continua meet and affect one another. This is not a false problem summoned from thin air; instead, such questions haunt our conception of the ultimate nature of reality.

Thixis

At its core, this book is an examination of why contact between two things is more difficult than we imagine. I'd like to propose a technical term for this problem, based on one of the many ancient Greek words for 'touch.' *Thixis* refers to the contact between surfaces. This makes it an appropriate term for OOO, which holds that contact between two objects is always a matter of surfaces. Later we will consider the notion of *thixis* in greater detail. Remembering that OOO recognizes two different kinds of objects (the real and the sensual) it will also turn out that objects can only touch objects of the opposite kind. Neither two real objects nor two sensual ones can make direct contact; they can only touch their opposite kind, just as the poles of a magnet repel each other if they are both north or both south. But this insight soon gives way to the deeper truth that discrete objects can only make contact with something continuous, and that two continua can only touch through something discrete. Let's coin the word *heterothixis* as a technical term for this principle. In due course we will return to this point.

Our philosophical journey into the paradoxical coexistence of the continuous and the discrete begins with Aristotle in ancient Greece, and will take us all the way to modern physics and mathematics. But to clarify this problem does not mean to dissolve it once and for all. Problems in philosophy, unlike those in mathematics, engineering or medicine, can neither be eliminated nor definitively solved. Once a new philosophical problem is detected, it tends to return intermittently in new variations across the centuries. The philosopher's job is not to solve it, but to propose new options for dealing with it;

philosophy is driven by the love of wisdom, not some knowledge safely in our possession once and for all. For instance, Kant never 'solved' the famous mind-body problem that emerged from René Descartes's philosophy, but in some sense made it more extreme than ever, with his new split between visible phenomena and never-visible noumena, or 'things-in-themselves'.

There is a tendency to think of all the branches of knowledge as modernizing in a single direction, like improved cancer treatments or the increasing speed of computers. But this incremental picture of progressive knowledge does not work well for everything. Politics does not get 'better and better' over time; in any nation it shifts between high points of human flourishing and low points of stagnation and civil conflict. We would not say that painting is 'better' today than it was during the Florentine Renaissance, even if certain pigments can now be produced more reliably, and even if the system for training vast numbers of painters is bureaucratically better organized than in fifteenth-century Italy. The same holds for philosophy, which generates new insights in every century, but does not produce irreversible forward progress. We cannot say that the philosopher Baruch Spinoza (1632–77) 'proved that God is nature' in the same way that the mathematician Carl Friedrich Gauss (1777–1855) proved the fundamental theorem of algebra. Mathematicians are bound by Gauss's result; philosophers are not bound by Spinoza's.

Stated differently, 'proof' is not really the medium in which philosophy is conducted; indeed, every attempted proof is met by an army of counterproofs extending over long stretches of time. This is not what happens in astronomy or chemistry, not to mention such technical fields as industrial engineering or the design of superior submarines. Yes, attempts are made in academic journals to 'refute' various statements by Plato or Aristotle on a weekly basis. Yet most of these attempts vanish forever into the mist, while the great ancient thinkers repeatedly return from the dead to guide our thinking. This is not because they show us perennial truths, as conservatives like to believe, but because classic works make basic decisions on fundamental issues and thus inspire us to make our own. They do not whittle every stick to the point where only dust remains, as academic discourse too often does.

In closing, let's return to the beach and its medley of waves and stones. If I skip a stone through an oncoming wave, they have clearly interacted. I have made a cut in the continuous wave by puncturing it at one discrete point instead of another. But in using the stone to make this cut, I see that the stone is not just a self-contained discrete unit, but belongs to a continuous space and time just like the wave itself. Without this shared citizenship in a continuum, nothing could interact. By gaining a better understanding of this simple image from the beach, we will learn something important about the ultimate nature of reality. At the close of this book we will learn the strange lesson that the continuum where objects meet lies on the interior of another, larger object.

The philosophers Aristotle (Chapter 2) and Henri Bergson (Chapter 4) are not normally viewed as allies, but we will see that they are joined in recognizing that neither the continuous nor the discrete is alone in the world; both principles play significant roles. By considering the Arab historian Ibn Khaldun's theory of generations and the medieval death match between nomadic hordes and civilized urban centres (Chapter 3) we will gain a sense of how the continuous and the discrete function in human history. Turning to the philosopher of science Thomas Kuhn (Chapter 5), we will examine the distinction he draws between the gradual progress of mainstream cumulative 'normal science' and the sudden jumps that he links with scientific revolutions. In evolutionary theory we will encounter the strife between Charles Darwin's gradualist model of species change and Niles Eldredge and Stephen Jay Gould's idea of 'punctuated equilibrium' (Chapter 6), a debate that provides one of the most famous intellectual battlegrounds between the continuous and the discrete. Zeroing in on a debate between architects that unfolded from 1988 onwards (Chapter 7), we will examine the key role played by the continuous and the discrete in aesthetics. We will then return to Gould's notion of 'Nonoverlapping Magisteria' (Chapter 8), which conceives of human knowledge as made up of multiple autonomous zones, as opposed to those critics who insist on a single authority called 'reason' or 'rationality' empowered to weigh in on everything that exists. Finally, we will turn to perhaps the most famous instance of conflict of the continuous and the discrete, the wave–particle duality of matter discovered by quantum physics, while briefly considering what continuity means in mathematics (Chapter 9). This will lead to the concluding Chapter 10, in which the philosophical stakes of waves and stones will be illuminated from a fresh angle.

The reader may have noticed that the chapters just summarized address the concerns of both natural science and disciplines concerned with human affairs. The modern tendency is to view these as two radically different kinds of fields. Natural science supposedly deals with reality as it is, while human affairs concern socially constructed concepts, and involves a subjective projection of meanings and values on to a cold, grey universe ruled by mechanical clockwork. It is certainly true that humans are a major *cause* of history, language, society and the arts in a way that they are not causes of the particles studied by physics. But this is beside the point. For even if humans are responsible for creating human history, we are not able to mould the laws of history to work as we see fit. The generational cycles described by Ibn Khaldun in Chapter 3 below cannot be altered by human willpower, and the cultural gap described by René Grousset between nomadic and sedentary civilizations is every bit as wide as the natural gap between protons and neutrons. To create something is not to control it, as every parent knows.

Physics and Metaphysics: Aristotle's Hidden Duality

'Whenever philosophy has established real contact with Aristotle, it has immediately become more precise and serious.'[1]

Julián Marías, *History of Philosophy*

To anchor this book in Aristotle (384–322 BCE) is to take a certain risk. Although he is widely acknowledged to be one of the most import- ant philosophers of all time, in some quarters he has the reputation of a stuffy old bore. Bertrand Russell (1872–1970) tells us that Aris- totle's *Nicomachean Ethics* 'appeals to the respectable middle-aged', that it is useful in 'repress[ing] the ardours and enthusiasms of the young', and that 'to a man with any depth of feeling it is likely to be repulsive'.[2] A more recent verdict can be found in Robert M. Pirsig's *Zen and the Art of Motorcycle Maintenance*, one of the great philosophy bestsellers of the twentieth century.[3] Published in 1974, Pirsig's book tells the story of a cross-country motorcycle trip with his son, interspersed with recollections of Pirsig's own academic failures, severe depression and forced electroshock treat- ment. It is a work of genuine intellectual and spiritual merit, but its closing chapters wrongly portray Aristotle as an enemy of bona fide philosophy. In Pirsig's eyes, Aristotle is the one who led West- ern thought down the fruitless path of arranging reality into sterile categories and dull oppositions, boring us with commonsensical lessons on matter and form, substance and accident, grammatical subjects and predicates, while predictably dividing all living crea- tures into plants, animals and humans. When viewed in this way,

Aristotle seems to be nothing more than an ancient bureaucrat of the obvious.

If this were an accurate picture, I would happily join in revolt against the Aristotelian tradition. But that will not be needed, since the picture is utterly inaccurate. In Aristotle we actually have one of the strangest and most intriguing thinkers in human history. Despite his familiar place in Western philosophy and his high status in the Catholic Church through the mediation of St Thomas Aquinas, there is a certain exoticism to Aristotle's historical path. His works were preserved mostly in the foreign tradition of Islamic philosophy before his rediscovery in Europe sparked one of the first intellectual controversies in Paris, that enduring capital of controversy. In Aristotle we also find a bizarre, endearingly morbid sense of humour, as when he says that touch is the most important sense because it is the only one that can destroy us if we do it too hard.[4] There are also the amusing examples he gives of good luck: 'such as if all of a man's brothers are ugly but he is good-looking', and 'if someone who regularly made a journey were the only one not to, while others did for the first time – and were killed.'[5] The scholar W. D. Ross describes Aristotle's 'mocking disposition' as being visible even in his face, which I do not find difficult to imagine.[6]

All of his twisted witticisms aside, what is most captivating about Aristotle is the deceptive strangeness of his ideas themselves. In fact, I have long dreamed of writing a book called *Weird Aristotle*, and may eventually do so. The apparent technical aridity of his prose masks unresolved tensions that often take us to unexpected places. For instance, in his *Metaphysics* he says it is impossible to define any thing exhaustively, since things are specific individuals but definitions use universals, referring to terms that can describe many different individuals at once: honest, heavy, strong, cruel or beautiful.[7] By way of example he notes that defining a particular person as a 'pale animal' falls well short of the truth, given that each of us is a specific person with a unique life story and a personality that is impossible to describe exactly in words. The same holds for any description of anything, from dancers to oceans to lightning-bolts. While this point may sound dull or obvious, it points to a permanent gap between language and things. And given that knowledge entails a description of things in language, this leads us in turn to wonder how knowledge is possible at all.

One of the Aristotelian ideas that intrigues me most is the distinction he draws between the continuous and the discrete, which is the central subject of this book. Since this topic proves to be one of the most fateful to have emerged from ancient Greek philosophy, let's begin by revisiting the first steps of that tradition.

The Pre-Socratic Thinkers

Western philosophy began with the pre-Socratic thinkers of ancient Greece, who flourished from around 600 BCE to the time of Socrates himself (470–399 BCE).[8] The pre-Socratics generally focused their attention on theories of physical nature; Socrates turned inward towards human life itself. But when it comes to pre-Socratic Greek philosophy, the terms 'pre-Socratic', 'Greek' and 'philosophy' are all potentially misleading. First of all, the pre-Socratic thinkers did not know themselves by that name, since most of them lived and died without knowing that Socrates would ever be born. In their own time the pre-Socratics were known simply as the *physikoi* (a forerunner of our word 'physicists'), meaning those who investigate nature. The speculative talent of these early figures is undeniable. They proposed numerous imaginative theories about the cosmos, whether by claiming that everything is made of water, air or numbers, or by treating it instead as a colossal shapeless mass.

Second, there is something potentially misleading about calling the pre-Socratics 'Greek'. They were certainly ethnic Greeks, and spoke and wrote in the ancient Greek language. However, the heart of pre-Socratic geography lay not in the territory of modern Greece, but in former Greek colonies located in what are today the west coast of Türkiye, Sicily and the southernmost portion of the

Italian peninsula. Only towards the end of the pre-Socratic period did Athens finally become an important site for speculative thought.

That brings us to our final point: it might even be deceptive to call these pre-Socratics 'philosophers', fascinating though they are. The *physikoi* are probably better understood as the first scientists of Western civilization rather than the first philosophers. Although renowned for their cosmic speculations, their theories aimed mainly at providing foundational knowledge about nature. Even their practical activities were those we would associate in our own time with mathematicians and scientists rather than philosophers. They predicted solar eclipses and built dams (Thales), discovered fossils on mountaintops and drew geological conclusions (Empedocles), designed the first known map of the world (Anaximander) and felt traumatized by the existence of irrational numbers (Pythagoras, who according to legend drowned himself as a result).

It is my view that Western philosophy proper begins with Socrates and his disciple Plato (Aristotle's teacher), through their transformative insight that *philosophia* means not wisdom, but rather the love of a wisdom that is never quite attainable. With Socrates and Plato we find an effort to account for the imprecision of knowledge itself, just as Aristotle later saw with his cautionary note about the inevitable failure of definitions. While the pre-Socratics made statements about reality that they held directly to be true ('everything is changing all the time', or 'everything is made of a mixture of air, earth, fire and water, joined by love and separated by hate'), with Socrates and Plato there is greater awareness of the inherent inadequacy of any statement about reality. Heraclitus may be the one pre-Socratic thinker who foreshadowed this later trend, with his famous remarks that nature loves to hide, and that the Oracle at Delphi (believed to be a messenger of the gods) neither speaks nor conceals but gives a sign. But *philosophia* in the strict sense begins with Socrates, who in Plato's dialogues routinely declares that he is ignorant and has never been anyone's teacher. Although he constantly asks others to share their definitions of love, friendship, virtue or justice, these discussions never lead to any conclusive definitions at all. In line with

Socrates' open-ended curiosity, when we speak of *philosophia* (love of wisdom) rather than *sophia* (wisdom), we commit ourselves to endorsing a gap between literal statements and reality itself. This will be one of the key themes of the present book.

As an illustration of the inadequacy of stating anything in direct prose, consider the case of the widely read German philosopher Friedrich Nietzsche (1844–1900). The respected *Stanford Encyclopedia of Philosophy* describes him as follows: 'He is famous for uncompromising criticisms of traditional European morality and religion, as well as of conventional philosophical ideas and social and political pieties associated with modernity.'⁹ While not inaccurate, these same words could be said about many different authors, which reminds us of Aristotle's remark about the difficulty of adequately defining individual people or things. More importantly, the passage above gives us no sense of Nietzsche's unusually powerful writing style: one of the principal factors that lures so many young devotees to his ranks. Although the encyclopaedia entry goes on to give more detail about his life and writings, no such write-up can claim to give us a perfectly accurate portrait of the German thinker. The failure of literal prose is even easier to see if we try to describe a specific sculpture or sunset. Which details would you include or exclude? Which aspects would be difficult or impossible to express as a literal statement? In fact, philosophy should not be considered a form of knowledge at all, but as something more like an eccentric younger cousin of knowledge. There will be more to say about the nature of philosophy in the chapters to come.

The first pre-Socratics dwelled in the harbour city of Miletus, now in Türkiye, but then a culturally Greek location under heavy Persian influence. Although nothing remains of ancient Miletus but ruins surrounded by farms, and while the age-old silting of rivers means it no longer sits directly on the sea, it is still an emotional experience for a philosopher to visit the place on a sweltering summer day. Staying with my wife's Turkish family at their small vacation house nearby, it is a pilgrimage I make on a nearly annual basis.

The famous trio of early Milesian thinkers were named Thales, Anaximander and Anaximenes. We know little about them or their mutual interactions. But considering the historical pattern of great Western thinkers appearing in clusters, it is likely that they were related in the manner of teacher and student. Although Thales was something of a celebrity, known as one of the Seven Wise Men of Ancient Greece, we have no complete writings from him or his compatriots. Their views are known only from fragments quoted by later authors. Generally speaking, these Milesians are considered the first Western thinkers to speculate on reality using the power of human thought rather than deferring to such religious poems as the *Theogony* of Hesiod (b. 776 BCE), which recounts the origin of the pagan Greek gods.[10] Already, these important early figures of Miletus staked out the two basic paths that all pre-Socratic thought would follow. The group beginning with Thales and Anaximenes thought that one or more ultimate physical elements must be the source of everything that exists. The other group, pioneered by Anaximander, sought a deeper ground of existence in the *apeiron*, a term referring to the cosmos as a unified whole that surpasses and contains all individual things.

Thales believed that the first principle of everything is water, by which he seems to have literally meant that everything is made of it. The younger Anaximenes later claimed priority for air, which he thought could be condensed or rarefied to produce all other materials: blood, wood, metal or bone (through condensation) and fire (through rarefaction). Between them came Anaximander, who didn't think that any physical element – be it water, air, or anything else – was fundamental enough to be the root of every-thing. As mentioned, the alternative he introduced was called the *apeiron*. Although this term is usually left in the original Greek rather than translated, the meaning in English is something like 'the unlimited' or 'the indefinite'. It can be visualized conveniently as an amorphous blob, or as a colossal whole deeper than any specific element or thing. This makes Anaximander the first West-ern thinker to offer a form of monism: the idea that everything is ultimately one. This view, of course, has been even more widespread among Asian thinkers.

Unlike later theorists of the *apeiron*, Anaximander seemed to think it would come to exist in the future, rather than in the past or the present. His teaching was that over an extremely long period of time, all the opposites in the cosmos would cancel each other out through the workings of justice, until nothing would be left but the indeterminate *apeiron* itself, devoid of any specific qualities or things. He seemed to regard this as the morally correct outcome, given his view of opposites as unjust excesses that must eventually pay a penalty. Since we have only a fragment of his work, and little biographical information, it's impossible to know what led Anaximander to this idea. But it bears a certain resemblance to the later Marxist theory of history, according to which the ongoing struggle of opposed economic classes will eventually lead to a classless, ideal society. This intellectual overlap is no accident: the young Karl Marx wrote his doctoral dissertation on pre-Socratic thought.[11]

Zeno's Shadow

The pre-Socratics of greatest interest to us in the present book are a fabled pair from Elea, Italy, roughly two hours by car south of Naples. In ancient times this was the location of the so-called Eleatic School, known for its view that change and motion are unreal, meaning literally that they do not happen. At first this may sound absurd: isn't everything on earth constantly changing? But the Eleatics insisted that this is only true if we follow the deceptive lure of the senses. If we listen to our reasoning minds, we will conclude instead that the cosmos is simply one; there are no specific beings, and nothing really moves or changes. The influence of Anaximander on the Eleatics is clear, although various Asian religions and philosophies had independently reached the same conclusion.

Two of the leading Eleatics were the naval officer Melissus and

34

the younger and better-known Parmenides, whose philosophical poem *On Nature* can still be read in fragments today.[12] Adding to the fame of Parmenides were the efforts of a charismatic, tall and handsome disciple, roughly twenty years his junior: Zeno of Elea. Parmenides and Zeno make an appearance in Plato's dialogue *Parmenides*, where these two Eleatics send the young Socrates, another main character, to his only recorded defeat in debate.[13] Their conversation is difficult, but of high philosophical quality; whether it really occurred remains unknown. Zeno is most famous as a creator of tantalizing paradoxes that readers may have encountered in school or elsewhere. By paradox I mean a situation in which we are led in thought to two apparently contradictory conclusions. For example, there is the so-called 'Cretan Liar's Paradox'. If someone says, 'I am lying,' there is an obvious paradox at work: if they are really lying, then they are lying when they say they are lying, which means they are actually telling the truth; if they are telling the truth, this means they are telling the truth about lying, which means they are actually lying. As for Zeno's own paradoxes, they lead us to an impasse where our senses tell us that things move and change while reason tells us they do not. Like the older Parmenides, Zeno urges us to follow reason rather than the senses, even if this leads us to seemingly unreasonable results. Let's consider two of Zeno's famous paradoxes:

Paradox 1: Imagine that you are standing in a room and wish to walk to the door. As easy as this may sound, it turns out to be impossible. The reason is that before reaching the door, you must first reach a point *halfway* to the door, which we will call Point A. Now that you have arrived there, in order to get from Point A to the door itself, you must first reach a point halfway between A and the door, which we will call Point B. But once you reach Point B, you cannot move to the door itself without first reaching a point halfway between B and the door, which we will call Point C. From this it is easy to see that there is no way to reach the door, since there will always be another halfway point between wherever you currently stand

and the door itself. No matter how close you come to the door, you can never reach it, since there will always be infinitely many points between you and the goal.

My guess is that few readers will accept Zeno's attempted proof that moving to the door is impossible. After all, everyone walks to doors many times per day and exits through them without difficulty, which seems like sufficient evidence that it's possible. But Zeno would caution us not to believe our lying eyes, since our powers of reasoning are more trustworthy. At this very moment I am looking at a cruise ship in the port of Long Beach, near Los Angeles. Since it is somewhat distant from where I am sitting, the ship takes up only about two centimetres of my field of vision. But naturally I do not conclude from this that the ship is only two centimetres tall. Instead, from earliest infancy we learn to *infer* the true size of distant objects rather than thinking that all objects are as tiny as they appear. In Zeno's case, he is simply asking us to push an analogous lesson to its logical conclusion. It might *look* as if a person has reached the door and left the room, just as it *looks* like the ship in the port is tiny. But in both cases we should trust reason as a more accurate authority than the senses. Many later philosophers also urge us to trust reason more than the senses; Zeno was simply one of the most extreme in this respect. Let's turn now to another of his famous paradoxes.

Paradox 2: Imagine that the ancient hero Achilles, renowned for his speed, enters a footrace against a typically slow and plodding turtle. We will assume arbitrarily that Achilles is exactly ten times faster than the turtle, though the exact amount is not important. In order to stage a fair race, it is agreed that the turtle will be given a head start of some agreed-upon length. Strangely, it turns out that Achilles will never be able to catch the turtle. To see why, let's imagine that the head start given to the turtle was exactly ten metres. Achilles begins the race by running ten metres ahead to where the turtle started. Since we have assumed that Achilles is exactly ten times faster than the turtle, that means that while Achilles was running the ten metres to catch up, the turtle ran one metre forward. So Achilles is now only one metre behind in the race, and surges forward again to catch up with the turtle at last. But during that very surge, the turtle (which is one tenth as fast as Achilles) moved forward an additional ten centimetres. Achilles is getting closer, but has not yet caught up. As he rushes forward another ten centimetres, the turtle was able to move another centimetre forward. Achilles now moves an extra centimetre himself, but the turtle has advanced by a millimetre, and hence the great Greek warrior is still one millimetre behind. From this it is easy to see that no matter how close Achilles comes to the turtle, he will never quite catch up, let alone pass into the lead, since the turtle will always advance forward ten per cent of whatever distance Achilles has just covered. The turtle cannot be passed.

The easiest way to respond to this paradox, just like the previous one, is to refute it with experience. People pass others in footraces on a regular basis; the same thing happens even while strolling down the pavement. As clever as Zeno's paradoxes may be, they seem to be contradicted by everything that happens in daily life. But once again, this response assumes that sense experience is a reliable guide to reality, which is precisely what Parmenides and Zeno deny. Yes, they will admit, it certainly *looks* as if people are reaching doorways and coming from behind to win footraces, but these are illusions, just as the two-centimetre cruise ship in the port is an illusion. The French philosopher René Descartes (1596–1650) would later make the same point about the sun, which seems to our eyes to be very small in diameter.[14] It takes an additional process of reasoning to convince ourselves that the sun is actually very large and looks small only because it is so far away. Yet we easily forget that we make such corrections to the senses. In earliest childhood we learned how to interpret visual data so as to estimate the actual size and distance of various objects, a process that is now mostly unconscious and automatic. Much like Descartes, the Eleatic thinkers trusted reason far more than the senses. Nonetheless, not all philosophers share this prejudice: the rationalists were opposed in history by the empiricists, who (unlike Descartes and the Eleatics) trust experience more than reason. Later we will meet some of these empiricist philosophers.

Once the time of the pre-Socratics had passed, a serious effort was finally made to refute Zeno. We cannot know if Socrates himself tried this, since he wrote nothing at all, other than some lost poems composed in his jail cell prior to being executed by the city of Athens.[15] We know Socrates primarily through the dialogues of Plato, along with some conversations recorded by the military commander Xenophon, and a biting caricature by the ancient playwright Aristophanes in his comedy *The Clouds*.[16] In Plato's dialogues the major character is Socrates in all cases but one.[17] Plato himself never speaks, though his presence is mentioned a few times. In the dialogues there is no serious effort to refute Zeno, perhaps in part because Plato shared the Eleatic distrust of the sensory world. The task of disputing Zeno's conclusions fell instead to Plato's greatest student, Aristotle. Since motion and change were central topics of Aristotle's philosophy, he was practically forced to demonstrate that Zeno's paradoxes were wrong; otherwise, his own theory of motion would never have got started. As we will see, this confrontation occurred in one of the most important Aristotelian works, the *Physics*.[18]

Introducing Aristotle

Since Socrates wrote nothing, and Plato wrote almost entirely through the mouth of his character Socrates, we can say that Aristotle is the first major Western philosopher to write unequivocally in his own voice. He was initially marginalized in Christian Europe, which preferred the legacy of Plato and his admirer St Augustine. But Aristotle's legacy flourished in the Arab world, where his writings were the subject of especially important commentaries by Ibn Rushd, known as Averroës in the West (we will meet him again later).[19] Through the influence of Averroës's commentaries, and Latin renderings made largely by Arabic-speaking Jewish translators, Aristotle eventually re-entered mainstream European philosophy. In the thirteenth century he became a pivotal influence on the Catholic philosophers Albertus Magnus and St Thomas Aquinas; later, he inspired such important thinkers as Francisco Suárez in what is now Portugal and G. W. Leibniz in Germany.

Although Aristotle is widely admired for his crucial contributions to logic, ethics, biology and the theories of poetry and rhetoric, the heart of his philosophical position can be found in two works in particular: the *Physics* and the *Metaphysics*. As its title indicates, the *Physics* gives us Aristotle's main discussion of physical nature. As for the *Metaphysics*, the traditional story is that the title was coined by one of his editors, probably a person named Andronicus of Rhodes.[20] Not knowing what to call the difficult writings that followed the *Physics* in Aristotle's collected works, the editor chose a disarmingly direct solution: *ta meta ta physika*, meaning literally 'that which comes after the *Physics*'. But due to the multiple meanings of the prefix 'meta-' in Greek, the title was interpreted by some as referring to that which is 'beyond' physics. As a result, metaphysics now refers to the branch of philosophy that deals with the ultimate character of reality, as opposed to philosophical inquiries focused on art, politics, law, language, the mind, nature or some other specific aspect of the world.

In modern Western civilization, physics has practically replaced religion for many as the source of ultimate truth. During the same period metaphysics has generally fallen into low repute, given the modern inclination towards visible and measurable happenings rather than seemingly abstract speculation on difficult philosophical themes. But for the philosophers and theologians of thirteenth-century Europe, Aristotle's *Physics* offered a better treatment of nature than any they had known before, while the *Metaphysics* engaged in broad philosophical speculation about reality as a whole.[21] This even included a discussion of God, who is described by Aristotle as the 'unmoved first mover'.

One of Aristotle's major themes is the meaning of *substance*. In particular, what he calls 'primary substance' is an individual entity that is the bearer of changing qualities. For instance, a person can be called a substance, since they remain the same person no matter what clothes they are wearing, what hairstyle they currently have, and no matter what age they are at any given moment. A horse can be considered a substance that endures from birth through death as one and the same horse, even though it shifts in size and colour as it ages, and also changes its state regularly from running to walking to eating to sleeping. What, then, *would not* count as a substance? One answer is that qualities such as 'white', 'strong' or 'impressive' are not primary substances, since they are not individual things but belong instead to many different things. Another answer is that Aristotle was mostly concerned with substances that exist by nature, and would not have been inclined to treat machines, corporations or sports teams as substances, even though these too are enduring objects that remain one and the same despite their qualities changing over time.

Many scholars have interpreted Aristotle's theory of substance as a reversal of the views of his teacher Plato, who was less interested in individual things than in the qualities that things share in common. Others have argued that Plato and Aristotle are philosophically closer than is usually believed.[22] But whatever the truth of the matter, there is no denying that the key to Aristotle's philosophy is the central role he grants to individual things. This emphasis on substance gave him strong motivation to answer such questions as the following: What counts or does not count as a substance? What is the difference between a substance and its accidental or temporary features? Although Aristotle talks about substance in both the *Physics* and the *Metaphysics*, the term is even more central to the latter work. In fact, the most important difference between these two books has seldom been noticed: the *Physics* deals primarily with the continuous and the *Metaphysics* with the discrete. In the terms of this book, the *Physics* is all about waves and the *Metaphysics* all about stones. We can say that for Aristotle space, time, matter and number do the work of continuity, while substance and qualitative change perform the work of discreteness.

Aristotle on the Continuous

Aristotle develops his idea of continuity through a critical assessment of Anaximander's *apeiron*, which we saw refers to the cosmos as a single indeterminate whole, as a pure oneness without qualities or parts. Aristotle rejects this by noting that nothing is ever just one: does the *apeiron* not even have a size or shape? If it does, these are specific properties, and that means that the *apeiron* isn't just one. Consider the following case. If I say there are five apples and five bananas sitting on the kitchen counter, the number five is the same in both cases, although the apples and bananas differ greatly from each other in qualitative terms. Many apples are round and red while bananas are yellow, curved and oblong. The quantity also tells us nothing about the condition of the fruit: whether it's underripe, ready to eat or already blackened and oversoft. If we turn our attention to a single banana, it's fine to say that the banana is one thing, but only if we recognize that this same unified banana also has hundreds, perhaps thousands, of different qualities. This means that while the banana is one thing, it is also many when viewed from the standpoint of its qualities. This is true of any object.

In the days before subatomic particles were discovered, it was thought that atoms were indivisible: pure units without parts. But in 1897 J. J. Thomson discovered the first subatomic particle; this was the electron, known for its tiny mass and negative electric charge. In 1911 Ernest Rutherford discovered further that atoms are not uniform from one end to the other, but have a tiny nucleus made up of protons, which are much larger than electrons and have a positive charge. In 1932 James Chadwick then discovered the neutron, which is roughly just a proton with a neutral rather than a positive electric charge. Much later, in 1964, it was theorized independently by Murray Gell-Mann and George Zweig that even protons and neutrons are made up of smaller particles, called 'quarks'. Their existence was experimentally demonstrated soon thereafter, in 1968. Hence we no longer say that atoms are the smallest unit of nature, but can say instead that physical matter is made of quarks and electrons.

This bit of recent scientific history would not have bothered Aristotle at all, since he didn't believe in the existence of uncuttable atoms anyway: the word *atom*, in ancient Greek, literally means 'that which cannot be cut'. As we will see, Aristotle believed that matter was continuous, meaning that it could be cut into smaller and smaller pieces as far as we wish, without ever reaching some ultimate, uncuttable unit. If he were alive today, he would probably assume that both quarks and electrons could still be divided into smaller units ad nauseam.

At any rate, Aristotle did not embrace the idea that the *apeiron* could be one in the sense of being purely simple and without parts; as we saw in the example of apples and bananas, nothing exists that does not have multiple qualities. Yet he did accept the idea of the continuum, referring to a continuous stretch of anything that can be divided infinitely downward as far as we please. In his own words: 'the continuous is divisible without limit'.[23] We can see this in the case of the number line, which contains as many numbers as we want it to have: there is no obstacle to putting hundreds or even millions of numbers on the line between the neighbouring integers 1 and 2. But this indicates that every continuum is always multiple, not one, and that seems contradictory. The multiplicity of the continuum already posed difficulties in the eyes of Zeno, since for him it entailed that there are infinitely many points between any two given points, leading him to reject the existence of motion. But for Aristotle the reality of motion had to be protected at all costs, given that his entire philosophy is in some sense designed to explain what motion is.

Be that as it may, Aristotle doesn't completely discount the idea of an unlimited *apeiron*. He does think this concept is sometimes useful, because if we decide that the unlimited doesn't exist, 'many impossibilities result', though this also happens when we say that it *does* exist.[24] If we deny the existence of the unlimited, Aristotle thinks that exactly three impossible consequences follow. First, there will be 'a start and an end of time'. Second, magnitudes, or quantities, will not always 'be divisible into magnitudes' in their own right. And third, 'number will not be unlimited'.[25] Since all of these points are important, let's take them one at a time.

The first purported impossibility is that without the unlimited, there will be 'a start and an end of time'. Aristotle's horror at this idea is somewhat difficult for us moderns to grasp. Both our monotheistic religions and our state-of-the-art Big Bang astrophysics tell us that the universe was created at some point in time, and will perhaps come to an end at some later point. But that is not Aristotle's conception of the world. As he sees it, time has always existed and always will. It is worth noting that he does not also think that space goes on infinitely. In his view, space only extends as far as the outermost sphere of the heavens: namely, the rotating shell where pre-modern physics believed that the stars were located, moving around the earth in unison at a greater distance than the planets.[26]

The second impossibility concerns magnitudes (quantities). Take any number: four, for instance. We can slice four into a pair of twos or into four separate ones, but we don't have to stop there. We can cut up all these ones into ever-tinier fractions, meaning that our original number four can be carved into as many smaller units as we wish. Given this ability to keep dividing things into ever-tinier components without end, something like the unlimited must exist. Without it we would eventually arrive at the ultimate smallest units of mathematics, which of course do not exist. Aristotle's view is that this is true not only of numbers, but also of time, space and matter. For him these are all examples of continua, meaning that they can be split into pieces as small as we wish without ever arriving at a tiniest part.

From this endless division of numbers into smaller parts, we advance to the third impossibility, which concerns the increase of numbers through addition. I was just a small child when my mother taught me how to count to 100; I remember becoming discouraged and starting to cry somewhere in the low seventies, thinking that we would never reach the goal. While for many children 100 symbolizes an immense quantity of anything, it's obviously a paltry sum compared with numbers used for everyday purposes, such as financial transactions, and that's to say nothing of higher mathematics. Indeed, just a few years after my traumatic march to 100, I took childish delight in imagining ever-larger numbers, simply adding zeros and other digits to an initial numeral, or engaging in mental exploits of repeated doubling or tripling. The possibility of doing so shows that the limitless must exist, since otherwise we would eventually arrive at a highest number and be unable to go beyond it. Note that this lack of a largest number cannot be refuted simply by saying 'infinity', as children also frequently do. For one thing, infinity is technically not a number. For another, the mathematician Georg Cantor (1845–1918) eventually proved that there are infinitely many different infinities, all of different sizes.[27]

At first glance Aristotle's points about division and addition might seem to be the same. We can divide any number into as many pieces as we like, or add to any number as much as we like; both procedures have no limit other than personal patience. But Aristotle notes a crucial difference between them.[28] In the case of division the total amount is already contained in the starting point, while in the case of unlimited increase it is not. To give an example, when we take the number four and begin cutting it indefinitely into tinier pieces, we never get beyond the four that we already had at the start; we are merely subdividing an already known quantity, no matter how minutely we explore its interior. Yet when we play the reverse game of increasing any number as far as our imagination allows, we will inevitably fail to reach a final, encompassing whole, since there will always be a number larger than whatever number comes to mind. This means that the unlimited cannot be some greatest cosmic whole, since for any given quantity there is always a larger one: there is no infinite container that holds everything. But every finite amount of matter, number, space and time *does* contain an infinite amount of increasingly tiny units inside itself.

Yet by admitting the existence of limitless divisibility, Aristotle seems to have fallen directly into Zeno's trap. We recall Zeno's argument that if space is infinitely divisible (as Aristotle also holds) then motion will be impossible, since we cannot possibly move through an infinite number of points. Aristotle does have a strategy for escaping this predicament, and it is one of the keys not only to his disagreement with Zeno, but to his philosophy as a whole. Namely, Aristotle draws a distinction between the *actual* and the *potential*, aleady familiar to us from everyday language and experience.

The ancient Greek word for potentiality is *dynamis*: the root of our English words 'dynamic', 'dynamo' and 'dynamite'. All these terms refer to the storage of vast reserves of power that are not currently unleashed, but which might be set loose under the right circumstances. Aristotle's word for actuality is *energeia*, the root of our English words 'energy' and 'energetic'. Although today we use 'dynamic' and 'energetic' almost as synonyms, for Aristotle there is a strict distinction between (a) the dynamic, as that which is not currently acting but might do so at some future point, and (b) the energetic, as that which is already acting right now. This is precisely why Aristotle introduced the idea of potentiality into philosophy: in order to make room for realities that exist without being expressed – or at least not fully so – in the present moment.

The classic example is an acorn, which is an actual acorn but also a potential oak tree, meaning that its acorn-qualities right now are not the full story of its reality. Although an acorn is not currently an oak tree, and a kitten is also not currently an oak tree, the status of 'not being an oak tree' is obviously different in the two cases. We can and do think of the acorn as a repository of forces not currently expressed that might someday shape it into a lofty, sturdy oak. But a kitten will presumably never grow into an oak tree without horrific genetic manipulations that lie beyond the scope of present-day science. A similar point holds for a human infant who is capable of very little at present, but might someday lead a large bank or artificial intelligence laboratory, or perhaps a street protest against these same institutions. There is also the infamous San Andreas fault, not far from where I write these words in the Los Angeles metropolitan area. Although quiet at this moment in 2025, the fault is likely to be the site of a major earthquake at some point in the next fifty years, presumably causing widespread damage.

Ultimately, what Aristotle is after with his discussion of the actual and the potential is an account of motion, which he famously defines as the activation of what is potential *insofar as it is potential*.[29] As head-scratching as this may sound, it's really rather simple. When an acorn grows into an oak tree, this involves the activation of what used to be *only* potential in the acorn: unlike the kitten, which never really had the potential to be an oak. As another example, consider a piece of bronze sitting on a worker's table in Athens. We can say that it is actually bronze and actually located in Athens rather than somewhere else. There is no way to *actualize* (to make actual) its bronzeness and its presence in Athens, since both of these are already accomplished facts. If we want the bronze to change in some way, we need to draw on its potentiality rather than its actuality. Hence we can say that the bronze is potentially a sword and potentially located in Miletus, and if we have the proper metal-working skill and means of transport we can actualize these potentials. For Aristotle the activation of potential is what motion means. However, we need to keep in mind that 'motion' for him refers to any sort of change, whether it be an acorn growing into an oak or a person becoming tired. It is not like modern physics, where 'motion' refers exclusively to the movement of objects in space. This difference between potential and actual has a good case to be called the centre of Aristotle's entire philosophy. At one point he even describes reality not as unified, but as 'twofold'. In his own words: 'Everything changes from what is potential to what is actively.'[30] Aristotle's fascination with movement was connected to his fascination with nature and its constant ongoing change, whereas his teacher Plato was more drawn to the unchanging and eternal, and hence towards mathematics rather than physics.

On that note we return to Zeno's paradoxes, which Aristotle addresses by way of his actual/potential distinction. Zeno is correct, Aristotle says, that it would be impossible to reach the doorway if we actually had to pass through an infinite number of points in between. Our movement to the door would not be continuous and ultimately successful, but would 'come to a standstill', since we could never even reach the next point on our way to the door, let alone the door itself.[31] There is no 'next' point, after all, since between where we are and even a close nearby point there are infinitely many other points. The reason motion can occur is that the path from our current position to the doorway is not *actually* made of an infinite number of points, but only *potentially* so.[32] Splitting any length of space into numerous separate points is something we can do as a kind of mental exercise, but this does not really govern our motion in space. That means that Zeno's examples fail.

The so-called paradox of the arrow is another of Zeno's most famous, and Aristotle challenges this one as well. Imagine that an archer fires an arrow directly at a target. Obviously, the fired arrow always appears to be moving. But if we consider any particular moment of its flight, the arrow must be in some definition position, which means that it remains motionless in that point. But if it is always motionless at any given moment, how can it also be moving? Predictably enough, this leads Zeno to claim that the arrow does not actually move, and that (once again) all motion is just an illusion created by our senses. Aristotle rightly notes that this argument relies on a questionable assumption. Namely, the arrow paradox 'results from assuming that time is composed of nows'.[33] In other words, Zeno reasons that in any distinct temporal instant, the arrow must occupy some definite spatial position. But Aristotle does not think that time is made of distinct instants any more than he thinks a line is made of indivisible points. Euclid's geometry would later define a point as 'that which has no part', and beyond this, we already know that no matter how closely we place two points together, an unlimited number of additional points can always be placed between them.[34] Aristotle contends that the same holds for time and space as well, not just magnitude.[35] In the case of the arrow paradox, there is no definite number of moments between the time the arrow is fired and the time it strikes the target. No two nows can be contiguous for the same reason that no two points on the number line can be: it is always possible to add another now between any two nows that seem at first to be consecutive.

Aristotle makes some additional remarks about time, in much the same spirit. Since he views time as a continuum, there can be no definite 'now' in which something begins to happen. The moment of falling in love or turning towards the dark side can never be pinpointed exactly, for the simple reason that nothing in time can ever be located in a single instant. He also makes the crucial point that our sense of time is rooted in our observation of change. As he notes, 'when we ourselves do not change in our thought or do not notice that we change, then it does not seem to us that time has passed'.[36] This is important because many philosophers continue to think of time as something like a wind tunnel, as if it were a disembodied force that caused things to change, when it is really just our way of measuring change.

We've seen that the distinction between the potential and the actual lies at the heart of Aristotle's way of thinking. If there were truly an infinite number of halfway points between me and the door, Zeno would be correct: not only could I never reach the doorway, I could never move at all. For there would always be an infinite number of points between my current position and even a nearby point. Aristotle's response was that only in a derivative way can we conceive of space as made up of distinct parts: the continuum is *actually* one, but *potentially* many. From an Aristotelian standpoint, Zeno's paradoxes pose no threat to the existence of motion. By introducing his concept of potentiality, Aristotle has no need to say that the continuum actually has an infinite number of parts. It is enough to say that we can potentially divide the continuum into as many pieces as we please: in our minds, that is. But the world in which motion takes place is continuous, unfolding as it does over undivided stretches of space. If I want to walk to the door, I simply walk to the door; there is no actual infinity of points I must cross in order to do so. The shift from Zeno's to Aristotle's understanding teaches us that motion is real. This means that Aristotle's extensive physics of motion (which includes all forms of change, not just physical movement) was able to get off the ground and take flight. It would become an extremely useful tool not only in the ancient world, but also in early Islamic civilization and then medieval Europe, before being rejected decisively by seventeenth-century physics. But that is a story for another time.

Aristotle on the Discrete

Looking back on the pre-Socratics, we can see that their *apeiron* counts as a continuum even though they never explicitly defined it that way. The *apeiron* like the number line is boundless, extending without limit in every direction, and also consists of no definite number of parts. How, then, can we move from the boundless *apeiron* to the multitude of individual beings that we see all around us? At least two pre-Socratic thinkers – Pythagoras and Anaxagoras – offered origin stories about how the *apeiron* was broken up into countless pieces. In any event, we have seen that there was a basic division of the pre-Socratics into two camps that amounted to an early dispute between the continuous and the discrete. Those who believed in the *apeiron* were adopting a basically continuous model of reality, while those who believed instead in ultimate physical elements (whether it was water, air, atoms or four separate elements) were advocates of a discrete conception of the world. Aristotle is the first person we know of who divided the pre-Socratics into these two factions, a decision surely motivated by his own joint fascination with the continuous (*Physics*) and the discrete (*Metaphysics*).

In Aristotle's *Metaphysics*, potentiality and actuality also continue to play an important role. He is quick to inform us that the question here is no longer Zeno's point about objects being unable to move physically through an actual number of intermediate points.

The main issue in the *Metaphysics* is qualitative change. Namely, if an object were nothing more than its current, actual state, it would never be able to change at all; potentiality is therefore needed as the engine of all transformation. This is not an imaginary concern aimed at a straw man opponent. Aristotle was engaged in dispute at the time with a rival philosophical school, the Megarians, who held that a thing is nothing more than its sum total of current activities. The Megarians were named after their small home city of Megara, which still exists to the west of Athens today. The leader of the school was one Euclides, a former student of Socrates. Aristotle summarizes the Megarian doctrine in the *Metaphysics* as follows: 'someone who is not building is not capable of building, but someone who is building is capable if and when he is building, and similarly in the other cases'.[37] In simpler terms, no one is a house-builder unless they are building a house at this very moment. Aristotle is unconvinced. He complains that according to the Megarian teaching, when someone stops building, 'he will not possess the craft', meaning that builders will no longer be builders once they stop or even while they take a brief lunch break. But this is absurd, because 'if the builder starts again immediately, how will he have learned again how to make use of this craft?'[38] In other words, there is a big difference between a master house-builder who currently happens to be asleep and those of us who are awake and may have tools in our hands but have no idea what to do with them.

Far from being an outmoded way of thinking, the Megarian approach is adopted even by some recent authors. A good example is the previously mentioned French philosopher Bruno Latour. The method he helped develop is called Actor-Network Theory; it maintains that everything is equal to its sum total of actions here and now.[39] As Latour puts it, objects (which he calls 'actors') can only be defined by their actions. There is no other way to define an actor than by asking what other actors it 'modifies, transforms, perturbs or creates'.[40] A thing is simply the effects it has on other things, and nothing more. In this way Latour and his colleagues devised a technique for analysing any situation by focusing on what things do, rather than getting lost in speculation about what their hidden properties might be. As a method in the social sciences, it has admittedly yielded many fruits.

One of these fruits is that Latour is able to deflate theories that try to explain anything in the world by using big, fuzzy abstractions like 'Science', 'Society' or 'Capitalism'. Instead of these terms, the Actor-Network theorist looks more concretely at what every human and non-human actor is doing in any given situation. Along similar lines, the French historian Fernand Braudel (1902–85), much admired by Latour, writes a history of capitalism that does not assume that we understand in advance what this economic system really means. Instead, he looks in detail at the European development of local markets into regional markets, regional markets into national ones, and national markets into international monopolies.[41] When one thing transforms into another, Latour insists that we explain each step, and not just say that the later ones existed in germ in the earlier ones. We cannot say that a cake is 'already there in the recipe', since from raw ingredients to the final cake there are a number of steps that must be properly executed. We could certainly call the ingredients a 'potential' cake, but for Latour the idea of potentiality risks intellectual laziness. He would agree with the Megarians that we should focus solely on actual things and what is needed to turn them into different things. And in the Megarians' case of the house-builder, Latour would surely have agreed that a house-builder is simply whoever is building a house at this very moment.[42]

Aristotle counters such ideas by insisting that if everything were completely actual, then nothing could ever change. At this moment I am actually typing these words on an Apple MacBook Air computer while sitting at my dining table in California. Yet this will not always be the case. In a few hours I may be at the supermarket instead. I frequently travel outside California, and some day might not be living here any more. Before long this book will be finished, and the need to type it will have ended. Aristotle would say that all this is possible only because I have the potential to change my current situation; without such potential,

I would forever be frozen in the act of sitting here typing. In similar fashion, a happy person is potentially sad, and a dusty little pearl-fishing village called Dubai was potentially a wealthy world metropolis, which is exactly what it became. These scenarios are not properly accounted for by either the Megarians or Latour. Some of you may have studied the difference between potential energy and kinetic energy in secondary-school physics. The idea is the same: whenever something occurs in the world, it is because one or more entities were previously in a position to make it happen before it actually did. That's what potential energy is all about. A vase was stationed precariously on a windowsill ten floors above the pavement (potential energy) and thus it was able to fall in a strong wind (kinetic energy); an athletic young girl had abundant energy stored in her muscles (potential) and for this reason was able to climb the tree in front of the school (kinetic).

I'd like to lay special stress on Aristotle's passion for movement and change since, in recent decades, his theory of enduring individual substances (which we will explore shortly) has unfairly earned him the reputation of a boring philosopher of stasis. He warns us against a tendency that is especially widespread today: the frequent proclamation that 'everything is in flux', all is changing constantly, nothing remains the same. Aristotle responds that this statement isn't quite true, since nature contains rest as well as movement. Just as importantly, change in nature is not usually gradual or continuous. Consider a stone being worn away in a river. It does not lose the same number of particles in every second; instead, sometimes nothing changes, but then all of a sudden a significant piece of the stone is broken away by the water. Likewise, when we read Edward Gibbon's great multi-volume work on the decline and fall of Rome, we do not find an equal amount of Roman decline on every page; there are specific moments in Gibbon's history where decay is unusually rapid or severe, and others in which it is even briefly reversed.[43] The renewed embrace of radical flux and flow in our own time has unjustly diminished Aristotle's reputation as an innovator. But he is nothing if not a philosopher of movement.

Substance

Along with his insights into motion, what Aristotle has to say about substance – discrete individual entities – is also of considerable interest. Of the two groups of pre-Socratic thinkers, those who focused on ultimate physical elements (water, air, atoms) were downward reductionists who dissolved everything into its smallest pieces. While it can be very useful to know what something is made of, this can lead us to overlook important features of the original substance. We know that water puts out fire and quenches thirst, but neither of these capacities is found in its smaller components, hydrogen and oxygen. As already seen, Object-Oriented Ontology (OOO) calls the reduction of a thing to its parts 'undermining'.[44] Later in ancient Greek philosophy we find a different sort of reduction prevalent among the Sophists, the enemies of Socrates and Plato. One of the Sophists, Protagoras, is famous for saying that 'man is the measure of all things'. This amounts to the upward reduction that we called 'overmining', since it shifts attention from the inherent constitution of things to the question of how these things affect the human mind. The easiest way to avoid these two extreme methods is to focus on things that are neither the smallest nor the largest components of the cosmos, and this is exactly what Aristotle tries to do. Hence he is often mocked by philosophers as being obsessed with 'mid-sized everyday objects', as if this were proof of a limited and overly commonsensical vision. But we can easily turn this trick against his critics by saying that they themselves are obsessed either with 'tiny-sized physical objects' or 'large-sized cosmic wholes'. Medium-sized things are real, and they have an important place in philosophy.

Like most trailblazers, Aristotle is sometimes inconsistent about his central topic. In the *Metaphysics* he not only asks the question 'What is substance?' at least five times, but gives slightly different answers in each case. In some passages he insists that nothing is more a substance than anything else: the sun is more powerful and durable than a dragonfly, but the dragonfly is just as much a dragon-fly as the sun is the sun, and hence they are equally substances. At other times he speaks of substance in the comparative terms of one thing being 'more' or 'less' substance than another. Later he stresses that whatever is *natural* is somehow more substance than a table or anything else put together by human artifice. It is true that for Aristotle, substance means many things. It is a unity over and beyond the sum of its parts; it is a somewhat inscrutable thing that can never be adequately defined in words; it has the capacity to have opposite qualities at different times; it has abundant reserves of currently unexpressed potential. But perhaps most importantly, the concept of substance is needed in order to ensure that the discreteness of reality is taken just as seriously as its continuous side. Let's examine these aspects of his theory of substance more closely.

According to Aristotle, everyone in his time agrees that earth, water, fire, plants and the sky are substances. So too are 'bodies in general and the things composed of them, both animals and divine beings, and the parts of these. Each of these is said to be a substance because it is not said of an underlying subject but rather the other things are said of it.'[45] The easiest way to understand this point is to think of the grammatical difference between nouns and adjectives. We say 'the car is green', not 'the green is car'; this shows that we recognize the car as the underlying substance and its colour as an accidental quality that can be modified if we wish. While some have criticized Aristotle for assuming that the structure of language is an accurate guide to reality, he is certainly correct that the car appears to us as a more robust thing than its green colour. Normally, we do not have the experience of colours taking on the form of different objects, though we often observe objects changing colour.

Just as we saw in the case of the *apeiron*, an object cannot simply be one, since it will always have multiple qualities, which for philosophers means something like traits, features or characteristics. Yet an object must also be one so as to be able to unify those many qualities. As Aristotle puts it: 'of all things that have several parts and where the totality of them is not like a heap, but the whole is something beyond the parts, there is some cause of it'.[46] By way of example, he notes that a 'syllable is not its phonetic elements, BA is not the same as B and A, nor is flesh fire and earth. For when they – for example, the flesh and the syllable – are dissolved they no longer exist, whereas the phonetic elements do exist, and so do the fire and the earth.'[47] Although we no longer share the ancient Greek view that fire and earth are indestructible physical elements, we can still take his point that the whole is something more than its sum total of parts.

Despite its unity, a substance must always be composite as well: not only in the qualities it possesses, but also in the parts of which it is formed. In this way Aristotle brushes against the later concept of emergence.[48] Although a substance has parts, it is not reducible to them, since otherwise it would be a mere collection rather than a unified substance. If someone tried to decree the existence of a new substance composed jointly of France, red pepper and the music of Mozart, we would quickly conclude that these three random objects do not form a unified substance with emergent properties of its own. Yet things are rather different if we list water, tomato paste, kidney beans and spices: in this case we should probably say that they are unified in the new substance 'chili'. Although Aristotle would likely not recognize chili as a substance given that we do not find it existing in nature, I am inclined to do so: after all, the chili has a certain something not found in its ingredients taken in isolation. Even so, I would also say that each of the ingredients in chili should also be treated as substances, since each of them has an independent reality not entirely transformed by the chili as a whole.

But here we have touched on an interesting problem. Namely, is it really the case that the various ingredients of a whole retain their independence while simultaneously belonging to the larger substance of which they are ingredients? For example, if a berry bush is a substance, then is each of the individual berries also a substance? In one passage Aristotle does say that the parts of animals and plants are substances just like animals and plants themselves.[49] But a more typically Aristotelian answer comes earlier in the *Metaphysics*, when he says 'it is only when the whole has been dissolved' that the parts can exist as substances in their own right.[50] In terms of our example, this means that individual berries are not substances unless they are picked from the bush. Until that occurs, they are only 'potentially' substances, since they remain subordinated to the unity of the plant as a whole, which would be considered the only substance on the scene until berry-picking happens.

Quite aside from the case of berries, it is easy to see why Aristotle might think that a liver or kidney should be treated as part of a unified human substance, and might also think that these organs only become independent in cases where overall bodily function breaks down. The best example of this was unknown in his time but familiar in our own: a healthy organ is removed from a deceased donor and shipped elsewhere to be transplanted into a needy recipient. During its period of transit from one body to another, the organ could certainly be viewed as an independent substance. Aristotle would make the same point for the parts of plants, given his view that the status of 'substance' is always temporary and contingent. Here he is different from his distant heir G. W. Leibniz, who claimed all substances existed from the dawn of time, with no new ones created later.

As mentioned, Aristotle has a marked tendency to favour nature as the ultimate authority on what counts as a substance and what does not. Berries may exist as substances once they are picked, but a raspberry pie could never count for Aristotle as a substance, given his general philosophical suspicion of artificially produced things. In our own time we are less likely to share this prejudice in favour of nature. Today's world is brimming with artificially produced objects that nonetheless behave like substances: chemical molecules synthesized in laboratories and used in medical drugs, important 'transuranian' elements such as plutonium that are rarely found in nature, along with aeroplanes, cars, drones, robots and medical prostheses, not to mention the European Union and the International Olympic Committee. The fact that complex machines do not occur naturally in forests does not mean that coffee grinders are less deserving of the name 'substance' than strawberries. The point is especially convincing once we recall the centuries of careful breeding and fattening needed to produce the strawberries we purchase in supermarkets today: which are plump and bright red, but often less flavourful than their smaller and wilder cousins.

In our own era, when natural science is usually taken to be the gold standard of knowledge, it might seem antiquated when Aristotle praises philosophy as the superior discipline. The reason he gives (despite being himself one of the greatest scientists of ancient Greece) is that only philosophy is concerned with substance. I take his point to be as follows. Given that substance is what endures despite all changes in its qualities, and given as well that we cannot define any substance in words, there is something elusive about it that makes it difficult to measure adequately or describe in exact prose language. That makes substance an excellent match for *philosophia*, the love of a wisdom that can never fully be had, rather than a clear and direct possession of wisdom. The natural sciences will always favour that which is more tangible and directly discernible, yet substance itself is neither of these things; it is something deeper than all its visible or testable features. The adoration of precise quantitative measurement in the sciences is one of the reasons modern thought dismisses Aristotelian substance as outmoded, or at least as not very useful. But such dismissal is easily reversed once we recall that knowledge is not the only form of human cognition. That is to say, exact knowledge is always supplemented in daily life with *philosophia*, aesthetic sensibility, practical know-how, political instincts and personal empathy, along with other sorts of vague intuitions and partly ineffable skills. The fact that some things exist mostly in shadow does not make them less real than those that sit in direct sunlight.

None of this means that we need to accept Aristotle's own hierarchy placing philosophy above science. I for one do not agree, just as I do not agree that a globe of the earth is 'better' than a street map of London; obviously, which of the two is better depends on what you are trying to do. But present-day society is so enamoured of the exactitude, the concrete achievements and even the monetizable industrial results of natural science that philosophy has reached something close to the historical low point of its prestige. This can be seen from the fact that so many present-day philosophers try to be recognized as 'scientific' in their thinking, while few scientists aspire any more to be called philosophers, a significant change from previous centuries.

Despite certain undeniably dated aspects of Aristotle's theory of substance, it is well worth reviving due to its many powerful features. Some philosophies treat substances (or discrete objects more generally) as nothing more than arbitrary chunks torn out of a primal whole by the human mind, as if the world itself were really just a continuum.[51] But this position contradicts itself by assuming in advance that the mind is independent enough from the whole to be able to perform the work of creating individual things from whole cloth. As we saw, the pre-Socratic philosopher Anaxagoras argued that reality was initially an *apeiron* but was broken into individual pieces when a powerful mind caused it to rotate, vibrate and shatter. Those who uphold such views – whether then or now – need to explain how, in a supposedly continuous cosmos, the mind was different enough from this unity to be able to produce discrete objects in the first place.

When it comes to substance, Aristotle is not shy in criticizing Plato, despite deep personal admiration for his former teacher. Plato's philosophy supposes a realm of perfect forms; scholars still dispute whether this realm was located in a different world from our own. In this higher realm we find the ideal versions of everything, such as 'blue' or 'justice', which exist in the sensory world only in the form of pale copies: blues that are not quite pure blue, or justice mixed with bits of nepotism and corruption. Aristotle rejects Plato's perfect forms or qualities, which he instead takes to be derivative of individual things. Instead of all cats on earth being shadows of the perfect form of cat-hood (as with Plato), Aristotle thinks we simply observe many imperfect cats and form the idea of 'cat' from everyday experience. Plato contends that we are born with the perfect forms already in our minds, and merely need to be reminded of them; for Aristotle our minds are blank at birth and we learn everything through the senses. As Aristotle puts it, each substance is one, and 'what is one cannot be in many places at the same time, but what is common does belong in many places at once'.[52] Restated more simply, Plato's forms are like adjectives that can apply to many different things at once: 'blue' to all the blue things or 'just' to all the just rulers in the world, even if none of them is a perfect example of these qualities. But a substance (an individual thing) is unique, and often has different qualities at different times, such as 'awake' or 'asleep', 'happy' or 'angry', 'young' or 'old'. For Aristotle, this makes substance better suited than Platonic forms to be the underlying stuff of reality.

The Problem of Interface

To summarize, Aristotle sees time, space, motion and number as continuous, but substance and qualitative change as discontinuous. This leaves us with the question of how these two domains could ever interact; otherwise, they would represent two separate and parallel worlds, each with its own set of rules. In a world of perfect continuity, individual substances would only exist as delusions of the senses, which is precisely what Parmenides and Zeno thought. In a world of perfect discreteness, nothing could interact with anything else: everything would remain isolated in its unique individuality. This is precisely why Leibniz in the early eighteenth century said that monads (his name for substances) have no windows, and why he called upon God's pre-established harmony as the only possible solution for how monads could communicate.[53]

If two things are continuous, this means they blend into one; if two things are separate, this means they do not touch at all. What interests us as much as Aristotle is the intermediate case where two objects touch without blending together. His term for this middle case is 'contact', an idea he never sufficiently develops. The problem is knowing exactly where such contact occurs. Common sense obviously sees no problem here: I am a substance, my friend is a substance, and it is easy for us to make contact simply by shaking hands. But difficulties soon arise. For one thing, I am not the same as my hand and my friend is not the same as his hand, since in both cases the hand is merely part of each of us. At best we could say that shaking hands is merely a contact of hands, not of people, and then we would need to know how a part of me (like my hand) can transmit its actions and capacities to me as a unified human being.

But beyond that, we cannot even say that two hands touch while shaking. After all, the hands do not meet in some sort of obscene total-surface contact, but only at a limited number of points along the outer layer of skin, so that we now need to know how those limited points on the skin are able to affect the hand as a whole, and so forth. In short, any contact between two separate things is more mysterious than the physical assumptions of everyday life would lead us to believe. Descartes faced an analogous problem when trying to determine where his two finite substances (mind and body) could possibly meet. His theory was that the soul is contained in the pineal gland of the brain, whose function was unknown in his time, though today we know it produces melatonin.[54] Although Descartes' solution was wrong, he was perfectly right to ask how two totally different kinds of things could possibly meet. What, in the end, is contact? That's what the problem of interface is about.

3.

Dynasty: From Ibn Khaldun to Mongolia

There is an age-old question posed repeatedly by historians and philosophers, one that goes something like this: is history made primarily of great events and sudden turning points, or does it consist instead of a gradual collective drift in which no specific moment is decisive? The relevance of this question to the present book should be obvious. Is history fundamentally continuous, or basically discrete? The gradualist approach to history often gives the impression of being more democratic and inclusive (more 'Left'), since it focuses on the dynamics of regular people rather than the heroic deeds of aristocrats and geniuses.

But caution is needed, since the supposed political alignments of these two ways of viewing history are easily reversible. Yes, the strategy of writing gradualist micro-histories of everyday life might seem more favourable to the cause of the masses: histories of square dancing in the American West, or the plight of low-level soldiers in Napoleon's army. Here the grind of everyday practices is treated as no less important than great events and the demigods who trigger them. Nonetheless, the political Left also pays homage to a more discrete model of history, given its commitment to revolutionary discourse and to hero-worshipping biographies of such figures as Lenin or Mao. In analogous fashion, postwar French philosophy introduced the notion of 'epistemological ruptures', in which previous long-standing obstacles to thought are overcome or abandoned with relative suddenness: think of the French Revolution's overthrow of the monarchy, or the invention of the steam engine. This idea of immediate and irreversible change is usually credited to the philosopher Gaston Bachelard (1884–1962).[1] Later there came the closely related idea of the 'epistemological break', found in the works of the hardcore communist Louis Althusser (1918–90) and the more moderate Leftist Michel Foucault (1926–84), the recent king of the social sciences.[2] This is best seen in Foucault's view that certain familiar institutions – insane asylums, prisons, medical clinics – were produced by a new link between knowledge and the desire for social control, one that he argues belongs to the modern world alone.[3] As Foucault sees it, the important things in history do not happen gradually.

We have long been accustomed to living through events that are said to mark a radical break with everything that came before: the French Revolution, the scientific revolution, the sexual revolution, the computer revolution and other such formulations. The Paris Student Revolution of May 1968 (the very month of my birth) is sometimes described as the 'dawn of postmodernity', however vague this phrase may seem. Bruno Latour claimed instead that 1989 was the most decisive year of recent history, due to the end of the Cold War and the first major climate conference also held in that year.[4] The 11 September 2001 terrorist attacks might have seemed at the time like another watershed event, though with the passage of more than two decades, debate continues as to whether the attacks transformed us deeply or only in transient fashion. The COVID pandemic that began in late 2019, another candidate for revolutionary transformation, brought unprecedented disruption to the daily lives of those who endured it.[5] Even more recently, 2022 was a breakthrough year for public access to artificial intelligence software with the first public availability of ChatGPT. There is also the lingering spectre of global warming, which could become the transformational catastrophe that nuclear weapons once promised to be, and still may be.[6] Again and again we are asked to imagine total upheaval in how the human race dwells on this planet. Are all these so-called revolutions nothing more than media hype, or is contemporary life more punctuated by revolutionary events than in the past? The question is important, but difficult to answer.

It may be easier to consider the question of sudden ruptures by looking to the past, to historical events on which we now have a longer perspective. For this reason I propose that we begin with the Arab Middle Ages, more specifically with *The Muqaddimah* of the famed historian Ibn Khaldun (1332–1406), a book that offers a detailed theory of intermittent and inevitable change in human affairs.[7] One need only browse the pages of this lengthy masterpiece to see why it was described by the British historian Arnold Toynbee (1889–1975) as 'undoubtedly the greatest work of its kind that has ever yet been created by any mind in any time or place'.[8] In fact, Ibn Khaldun deals with the question of the continuous and the discrete in two separate and equally important ways. First, he argues that history is discrete in the sense that it is structured by the differing positions of successive generations. Second, he sees the conflict between the continuous and the discrete as embodied in a rivalry between (a) the dwellers of walled cities separated cleanly from the surrounding landscape, and (b) the nomadic inhabitants of continuous wastelands who destroy or conquer these cities before later becoming urbanized in their own right. Although Ibn Khaldun's focus is on the disruptive role of the desert-dwelling Bedouins and Berbers, his theory effectively has universal scope in explaining the dynamics of human history.

Generations and Corruptions

One of the most picturesque scenes of modern European history dates to October 1806. The German philosopher G. W. F. Hegel (1770–1831) was completing his first great book, the *Phenomenology of Spirit*, at the very moment when Napoleon was attacking his home city of Jena.[9] What makes this crossing of paths so fascinating is that the French commander embodied a key facet of Hegel's philosophy: Napoleon represented 'the world-spirit on horseback' for Hegel, as we read in a famous letter to his friend F. I. Niethammer.[10] In Hegel's eyes, Napoleon was not just a selfish conqueror pursuing individual interests, even if vainglory was certainly part of the picture. By pushing the ideals of the French Revolution beyond France itself – including the idea that careers should be 'open to talent' rather than handed automatically to the aristocracy – Napoleon embodied an advance of civilization for Hegel, despite the often horrific violence of his campaigns.

Unfortunately, the still-unknown Hegel and the well-established Napoleon never actually met. But there is an earlier scene from history that did involve a personal meeting between thinker and conqueror, and it happened in Damascus in 1401. At that time the city was under siege by the ravaging commander Timur the Lame (or Tamerlane), a Muslim born to a Turkic father and Mongol mother. His capital was the city of Samarkand, located in what is now the nation of Uzbekistan. For all his destructive assaults, which spread in every direction of the compass, Timur also had a passion for intellectual debate: the historian René Grousset speaks with dark humour of 'his dual aspect as sophistical *littérateur* and mass murderer'.[11] Given his intellectual interests, Timur could hardly resist the chance to meet a renowned scholar trapped within the walls of Damascus. The captive scholar in question was none other than Ibn Khaldun.[12] The meeting seems to have gone well, at least by the bloody standards of Timur. He did release some prisoners afterwards as a goodwill gesture, though Damascus was not spared the atrocities that (as we will see) Timur so often inflicted on conquered cities.

Just as Napoleon's actions expressed an important aspect of Hegel's philosophy, the career of Timur embodied some of the central ideas of Ibn Khaldun's theory of history. Although Timur was an urban commander who led the army of an ostensibly civilized empire, he and that army matched or even surpassed the murderous record of the nomadic hordes (flowing mainly from Mongolia) who preceded them in history. Timur's unusual status as a civilian who annihilated cities in the manner of nomadic conquerors makes him the perfect test case for the theories of Ibn Khaldun, who laid so much stress on the eternal tension between sedentary and nomadic peoples. For thousands of years, cities lived in terror of being laid waste by pastoral hordes sweeping in from the vast grasslands of Eurasia.

But let's begin with the theory of generations. In the eyes of Ibn Khaldun, all natural and historical existence has a cyclical character; the rise and fall of human destinies is no different from the births and deaths of animals or the changing of the seasons. 'The world of the elements and all it contains comes into being and decays,' he writes, and in similar fashion, 'human prestige comes into being and decays inevitably'.[13] He was certainly not the first to notice the cyclical rise and fall of human power, something that must have been clear already to our Stone Age ancestors. Ibn Khaldun's originality lies, instead, in his close examination of how the internal logic of human generations governs this process. For instance, a ruling family does not rise to greatness and fade away gradually like sunlight at dusk. Instead, it jumps from parent to child and child to grandchild; each generation in the series has a different internal character. The process generally runs as follows. An initial leader, toughened by hardships and marked by relatively pure moral standards, establishes a new regime. But with each succeeding generation the dynasty's heirs sink deeper into luxury and perversion, until by the fourth generation the dynasty is doomed to collapse. Ibn Khaldun shares numerous stories about great empires rotted by the internal decay of spoiled and lazy heirs. For the most part, he conceives of this cycle of inheritance as lasting four (sometimes five) generations: 'The builder of the family's glory knows what it costs him to do the work, and he keeps the qualities that created his glory and made it last. The son who comes after him had personal contact with his father and learned these things from him.'[14]

So far so good, though Ibn Khaldun thinks the son is necessarily inferior, since his knowledge comes solely from study rather than from rising to power through struggle like his father. Still, it is only with the third generation that things really start to go badly: 'The third generation must be content with imitation and, in particular, with reliance upon tradition.'[15] A few leading members of the third generation may have vague childhood memories of the dynasty's first conquests, but for the most part they learn from the oral advice of their elders and from recorded history. It is with the fourth generation that the family's ultimate decay sets in, since this generation is 'inferior to the preceding ones in every respect'. More specifically, its members have 'lost the qualities that preserved the edifice of glory'.[16] With their spoiled sense of entitlement, the now distant heirs of the founder take their political success and their people's obedience as something existing by nature, rather than earned through difficult labour. Eventually the people rebel; power is transferred to a different family line, generally 'in another group of the same descent'.[17] In most cases, then, it is simply a matter of passing from the main branch of a family or group to a parallel one. Consider the case of France, where the Bourbon dynasty led the country from 1589 until King Louis XVI was guillotined in 1793 during the French Revolution. After the fall of Napoleon, the Bourbons were restored to power from 1815 to 1830. Yet their prestige had fallen so far that they were toppled again in a lesser revolution and replaced by their cousins, who founded the Orléans dynasty.

Ibn Khaldun sees this as a common historical pattern. Later we will turn to the Mongol Empire founded by Jenghiz Khan (often spelled Genghis Khan or Chinggis Khan), who had foreseen the fate of his grandchildren in terms similar to those of the Arab historian: 'After us, the people of our race will wear garments of gold; they will eat sweet, greasy food, ride splendid steeds, and hold in their arms the loveliest of women, and they will forget that they owe these things to us.'[18] This sounds like something Ibn Khaldun might easily have said himself. If Jenghiz Khan had been blessed with literary talent, he might well have written an early treatise similar in spirit to *The Muqadimah*.

The same phenomenon can be seen in billionaire families today, as the stern workaholic founders of business empires are followed in quick succession by scandal-seeking playboys and sex-tape starlets: Dodi Fayed and Paris Hilton come quickly to mind. For Ibn Khaldun, the key to any empire is a feeling of group loyalty; once this is gone, younger generations tend to care only about themselves and their pleasures. In this way the game is lost. Especially poignant is Ibn Khaldun's account of how the early Islamic Empire degenerated before falling to a series of Muslim but non-Arab peoples.[19] Islam was introduced and spread by Arabs, but through what Ibn Khaldun saw as their growing decadence they eventually lost power to such outside groups as the Turks and the Berbers, each of them taking up the torch of the faith in turn.[20] Later the declining Seljuq Turks (who ruled before the Ottomans) gave way to the Mamelukes, a social caste of military slaves; the Mamelukes in their own degenerate phase were massacred in Cairo by Muhammad Ali in 1811.[21] The religion and language of the Arabs endure in importance to this day, but at the time of writing in 2025, only three of the ten countries with the most Muslim residents are predominantly Arab in ethnic terms: Egypt, Algeria and Iraq.

Ibn Khaldun does allow for exceptions to the four-generation cycle, though it never varies by much. Sometimes a ruling house is destroyed faster than this, and at other times 'it may continue unto the fifth and sixth generation, though in a state of decline and decay'.[22] The number four is not chosen arbitrarily, but refers to four specific stations that each human generation occupies in turn:

1. The builder.
2. The one who has personal contact with the builder.
3. The one who relies on tradition.
4. The destroyer.[23]

The builder creates an empire against all odds, succeeding through unusual personal talent and ambition. The next generation remembers this struggle from their youth and is inspired by the direct example of their elders. Then comes the third generation, supported by cultural memories of conquest rather than personal experience. Finally there are those who indulge in the luxuries of the dynasty while having no ability to maintain its greatness. The underlying idea of this schema is that education without direct personal effort is inevitably tragic.[24] The discoverer of a truth, the founder of a power, must be in sufficient contact with reality to attain greatness in the first place. But as this person passes their hard-earned lessons to the young, personal experience is replaced by a kind of student hearsay, and this only grows worse over time.

The Spectre of Decay

But what exactly does Ibn Khaldun mean by a 'generation'? At first he is imprecise, treating it as a self-explanatory matter of parents having children, then grandchildren, then great-grandchildren, and so forth. But since not all members of a generation have children at the same age, a more exact accounting of generations is needed. Eventually, Ibn Khaldun tries to provide a definite mathematical answer as to how long it lasts. Arguing partly on the basis of holy scripture, he concludes that 'forty years is the shortest period in which one generation can disappear and a new generation can arise'.[25] In the twentieth-century School of Madrid lineage of José Ortega y Gasset (1883–1955) and Julián Marías (1914–2005), we have later philosophers who take Ibn Khaldun's theory seriously indeed. Their own tendency was to insist that a generation lasts exactly fifteen years, for reasons that need not concern us here.[26]

A prime modern example of a dynasty is of course the British Royalty; it sheds light on the problems with a rigid model of generational shifts. Queen Elizabeth II was on the throne for seventy years, while her son King Charles III is already old enough that a comparable reign in his case lies beyond the realm of mortal probability. It would beggar belief to say that all British citizens born during Elizabeth's lengthy reign from 1952 to 2022 belong to a unified 'Elizabethan Generation'. In a case like the United States, whose elected monarchs (the Presidents) rule for only a few years at a time, generations cannot possibly be defined by who was in office at the moment of one's birth: never once have I considered myself a member of the 'Lyndon B. Johnson Generation'. In modern mass democracies in particular, a generation is something more ambient, referring to the pace of cultural shifts: including technological breakthroughs and even athletic and musical trends, not to mention television series familiar to one generation but not the next. For our purposes, the exact length of a human generation is not so important. It is enough to be aware of Ibn Khaldun's basic idea that human civilization moves by steady but intermittent generational drumbeats rather than along a slow-moving slope. This great Arab historian is primarily a thinker of the discrete rather than the continuous.

Ibn Khaldun often stresses the hardships faced by the first, conquering generation of a dynasty. For instance, Muslims revere the pious purity of the Prophet Muhammad and his Companions. But the reader of Ibn Khaldun is likely to be shocked by his depiction of their poverty. He reports that these earlier Arabs were sometimes reduced to eating such repellent dishes as scorpions, beetles or crushed camel hair.[27] Slightly later in Islamic history, even the mighty Caliph Umar would repeatedly mend his garments with leather patches, a stark contrast with his string of victories over wealthy and powerful kingdoms.[28] The early Muslims were also largely illiterate, which explains why the Qur'an initially took the form of an oral recitation learned by heart.[29] But however awe-inspiring the piety and military prowess of these early generations, when we think of the golden age of Islam we probably think of the literate and cosmopolitan Baghdad of *The Book of the Thousand and One Nights*, of the breakthroughs in optics and algebra that came with a more developed civilization, and of the sublime Islamic architecture that required economic investment well beyond the means of the earliest Muslims.[30] All this indicates that there are multiple generational cycles at work within the same civilization, which may be in political decline at the same moment that it achieves artistic or intellectual excellence.

We have seen that one of the basic mechanisms Ibn Khaldun has in mind for the inevitable decay of dynasties is increasing generational distance from the simplicity and enterprising spirit of the founders. As mentioned, degeneration occurs whenever group feeling declines, and there are a number of ways this can happen. One of the most blameworthy is the increased luxury in which each succeeding generation is raised, a point made repeatedly throughout *The Muqaddimah*. For instance: 'Luxury wears out royal authority and overthrows it.'[31] Or again: 'The new generations grow up in comfort and the ease of luxury and tranquility. They forget the customs of desert life that enabled them to achieve authority, such as great energy, the habit of rapacity, and the ability to travel in the wilderness and find one's way in waste regions.'[32] Mincing no words, Ibn Khaldun concludes that the dynasty's heirs eventually turn into 'urban weaklings' who are incapable of taking care of themselves.[33] They are now 'cowards and lazy fellows' who have 'shed the characteristics of courage and manliness' and are entirely preoccupied with seeking power through 'assiduous competition for leadership'.[34] The dynasty's territories become overpopulated, leading to 'the growth of putrefaction and evil moistures' in the most crowded places.[35] Here he bemoans the frequent plagues in Cairo and Fez, two especially populous Muslim cities of his day.

There are other signs of dynastic decay. One is when rulers become progressively cut off from their people, surrounded instead by crafty ministers, cynical courtiers and opportunistic cousins, all of them having every incentive to prevent the ruler from coming to grips with the true strengths and weaknesses of the empire.[36] With the original vigour of the dynasty having long since passed, the ruler is no longer in a position to use military force to maintain order, and must increasingly resort to buying off enemies and relying on foreign mercenaries rather than citizen armies (which Niccolò Machiavelli in Italy would famously warn against two centuries later in *The Prince*).[37] There is also the dangerous temptation for rulers to operate businesses of their own, which undercuts those who pursue these endeavours without the benefit of royal connections.[38] As the ruling class grows fat on corruption and idleness, the growth of economic injustice in society destroys the morale of the people. Along with higher taxation and unfair competition comes unchecked increase in crime.[39]

Ibn Khaldun expresses a profound fatalism; he sees the cycle of dynastic decay as both inevitable and recurrent throughout human history. As he puts it in an understated gem of a passage: 'Things repeat themselves. One thing contains the clue to another.'[40] For everything that exists, rise is followed inevitably by decline. When he ends his chapters with such phrases as 'God gives success and support', this is only partly a mark of religious fervour. It also expresses his understanding that the outcomes of human effort are always somewhat opaque, obscure and ultimately not in human hands.[41] One of the deceptive things about a dynasty in decline, he says, is that such decay is often concealed by a final burst of activity, much like a terminally ill animal that seems to regain full vigour just before the end of life. As he puts it in a memorable image, a dynasty 'lights up brilliantly just before it is extinguished, like a burning wick the flame of which leaps up brilliantly a moment before it goes out, giving the impression it is just starting to burn, when in fact it is going out'.[42] The downfall of a dynasty is likely to begin in its peripheral regions: 'it begins to crumble at its extremities. The centre remains intact until God permits the destruction of the whole dynasty. Then, the centre is destroyed.'[43] Think of how the downfall of the British Empire was first noticed through unrest and rebellion in the colonies: whether non-violently as in India, or violently as in Kenya. In a different case that unfolded during the final editing of this book, the half-century-long Assad regime in Syria suddenly crumbled, as rebels from the periphery moved on the capital city.

Another possible downward path is through dynastic split, familiar from such cases as the posthumous fragmenting of Alexander the Great's domain, the severing in two of the later Roman Empire or more recent cases following the deaths of mafia or cartel kingpins. In the Mexican state of Sinaloa, such a conflict is now underway between the respective factions of the imprisoned drug lords Joaquín 'El Chapo' Guzmán and Ismael 'El Mayo' Zambada, with bodies littering the streets of Culiacán. To be more precise, Ibn Khaldun recognizes two possible kinds of split. The first, which is generally the less destructive sort, occurs when a governor or some other regional leader consolidates power and eventually takes command of the whole.[44] The classic example (not mentioned by

Ibn Khaldun) is that of Julius Caesar returning from Gaul to over-throw the crumbling Roman Republic, becoming dictator of a new Imperial Rome. The second type of split, potentially far more dangerous, happens when 'someone from a neighbouring nation or group revolts against the dynasty'.[45] An example of this was the initially petty kingdom of the Ottomans in Bursa, one of the weakest among the Turkish polities in Asia Minor. These Ottomans eventually rose and captured Istanbul in 1453, putting an end to the previously durable Byzantine Empire. In this case it was not only a particular group of rulers who were swept away, but all of the Byzantines, who no longer exist as a unified people. Although the Greeks and the Russians both claim to be the true heirs of Byzantium, any continuity is arguably more religious than demographic in character. Another well-known example is how the decay of the Roman Empire eventually led to incursions by Goths, Vandals and Huns, to the point that nothing like a Roman people can be said to exist any more, except in a limited sense referring to the residents of present-day Rome. Whether it happens through decadence or dynastic split, we have seen that Ibn Khaldun's model of history is punctuated rather than gradual. Decline does not occur gently, but in almost stroboscopic fashion.

Nomads Versus Civilians

As mentioned earlier, Ibn Khaldun's theory of history consists of two distinct but interrelated mechanisms, both of them linked to the problem of the continuous and the discrete. So far we have focused on the first: the punctuation of history into discrete generations, each of them marking a stage of decay by contrast with its predecessor. The second element of his theory involves the eternal rivalry between luxurious sedentary civilizations and the primitive but vigorous nomadic outsiders who threaten them. The primary difference between these groups is that nomads are animal herders who live on horseback (or camelback) and sleep in tents, while sedentary peoples reside in cities or on their surrounding farm-land. This conflict did not occur in the Muslim world alone, but dominated Eurasian history from ancient times until the sixteenth century, and we cannot assume it will not re-emerge in modified form. The previous cycle of urban/nomadic conflict is usually said to have ended with the invention of modern artillery, for which the nomadic armies were no match. This enabled Ivan the Terrible (1530–84) to shatter the heirs of the tent-dwelling Golden Horde, a group of Mongols who had dominated Russia for centuries.[46] The primary site of the centuries-long dispute between cities and nomads was the vast steppe: the seemingly endless stretch of unforested grassland running from Mongolia all the way into Eastern Europe, which provided a historic runway for invading Huns, Avars, Turks, Mongols and others. But as a Muslim resident of what is now called the Middle East, Ibn Khaldun was more directly concerned with the desert than the steppe, and especially with those traditional desert nomads who are still with us today: the Bedouins in the East and Berbers in the West.

Ibn Khaldun admired these nomadic peoples for their vitality while remaining firmly opposed to what he describes as their violence and rapacity. He speaks of them in terms that may offend our sensitive modern ears, but which need to be quoted for an insight into his way of thinking. Nomads riding camels, he tells us bluntly, 'are the most savage human beings that exist. Compared with sedentary people, they are on a level with wild, untamable animals and dumb beasts of prey. Such people are the Bedouins. In the West, the nomadic Berbers and the Zanâtah are their counterparts, and in the East, the Kurds, the Turkomans, and the Turks.'[47] But alongside his contempt for the lifestyles of these nomadic peoples, Ibn Khaldun has ungenerous things to say about urban civilians as well, beginning with the remark that 'sedentary peoples are much concerned with all kinds of pleasures'. Their souls, therefore, are 'colored with all kinds of blameworthy and evil qualities'.[48] By contrast, he admired the cultural simplicity of nomadic people, which in his eyes made them less prone to corruption.

In some respects Ibn Khaldun seems like a herald of modern political thought. Traditionally, writers on political philosophy in both the European and Islamic traditions tried to theorize what the 'ideal city' would look like, as seen in such classics as Plato's *Republic* and al-Farabi's *On the Perfect State*.[49] Ibn Khaldun was among the first to avoid such speculation, focusing instead on the ways in which politics actually works in a world of less-than-perfect humans. This makes him a forerunner not only of Machiavelli, but also of Thomas Hobbes (1588–1679), who wrote about the nasty, brutish and short lives of humans in the pre-civilized 'state of nature', thoughts inspired by the ruins of the English Civil War that ravaged the country from 1642 to 1651.[50] But Ibn Khaldun was a city-dwelling Arab who spent his prime years in the advanced civilization of Cairo. This meant he was most concerned with the primitive tribes of the desert as sources of possible turmoil.

As he sees it, the conflict between nomads and sedentary people is a natural one.[51] Throughout history, nomads have generally been the inhabitants of some continuous wasteland, whether the deserts of the Arab world or the seemingly endless stretch of Eurasian steppe. The Vikings are a different sort of example, since the relative isolation of Scandinavia made it the perfect incubator for a different sort of marauder: one that travelled by longboat rather than on horseback. But in world-historical terms, the steppe was the largest preserve of such destroyers. Consider for instance the Asian Seljuqs, who first invaded and Turkified the Islamic world. A more famous example is the Mongol hordes who deluged the Russian grasslands, the Baghdad-based Islamic Caliphate and eventually the whole of China, not to mention their genocidal stampedes through Central Asia and Eastern Iran. As seen earlier, when historical catastrophes of this magnitude occur, entire peoples sometimes disappear from the face of the earth. We know that Chinese civilization survived the Mongol and Manchu periods intact, if not unaltered. But Ibn Khaldun speaks movingly of the pre-Islamic Persian population, who were once rather numerous: 'when the Persians came under the rule of the Arabs and were subjugated, they lasted only a short while and were wiped out as if they had never been'.[52] He is speaking here of the older Zoroastrian rulers of Persia, builders of empires who eventually vanished amidst the Islamic wave that washed across the Middle East, forming the basis for the Shi'a Islamic Iran we know today.

Along with Ibn Khaldun's belief in the inevitable mortality of dynasties, he also speaks of a natural geographical limit. The Roman Empire at its peak, under Emperor Trajan (who ruled 98–117), extended from the Scottish border all the way to present-day Iraq, but the Romans were never able to penetrate far into Germany. Debates persist as to whether this failure was structurally inevitable, as Ibn Khaldun might have said, or whether the German destruction by treachery of General Varus' two legions in the year 9 was a fluke that obstructed Rome's natural growth. But even if we imagine scenarios under which Rome subjugated the German forests, it is hard to imagine its armies crossing vast deserts to conquer Sub-Saharan Africa, despite the surprising fact of Roman explorations as far as present-day Mali, Chad and Lake Victoria. Keeping in mind Gibbon's remark that 'to the Romans the ocean remained an object of terror rather than curiosity', it is still harder to imagine a Roman Columbus or Cortés carrying the Imperial eagle beyond the Atlantic.[53]

One key difference between nomads and city-dwellers can be seen in their varying styles of warfare, a crucial fact of pre-modern history. Civilized empires such as the Romans and Chinese would arrange their armies in battle order and advance in closed formation. By contrast, nomadic powers such as the Jenghiz-Khanite Mongols would approach the enemy stealthily, fire off a lethal flurry of arrows, then rapidly retreat.[54] While this form of warfare appeared to be obsolete throughout the modern period, except in the course of occasional guerrilla conflicts, it seems to be on the rise again today. Despite the ongoing traditional land war in Ukraine, the number of large battles between conventional armies has been decreasing, giving way to battles that often seem pre-modern. In 2023 the mercenary Wagner Group dared the first hostile march on Moscow in over eighty years, and for the better part of a day it looked like a genuine threat to Vladimir Putin's Russian regime. Major set-piece battles are increasingly replaced by a twilight conflict of the special forces of nation states, transnational terrorist outfits and assorted independent mercenaries or well-armed criminal cartels.[55] When in 2011 the United States attacked Osama bin Laden in a small helicopter raid in Pakistan, killing bin Laden himself and a handful of followers, it was considered a bigger military event than many past battles in which tens of thousands were deployed. Large airborne raids with dumb bombs are giving way to drones that harass and target individual soldiers, not to mention the cyberwarfare potentially able to shut off heat and water in distant individual homes.

With their sudden and demoralizing assaults, today's ISIS, Boko Haram, al-Qaeda, al-Shabaab, Mexican cartels and Somali or Indonesian pirates are more reminiscent of the nomads described by Ibn Khaldun than of the massive armies of World War II, whose continued dominance was assumed throughout the twentieth century. Once again, Ibn Khaldun speaks in terms that may offend modern readers, but that help illuminate his vision of history. In his own words: 'savage nations living in the desert' can 'earn their sustenance with their lances and their livelihood by depriving other people of their possessions'.[56] Turning to the Bedouins in particular, he bluntly describes them as 'fully accustomed to savagery and the things that cause it. Savagery has become their character and nature. They enjoy it . . .'[57] This is because they represent 'the negation and antithesis of civilization'.[58] They tear stones from buildings to create bases for their cooking pots, and rip wood from elegant cities to use for tent poles. Yet herein lies the nomadic paradox, another of the chief engines of human history. For as ferocious as he appears to be, 'urbanization is found to be the goal to which the Bedouin aspires'.[59] Ibn Khaldun's evidence is that whenever these nomads conquer farmland or cities, they soon settle down and become civilians in their own right. Let's now turn our attention from the Middle East to East Asia: the homeland of a horde of nomads who plagued civilization for centuries.

Barbarians at the Gates

If we think of cities and their agricultural hinterlands as discrete pockets of culture and education, nomadic armies have generally resented these urban islands, laying waste to their riches in the name of restoring a continuous steppe. No example from history is more famous, or more dreaded, than that of the great Mongol hordes. Shortly before World War II the French historian, curator and Académie Française member Grousset published his massive work *The Empire of the Steppes*. It appeared in English translation in 1970, and remains a widely read source on its topic.[60] Any reader who tackles this book will be much enlightened, though I would recommend having a historical atlas of China and Central Asia at hand. Although Grousset's book does contain helpful maps, there are not nearly enough to follow the narrative clearly at all points, given the many unfamiliar place names that riddle his pages.[61] In Grousset's hands, the same Bedouin desert patterns described by Ibn Khaldun are deftly illustrated through the nomads of Asia.

As mentioned, the Eurasian steppe is a largely treeless grassland extending for thousands of miles from Bulgaria and Romania in the West all the way to Manchuria in the East. There are additional, smaller pockets of steppe in Hungary and Anatolia (the modern Turkish heartland). As I write these words, Russian and Ukrainian armies continue to battle in the same western portion of the steppe that Adolf Hitler once used to invade the Soviet Union in the opposite direction. This sweeping geographical continuum provided the stage for events of such importance that Grousset calls the region 'the paramount fact in human history'.[62] The ancient Greek historian Herodotus writes about the nomadic Scythians, while the Hellenistic author Polybius informs us about the Sarmatians, both groups probably hailing from the region of modern Iran.[63] These tribes were important carriers of Babylonian art forms as far away as China. Yet beginning in the seventh century BCE they also periodically ransacked the Middle East, actions that Grousset calls the beginning of 2,000 years of terror descending from the steppe on the more developed southern empires.[64]

Soon these movements would reverse, running instead from East to West. The Huns and later the Turks drove along different paths towards the Mediterranean world, until this pattern of invasion reached its peak with the Mongols of Jenghiz Khan and his descendants. The Mongols would eventually conquer China and establish the Yüan Dynasty in 1271 under Kublai Khan. A third group alongside the Turkic and Mongolian peoples were the Tungusic nomads. While they are less well known, and never ventured far to the west, under the names of the Jurchids and Manchus they often struck fear into the hearts of the Chinese people. With their famous pig-tailed hairdos, the Manchus even conquered China and founded the Qing Dynasty there, which ran from 1644 to the relatively recent year of 1912.

One of the merits of recent scholarship has been to offer a more balanced perspective on cultures that were once dismissed as primitive or barbaric. While such reservations provide a useful cautionary note, they can sometimes lead to a disingenuous form of neutrality. If the conflicts described in this chapter were occurring today, every reasonable reader of this book would quickly back the farmers, artisans and literate city-dwellers over the savage raiders who repeatedly swept in from the steppe, burning cities to the ground, committing mass rape and slaughtering the men before enslaving the women and children. Whatever the merits of these nomadic cultures may have been, none of us today would view their activities with anything but horror. They would eat raw meat and sacrifice virgins at the funerals of heroes: activities that fully deserve the title 'barbaric', whatever skeletons may be hidden (or not so hidden) in our own civilized closets. Today's world of educated and cosmopolitan liberal democracy, with all its civil rights and civil liberties, would not be possible if we were still at war with these hordes. It is also worth noting that the conflict in question was not a permanent strife between distinct races. Instead, it was an ethnically mutable battle between civilization and those who flattened its cities, exterminated its residents and aspired to return the unforested portion of the world to its former condition as grazing land. Today's barbarian people is often tomorrow's builder of golden cities and wonders of the world, but it doesn't start out that way.

Moreover, the steppe was not populated by mutually peaceable herders who joined in harmony to attack farmers and urbanites; they were invariably cruel to each other as well. They battled for the best land, and conditions of drought or overpopulation would provoke violent invasions of one group's territory by another. Since all of these peoples were nomads, if they felt threatened it was usually enough to move somewhere further out in the steppe, beyond the range of their current enemies. Despite the immense geographical scale of this region, its relative homogeneity made it easy for the nomads to relocate over hundreds or thousands of miles when circumstances required. This meant that conflicts on the steppe could have profound effects on spatially distant kingdoms. For instance, clashes between nomads and Chinese armies often had repercussions for India, Persia, Asia Minor or even Europe at the other edge of the region. In Grousset's words: 'The slightest impulse at one end of the steppe' would inevitably start 'a chain of quite unexpected consequences in all four corners of this immense zone of migrations'.[65]

Despite their often violent histories, many of these nomadic peoples grew tamer and more sedentary over time. The originally savage and shamanistic Turks of the Altai Mountains in Mongolia eventually became bulwarks of urbane Persian civilization. Later these Turks assumed rule over the Islamic Caliphate, and in their Ottoman form would develop a modern army envied by the princes of Europe. More recently, they developed the modern economy and democracy of Atatürk's post-Great War Turkish Republic. Likewise, the Jenghiz-Khanite Mongols began their reign amidst piles of headless corpses. But within a few generations they had produced the highly cultured Kublai Khan, who promoted paper currency and discussed the finer points of religion with Buddhist monks and Papal emissaries from Rome. Along with the internal punctuated rhythms of generational change discussed by Ibn Khaldun, there was also this basic conflict between the discrete walled cities of civilized empires and the mass of illiterate steppe-dwellers firing arrows from horseback. In this respect, the nomads formed a kind of historical *apeiron* (recalling Anaximander's concept), erasing the efforts of discrete civilizations by burning their cities to the ground and leading survivors away to servitude in foreign lands.

Of all the nations that bordered the steppe, China was stationed in more dangerous proximity to the nomads than any other. This threatening dynamic took on permanent physical form in one of humanity's most awe-inspiring monuments: the Great Wall of China, whose earliest sections predate ancient Greece even if much of it dates to the later Ming Dynasty.[66] Dozens of nomadic tribes flow through Chinese history, accompanied by countless battles, negotiations, plots, extortion payments and strategic intermarriages. Indeed, the very location of the Chinese capital often shifted in response to evolving nomadic threats. China's security was tormented not just from the north but also from the near west, since nomads could be found in both directions. We must now pass over many centuries of this conflict in silence, and shift our attention forward to the most famous Mongol period in world history.

By the late 1100s, the centre of gravity of Turkic peoples from Mongolia had moved far to the west, closer to where the nation of Türkiye is found today. Following their departure, the Mongols themselves had emerged from subjection to form a stratified society led by a falcon-assisted steppe hunting aristocracy.[67] But a more impoverished group remained in the wooded hills, led by self-declared sorcerers while hunting 'marten and Siberian squirrel, which they traded'.[68] These hardscrabble Mongols of the hills were the sturdy stock from which great empires would emerge, though there was still a long road ahead.[69] Among them there eventually arose a famed warrior named Yesugei, whom later tradition linked with ancient royalty, though he seems to have been just another clan chieftain engaged in constant warfare with the Tatars (of

Turkic blood), 'hereditary enemies of the Mongols'.[70] This violent adventurer abducted and married the wife of a rival chief, before the Tatars finally poisoned him at a meal in around 1167.[71] But no matter, since Yesugei had already performed his great historical deed: fathering a seemingly unremarkable boy named Temujin. This overlooked youngster would be known to posterity under the ominous title Jenghiz Khan, a name with the probable meaning 'Ruler of the Universe'.

In retelling the astonishing story of this future conqueror and his various heirs, Grousset draws heavily on the writings of the medieval Persian historian Rashid al-Din, along with an anonymous work entitled *The Secret History of the Mongols*.[72] Other scholars have been less trusting of the latter source than Grousset, dismissing it as little more than myths and hearsay. Yet it tells a riveting tale in which the young Temujin is deprived of his inheritance and endures a rough childhood in the company of his brothers and half-brothers, one of whom he kills for stealing his food.[73] After Temujin reaches adulthood and finds moderate political success, there follows a lengthy series of alliances and betrayals; these conflicts are initially local, or regional at most, with no hint of the vast geographical stage he will one day roam. Allied with his uncle Toghrul (known in China as the Wang-Khan), Temujin faces the expected challenges from his ancestral enemies the Tatars, along with less credible struggles with sorcerers, who were said to obstruct his rise with rainstorms and blizzards.[74] Be that as it may, Grousset reports that young Temujin succeeded 'not because he had formed any rigid plan of his future conquests', but because of his strong personality, which enabled him to put the 'perpetual state of guerrilla fighting to his advantage'.[75] Nonetheless, lingering tensions between Temujin and his uncle were exploited by the latter's sons and other ambitious Mongols, and by 1203 the future Jenghiz Khan was driven into exile. Without allies, he was back on his own at roughly thirty-six years old. This proved to be a blessing in disguise: 'having up to then played the part of the Wang-Khan's brilliant second in command, he was now to fight on his own account and for first place'.[76]

Temujin endured numerous reverses, but managed to elude a deadly trap set by the Wang-Khan's armies. He withdrew deep into the Asian continent, north of Manchuria, suffering with his loyal army throughout the summer of 1203 (much like George Washington's winter at Valley Forge half a millennium later).[77] But in a matter of months Temujin's situation had improved dramatically, as the coalition against him fell apart amidst squabbling. By means of deceit and secret marches, he forced a decisive battle and emerged victorious. The Wang-Khan and his son fled westward, where both would meet death in short order. Meanwhile, the people they left behind shifted their loyalty to Temujin himself.

From this point forward we can call him by his more famous name, Jenghiz Khan, since he now controlled all of eastern Mongolia. In the west, his numerous enemies gathered under the banner of the independent Naiman people.[78] But he smashed them in battle in 1203, not far from present-day Kharkhorin (Karakorum), Mongolia, which would become the Mongol empire's traditional capital. Within three or four years, Jenghiz had mopped up the last internal resistance to his rule: 'All Mongolia was now subjugated. Jenghiz Khan's standard – the white standard with its nine flames – was to become the flag of all Turko-Mongols.'[79] As mentioned, however, most of the Turks had passed much further west than Jenghiz Khan's current zone of control. New alliances were concluded and marriages made, and the cities of Eurasia were destined for a season of terror unlike any previously known.

Grousset does his best to avoid any lazy orientalist dismay in the face of this nomad king, often bending over backwards to depict the bright side of the man: 'In the framework of his way of life, his milieu, and his race, Jenghiz Khan appears as a man of a reflective cast of mind and sturdy common sense, remarkably well-balanced and a good listener.'[80] Yet however admirable his listening skills, he was very much 'the exterminator of peoples' that Grousset also calls him.[81] The fact that Jenghiz Khan was naturally courteous and noble in his bearing, or that literate urban societies are guilty of their own atrocities to this day, does not spare us the duty of taking sides against the continent-wide inhumanity of his Mongol hordes. If defeated and captured in our time, Jenghiz Khan's fate would be not that of The Hague, but of the Nuremberg Trials.

Systematic rape, massacre and enslavement are only the first and expected part of the story; the rest is more surprising and horrific. Piles of heads were amassed outside conquered cities, as if in the name of sport. Peasants were captured en masse and forced to take heavy casualties in storming their own nation's cities, as Mongol riders laughed behind them and killed any who tried to flee. With disturbing duplicity, the Mongols would feign departure from an already massacred place, then reappear suddenly to kill any survivors who had crept pathetically from the ruins. City-dwellers were made to bind each other's hands behind their backs before being slain with a tempest of arrows, and urban pets were often put to the sword. The Mongols even ghoulishly violated the tombs of the Seljuq hero Sanjar and the great Baghdad Caliph Harun ar-Rashid. Surrender before battle was sometimes rewarded by the Jenghiz-Khanites with mercy, though often enough a surrendering army was cut down on the spot regardless.

The emir of Diyarbakir (now in south-eastern Türkiye), one al-Kamal Muhammad, met with the following deplorable end: 'Mongols tore off pieces of his flesh and crammed them into his mouth until he died. His head, impaled on a lance, was carried through the great cities of Muslim Syria, from Aleppo to Damascus, preceded by singers and tabor players.'[82] The best political and military case to be made for these outrages was that the early Jenghiz-Khanites remained dwellers of the continuous steppe, and thus were unable to conceive of cities and farms as anything other than threats. As Grousset writes: 'if in eastern Iran Jenghiz Khan destroyed the brilliant urban civilization produced by a Firdausi or an Avicenna' – referring respectively to Persia's great poet and great philosopher – 'it was because he meant to create a sort of no man's land or artificial steppe in the southwestern marches to serve as a glacis or protective barrier to his empire'.[83]

In that era China was split between two separate dynasties. In the north were the Kin (or Qin or Jin), themselves of nomadic Tungusic stock, based in what is now Beijing. In the south was the traditional Chinese civilization of the Song Dynasty: masters of a vast network of cities, farms and rivers. The Kin initially served as a buffer preventing any direct contact between the Mongols and the southern Chinese, and it was against the Kin that Jenghiz Khan first moved. At first the Mongols struggled with the complexities of combat engineering and siege warfare, at which Chinese generals were so skilled.[84] Perhaps frustrated at the slow pace of success, Jenghiz Khan left his subordinates in charge of this northern China campaign and rode westward, where his army had better luck in decimating a series of great empires. They began by capturing land in what is now the western part of China.[85] From there they entered the territory of the once-mighty Khwarazmian Empire, where the Mongols laid waste to many of the great citadels of medieval Islam: such illustrious Central Asian places as Merv, Balkh, Bokhara and Samarkand became scenes of historic atrocity.[86] Often enough everyone in these cities was killed. An exception was skilled craftsmen, who were swiftly deported into the depths of Mongolia, facing lives of servitude to those who had murdered their families and friends. In 1221, the city of Nishapur in what is now north-eastern Iran was horribly punished after the Mongol commander Toquchar fell in combat there.[87] His widow happened to be Jenghiz Khan's daughter, and she saw to it that no living creature in the city breathed again. Afghanistan too was largely depopulated by the forays of Jenghiz Khan's army, and would remain so for years to come.

The conqueror returned to Mongolia in 1225, where he would die two years later. But during his lifetime his generals made gruesome assaults near the Caspian Sea, in present-day Russia, and as far removed as Kyiv, whose ruler was killed along with his entourage just days after surrendering peacefully. Under such later commanders as the famed Batu Khan (d. 1255), the Mongols would soak the fields of Hungary in blood while waging war as far west as Poland, Czechia and Croatia.[88] In 1243 Jenghiz Khan's successors penetrated Asia Minor as far as Konya (now in central Türkiye), reducing the formerly powerful Seljuq Turks to the status of vassals. In 1258 the great city of Baghdad was set aflame by the Mongol commander Hulägu, who exterminated hundreds of thousands of residents. The Caliph himself was trampled to death by horses after first being sewn up in a sack. Many historians say that the destruction of Baghdad was a blow from which Islamic civilization never fully recovered, meaning that it has yet to regain the military and cultural power it possessed during the early Middle Ages.[89] Only when the Egyptian Mamelukes defeated the Mongols in battle in 1260, in Ain Jalud to the north of Jerusalem, did the tide begin to turn against the Asian horde.[90] All of these grisly examples are cited to drive home the main point of our discussion. Namely, Mongol hatred of the discrete civilizations embodied in cities was so intense that it almost deserves to be called metaphysical rather than political. It was the greatest historical effort by the forces of the continuous steppe to crush urban civilization into powder. But we must also remember the equally powerful contrary trend in the hearts of these nomads. For they did not just destroy these cities and farms: after a time, they came to inhabit them, becoming civilized and then corrupted in turn. Here the later history of the eastern Mongols will prove instructive.

Kublai and Timur

We will now consider a strange pair of historical contraries: Kublai Khan and the aforementioned Timur the Lame (or Tamerlane). The first is a perfect example of the civilian aspirations of the barbarian: an heir of the ruthless Mongols, Kublai was nonetheless an enthusiastic admirer of Chinese culture who settled in China like a native dynastic ruler. By contrast, the later Timur was the head of an ostensibly civilized empire in the city of Samarkand, though his sadistic destruction of humans and their cities rivalled or exceeded the depredations of the Mongols themselves. This contrast will lead us to a closing discussion of the 'dialectical' character of reality.

While the Mongols would long enjoy political hegemony in Central Asia and even further to the west, our attention returns to China, which had fought these nomads for centuries. From Jenghiz Khan's death in 1227 to Kublai Khan's assumption of sole authority in 1264, the various battles of succession in the empire were paralleled by ongoing conquests. The Kin Dynasty of northern China had fallen at last to the Mongols in 1234, making the Jenghiz-Khanites neighbours of the southern Song Dynasty for the first time. The efficient Mongol ruler Möngke Khan established his brother Kublai as viceroy of China, a task for which he was well suited due to his intense fascination with Chinese culture. Following Möngke's death, Kublai battled for power with another brother, Ariq-Böga; after securing victory, Kublai kept Ariq-Böga in protective custody for the final two years of his unlucky life. Kublai was now supreme leader of the empire, though he kept his seat of government in Beijing rather than returning to the traditional capital in Mongolia itself.

After nearly two decades of struggle against the Song Chinese in the south, Kublai eventually conquered the whole of China in the name of his Mongol people. But this conquest was a double-edged sword, for as Grousset puts it: 'though Kublai, the offspring of nomads, may have conquered China, he himself had been conquered already by Chinese civilization'.[91] The Mongols had shifted gears. Originally the uncivilized lords of the continuous steppe, they had entered the labyrinth of Chinese urbanism and become subject to what Ibn Khaldun saw as historic law: the dynastic cycle with its phases of ripeness and decline. Beginning as the ultimate outsiders of Chinese society, the Mongols had risen to become the Yüan Dynasty of China, but were destined to rule for less than a century.

Kublai's troubles never quite ended. He failed at expansionist campaigns in Vietnam and as far south as the island of Java (in today's Indonesia), and his invading fleet was famously destroyed by a typhoon off the coast of Japan in 1281. He also continued to face challenges from other branches of his family. Although he came from hardy nomadic stock, Kublai was essentially a man of literate Chinese culture, which he thoroughly absorbed and in some ways even improved. His impressive postal service, with its galloping relays of fresh horses, reached an unprecedented level of quality across Chinese territory. Paper currency was promoted and respected, even if it was managed in inflationary ways. Historians may quibble with Marco Polo's account of his lengthy visit to Kublai's realm, but from his reports we gain the sense of an enlightened and boundlessly curious ruler.[92] Kublai remained savage enough to smother a rival to death under piles of carpets, yet he also hosted tolerant debate between Buddhists, Taoists, Nestorian Christians and Muslims.[93]

But even as Kublai continued to develop Chinese civilization in his own way, his indulgence in urbane luxuries set the stage for the degeneration of his own, less substantive heirs. Jenghiz Khan was right to predict his descendants would one day be softened by the culinary and erotic delights of city life. The elderly Kublai died in 1294. His later heir Toghan Temür rose to power in 1340, and Grousset's words about him are a fine confirmation of Ibn Khaldun's worries about fourth and fifth generations: 'Toghan Temür, a weak, vacillating person, found delight only in the company of his favorites and of Tibetan lamas. Dulled by debauchery, he took no interest in affairs of state and ignored the Chinese national rebellion now rumbling in the south.'[94]

That brewing rebellion marked the renewal of patriotic sentiment in China, which would lead in due course to the founding of the great Ming Dynasty. Ming rule would last from 1368 to 1644, when a new group of steppe outsiders (the Manchus) would establish the final pre-republican, pre-communist dynasty in Chinese history: the Qing. The story of the rise and decline of the Jenghiz-Khanite Mongols offers further confirmation of Ibn Khaldun's thoughts on both the historic role of nomads and the progressive corruption of dynasties. For as we already heard the great Arab scholar say, even though the nomad is 'fully accustomed to savagery and the things that cause it', it turns out that 'urbanization is found to be the goal to which the nomad aspires'.[95] It seems that not even the hardy nomads of the steppe can resist the luxurious and carnal lure of city life, which seduces them as soon as they try to destroy it. In Kublai Khan, we have the emblematic figure of the nomad turned civilized.

In 1336, just forty-two years after Kublai's death, a child was born whose life would embody the opposite movement: from civilian to nomadic pillager. Timur the Lame's relation to Kublai is comparable to that between the two famous canine heroes of the writer Jack London (1876–1916). In London's novel *The Call of the Wild*, a domestic dog tears out the throat of a Canadian and kills several Indigenous people before joining a wolf pack full-time (Timur). In *White Fang* we see the reverse movement, with a wolf-dog becoming domesticated (Kublai).[96]

Timur was born to an aristocratic Central Asian family in what is now Uzbekistan. Though hailing from Shahrisabz, he was closely associated with the nearby city of Samarkand (previously destroyed by the Mongols), where he established his capital and would occasionally pause to recuperate from his ceaseless military campaigns. His early career shows evidence both of great physical courage and of cynical deceit in his dealings with strangers, friends and family alike. Timur's own uncle Hajji Barlas was one of the many unwitting dupes outmanoeuvered by the future tyrant, as his nephew betrayed him and joined with his enemies. Another victim was his early adventure companion and brother-in-law Mir Husain. Their alliance was purposely eroded by Timur under various pretexts, until eventually Mir Husain was besieged in the city of Balkh and forced to surrender. Although Timur allowed him to depart unharmed on pilgrimage to Mecca, Mir Husain was swiftly killed by Timur's men, supposedly (though improbably) without their commander's knowledge.[97]

Equally complex was his religious practice, which somehow combined sincere Muslim piety with a staggering number of Muslim victims, running well into the millions. Sometimes he gave the excuse that those Muslims he killed were excessively tolerant of heretics. That was the case in the Sultanate of Delhi, where Timur would massacre roughly 100,000 Hindu prisoners before making towers of Muslim heads.[98] Clearly, he was an equal opportunity oppressor. Upon capturing the Mongol princess Dilshad Agha, he did not treat her with the expected dignities, but added her unceremoniously to his harem.[99] Grousset captures the spirit of the man perfectly: 'a farsighted Machiavellianism', capable of 'a consistent hypocrisy based on and identified with reasons of state'.[100] All in all, Timur's death toll may have reached as many as 20 million.

Before summarizing his path of destruction in greater detail, it is worth noting some of the typical features of his reign. One was the care he took to present his empire as the legitimate heir of Jenghiz Khan's. For years he did this by hiding in the shadows, pretending that some blood relative or other of Jenghiz was the actual ruler of the kingdom. Yet these public-facing figures were really just Timur's puppets: 'mere shadow kings', Grousset writes, 'poor, obscure men of straw whom no one ever thought or cared about'.[101] Timur kept up this game by not giving himself any definite title of leadership for many years. It was only in 1388 – at the age of fifty-two – that he finally proclaimed himself Sultan. Whereas the Jenghiz-Khanite empire continued to flourish through several successors, Timur's domain was held together by his forceful personality alone, and would fall apart rapidly following his death. He was shrewd in manipulating opponents on the political chessboard, though his physical courage sometimes spurred him to recklessness. For instance, he needlessly risked his life by challenging one Yusuf Sufi of Urgench to hand-to-hand combat (which Yusuf Sufi refused) and would sometimes enter battle on the Russian steppe like a common solider.[102]

There is something unsettling even in the pattern of Timur's military movements. Jenghiz Khan was guilty of sweeping crimes against humanity, but there was an underlying logic to his expeditions. After his abortive assault on northern China he moved westward, destroying the cities of Central Asia to create a wasteland barrier on the south-western flank of his empire. However grotesque his atrocities, his military thinking was sound. Timur's policy was different, and involved repeated marches to annihilate places he had already destroyed in the past, leading him to 'conquer Khwarazm three times, the Ili six or seven times', then 'eastern Persia twice, western Persia at least three times, in addition to waging two campaigns in Russia'.[103] In short, Timur presents the sadistic picture of a man sitting close to a fire but repeatedly leaving to stomp on beetles at the periphery of his campsite before briefly returning to the fire. This is so much the case that Grousset thinks it a waste of time to treat Timur's career in chronological order, and organizes it geographically instead.

It is also worth noting the counterproductive nature of many of Timur's conquests. As mentioned, some of his most ruthless victories came against political pillars of the same Islamic world to which he himself belonged: the Sultanate of Delhi, the Baghdad Caliphate and the early pre-Istanbul Ottoman state. It is true that Timur would have greatly expanded the map of Islam had he succeeded in conquering China, but he died in 1405 before launching this planned campaign, and his successors lacked even an ounce of the energy such an ambitious project would have required.[104] On the whole, Islamic civilization would surely have been better off if Timur had never lived, despite his scattered acts of vengeance against enemies of the religion. Ostensibly a well-read aristocrat of Samarkand, his legacy of annihilation rivals that of the Mongol nomads themselves.

A brief survey of examples will demonstrate the point. After a siege of three months, the inhabitants of the city of Urgench were massacred in 1379.[105] In 1382, lord 'Ali beg of Kelat surrendered during a siege, but was cruelly executed anyway.[106] The next year, rebels in the Persian city of Sebzawar were piled together alive and mixed with mud and bricks to form towers, in which they presumably died of thirst as the materials hardened.[107] In nearby Seistan everyone of all ages was massacred, though in this case only their heads were used to construct towers. Timur also took care to destroy the irrigation network there, and Grousset remarks that the 'desolation that strikes the traveler in these regions even today is the result of these acts of destruction and massacre'.[108] Next in Persia came Esterabad, where in 1384 even nursing infants were killed.[109] In 1386, some bandits accused of robbing the caravan to Mecca were thrown to their deaths from mountaintops.[110] The same fate awaited the residents of Van, now in Türkiye but then controlled by Armenia, whose inhabitants were also tossed from the mountains.[111]

When the residents of Isfahan in Persia rashly murdered Timur's tax collectors, he ordered the extermination of the entire city, resulting in a pile of 70,000 heads.[112] Thereafter the city of Shiraz surrendered in advance, though Timur still massacred most of its ruling family despite their sycophantic submission; like Jenghiz Khan, he forcibly deported the best craftspeople and intellectuals of that city to his own capital.[113] In 1395, Timur sold the many Genoese and Venetian merchants of the Crimean peninsula into slavery, despite the lavish gifts they had sent him.[114] In 1398, he disfigured Delhi by placing pyramids of heads at each of its four corners, before moving in early 1399 to the city of Miraj, whose Hindus were skinned alive.[115] As a sickening comical postscript he seized the Raja of Jammu, forced his conversion from Hinduism to Islam, then tested his sincerity by having him eat beef on the spot.[116]

General massacres were staged in Aleppo and Damascus (where Timur met Ibn Khaldun) between 1400 and 1401, and the next year Baghdad suffered an estimated 90,000 deaths at the hands of the Timurid army: fewer than had been killed by the Mongol conqueror Hulägu, though still a genocidal figure. In the same year, Timur killed many of the residents of present-day Tbilisi, Georgia, and destroyed all the churches in that city.[117] After defeating the Ottomans at Ankara in 1402 (at the site of what is now Ankara's airport) and capturing their Sultan Bayezid, he sent his raiders to burn down the beautiful Ottoman capital of Bursa. Grousset notes that this hampered the spread of Islam by delaying Turkish conquest of Constantinople (now Istanbul) for another half-century.[118] Timur then proceeded to Smyrna (now Izmir, on the Aegean Coast of Türkiye) where he battled then massacred the Christian Knights of Rhodes.[119] This is merely a partial list of Timur's record of destruction, most of it leading to nothing that outlasted him. Although

Samarkand briefly became the glittering capital of Asia, his empire would posthumously crumble. The Timurid heartland fell to the Uzbek people, of Turkic origin, who still live there today.

The Problem of Dialectic

One of the most renowned concepts in Western philosophy is the dialectic. It concerns the interplay of opposite terms, though not in the manner of balancing two complementary poles, as with Yin and Yang in Chinese philosophy.[120] Instead, the dialectic pertains to the way in which any idea (such as 'being') is at first simply negated by its opposite (in this case 'nothing'), before passing away into a higher unity that incorporates both (in this case 'becoming'). While the dialectic dates back to at least as early as Plato's dialogues in ancient Greece, its modern form is typified by the German philosopher Hegel, whom we met earlier as Napoleon attacked his home city of Jena.

Hegel was an heir to the philosophical legacy of Immanuel Kant (1724–1804), the German philosopher who declared that we cannot know reality directly, since humans are finite and therefore only encounter the world in a limited away. We cannot gain access to what he calls the 'thing-in-itself'. All human experience occurs in space and time, and is structured further by what he dubs the twelve categories of understanding, which include cause-and-effect relationships as one of the categories. We have no way of knowing whether these seemingly basic features of reality exist outside the mind; therefore, philosophy must limit itself to discussing the basic underlying features of human experience.

Hegel never cared for the idea of a world-in-itself beyond human access, believing as he did that nothing lies beyond the scope of reason. But he also took a dislike to another aspect of Kant's philosophy: his tendency to treat opposite terms (such as 'thing in itself' and 'thing as appearance') as if they were permanently fixed.[121] Hegel believed instead that when dealing with any pair of seeming opposites, it is often impossible to ascribe particular qualities solely to one or the other, since opposite terms implicate each other in increasingly complex ways. Let's say that we try to distinguish between the obvious opposites 'being' and 'nothing', as Hegel does at the beginning of his difficult book *Science of Logic*.[122] It turns out that we cannot say anything definite about either being or nothing, since that would be too specific, and would prevent them from having the status of pure being or pure nothingness that they were stipulated to have. As a result, we move back and forth mentally between being and nothing in a state of perplexity, which Hegel calls 'becoming'. But every case of becoming is different, and is not just becoming in some vague or neutral sense: it is always a *specific* becoming, which Hegel in turn calls 'determinate being'. And so the dialectic continues, running for hundreds of pages through a series of increasingly complex ideas.

Perhaps the most popular example is Hegel's famous discussion of the master and the slave, found early in his *Phenomenology of Spirit*. An influential account of Hegel's master and slave was given in 1930s Paris by the Russian émigré Alexandre Kojève (1902–68), whose lectures on the topic had extensive influence on the ensuing decades of French philosophy.[123] Two humans struggle with each other for recognition, with the result that one is victorious and becomes the master, while the other is reduced to a slave expected to acknowledge the master without being recognized in turn. At first it looks as if the master has all the power and the slave none at all. Yet the dialectic shows that the situation is more intricate than it looks. For one thing, the master is strangely dependent on recognition from a supposedly inferior person, the slave, which might make us question the supposed mastery of the master. For another, the master depends on the slave's labour, which makes the slave more aware of the demands placed on humans in the struggle for existence, while the master lives in an entitled dreamworld that blinds them to reality. Moreover, the slave comes to grasp how deeply the master depends on recognition for maintaining their superior sense of self. In this way a reversal of roles comes about, and the slave is revealed as having more strength, patience and insight than the pampered and abusive master. In other words, the slave begins to look more masterful than the official master, who in turn seems more weak and dependent than the so-called slave. Needless to say, this aspect of Hegel left a profound influence on the young Karl Marx, whose impact on world events is hardly in need of proof.[124]

Speaking of world events, one problem with the dialectic is the way it glosses over the gap between thought and the world. It is one thing to say that logic itself shows an intricate entanglement between master and slave, and quite another to hold that their positions in the world must soon reverse. More generally, in Hegel's books the dialectic moves from one stage to another as fast as Hegel can describe these movements in prose, although history itself takes time. For instance, American slavery lasted from 1619 to the end of the Civil War in 1865. However insightful the dialectic of master and slave might be, what is the reason this institution lasted for 246 years rather than 46 or 546, or rather than never existing at all? The world does not move at the speed of thought. Perhaps someone who believes in an abstract concept of being will need, say, only four months to realize that being is negated by nothing and that both pass into a higher becoming. But during that brief interval of intellectual failure they may spread their initially primitive philosophy of being via worldwide television broadcast, with significant effects on global thought. Dialectic is less concerned with the fact that one-sided ideas tend to prevail for a certain period, and more concerned with the ultimate philosophical judgement that being and nothing are both too abstract to provide a satisfying picture of reality. If a dialectician can show that monarchy and democracy negate each other as ideas, it is somehow implied that they will negate and subvert each other in the course of history as well. But little or no attention is paid to why this doesn't happen instantly, which means that the undeniable staying power of one-sided ideas and institutions is too little examined. For it is not the case that everything passes beyond itself into some sort of higher unity: a dog remains a dog and a hammer stays a hammer, whatever logical acrobatics we may perform.

In any case, there are countless further examples of dialectical movement to be found in the writings of recent philosophers. Slavoj Žižek (b. 1949), the prominent Slovenian thinker who owes so much to Hegel, has a particular talent for using the dialectic to reverse our most common thoughts at an almost industrial scale. Let's take just one of them. Žižek puts a particular spin on the career of Cold War CIA counterintelligence official James Jesus Angleton (1917–87).[125] Angleton's suspicions of a KGB 'monster plot' to infiltrate and deceive the CIA led to his becoming so paranoid that his search for Soviet spies was frozen: he never actually caught any spies, since he assumed that any supposed Soviet 'defectors' had merely been planted by Moscow as part of the plot. In view of his minimal results, Angleton might just as well have been a KGB spy himself, and some observers have actually suggested he was. Stated in dialectical terms, Angleton appears at first like a rabid defender of America, when in fact he was possibly just as harmful to Washington as a subversive KGB plot would have been. Perhaps he was even worse, since an actual plot would at least have run the risk of being detected and countered. Here Angleton is dialectically exposed as possibly more helpful to the KGB than to his own employer, the CIA.

A similar dialectical logic can be applied to Kublai, Timur and the closely related opposites they represent. As we saw, figures as different as Jenghiz Khan and Ibn Khaldun both noticed the logic by which barbarian nomads of the steppe or desert are enticed and finally corrupted by the delights of civilized life. In this respect Kublai Khan was just the sort of urban heir prophesied by his ancestor Jenghiz, though it was only Kublai's later descendants who took things to the point of full-blown degeneracy. As concerns Ibn Khaldun, though he spoke of inevitable decay in a dynasty, it is strange that his theory makes no obvious place for someone like Timur, even though he met the conqueror in person before surviving the massacre of Damascus. Ibn Khaldun clearly saw the magnetic pull of civilization on the barbarian, but never theorized the countervailing tendency of the civilized person towards barbarism, even while describing generational decay in matchless detail.

In Timur's case we can identify many possible, even contradictory, motivations for his descent into brutality: avarice, the lust for conquest and adventure, religious fervour, sadism, emulation of the deeds of Jenghiz Khan and the wish to glorify his capital of Samarkand with the riches of annihilated peoples. But there is a further possible motive, one initially hidden by Timur's immense physical courage. That motivation is fear, which need not take the form of conscious dread. It is enough to worry that everything one has might suddenly be stripped away. If anything in the human heart impels us towards inhumanity, it is no doubt the fear that our way of life might be stolen or subverted by some external force. To prevent our spouse from cheating, we may fanatically control every least thing that they do. We may worry that our existing social class will be levelled by radicals, who can only be stopped if thrown from helicopters into the sea (as in General Pinochet's Chile); that our own religion is threatened and must be preserved by faith-based massacre; that we may lose 1 million soldiers if we invade the enemy island, and can only avoid it by dropping atomic bombs on two cities. By definition the civilized person has more to lose than the rootless nomad, and whoever has much to lose has much to fear, and therefore much to annihilate pre-emptively.

The reason for taking this detour is that we might well wonder whether the continuous (barbarians on the steppe) and the discrete (civilians lodged behind particular walls) are fixed opposites in the Kantian sense, or whether each contains the seeds of its opposite as in Hegel's philosophy. Consider the sense in which the continuum potentially contains infinitely many discrete points, as in the case of the number line or of Zeno trying to walk to the door. Consider too that Aristotelian substance, that paragon of discreteness, is also a continuum made up of many potential substances: as with berries still on the bush, or organs currently unified in a human body. In similar fashion, we have seen that the nomads who rule the continuous steppe aspire to capture and enjoy discrete urban sites. Yet by contrast, at least one lord of a capital city (Timur) wished to level all rival cities to give rise to an artificial protective steppe of his own. In similar fashion, it may not be possible to assign one set of qualities exclusively to continua and another set solely to the discrete. Here we work in the shadow of Hegel's great insight, whatever its limitations may be.

4.

Flows and Instants: Bergson and the Occasionalists

The galloping progress of the natural sciences in late nineteenth-century Europe was met with what might be called a spiritualist reaction. Alongside the rapid spread of experimental laboratories we find the growing popularity of séances and other tokens of interest in parapsychology. Madame Helena Blavatsky (1831–91), an esoteric figure who claimed – perhaps falsely – to have visited Tibet, gained wide influence through her Theosophical Society even as she was mocked by critics as a charlatan. In that era of rapid scientific development, most mainstream scientists and intellectuals steered clear of these apparent regressions into what they dismissed as superstitious nonsense. A notable exception was the prominent American philosopher and psychologist William James (1842–1910), who engaged in dialogue with spiritualist figures and did not rule out the possibility that evidence might someday support their claims.

Across the ocean from James was a younger thinker he greatly admired, the French philosopher Henri Bergson (1859–1941). Bergson belonged to a serious lineage of French spiritualism that had developed during the nineteenth century, and which included such noteworthy figures as Victor Cousin (1792–1867), Maine de Biran (1766–1824), Jules Simon (1814–96) and the increasingly popular Félix Ravaisson (1813–1900). Bergson continued their legacy of interest in the non-quantifiable aspects of human experience, and also shared their lack of hostility to religion; in Bergson's case, this led to a growing closeness to Catholicism, to which he almost but never quite converted. These factors continue to make Bergson *persona non grata* among hardcore rationalist philosophers, committed as they are to cognitive exactitude, reverence for natural science and an unyielding atheism. Bergson's relevance to the present book comes from his peculiar solution to the gap between the continuous and the discrete, which differs from Aristotle's own.

Occasionalism

One of Bergson's most prominent admirers was the Lithuanian-French philosopher Emmanuel Levinas (1906–95), who shared Bergson's Jewish origins. While working as a Russian translator in the French Army during World War II, Levinas was captured when his unit was surrounded by a German offensive. Although forced to wear a yellow star in captivity, he was otherwise shielded by the Geneva Convention from the more extreme forms of Nazi persecution. The same did not hold for his family members, some of whom were executed in their own front yard in Kaunas, Lithuania, by SS soldiers. Despite his wood-chopping duties as a prisoner of war, Levinas was intellectually productive in captivity. The insights jotted down in his notebook led, after the war, to the publication of a short book called *Existence and Existents*.[1] The central conception of this work is that existence is something anonymous, indefinite and without parts; there are no real individual things in their own right. What we take to be independent objects are really just 'hypostatized' into existence through the work of the human mind. This is reminiscent of Anaximander's indefinite and continuous *apeiron*, as covered in our earlier discussion of Aristotle. In his own terminology, Levinas calls this continuous and anonymous whole the '*il y a*', which in French means simply 'there is'. This is not just an idea reached through abstract reasoning. Instead, Levinas thinks we experience the *il y a* directly in cases of insomnia, where everything seems to blend together and cannot be turned off like a light. This is a good example of a philosophy that sees a high degree of continuity in the world.

At the opposite extreme is the view that reality is radically plural, made up of countless entities that are so discrete that they cannot even make direct contact with each other. Such philosophies have existed frequently, and were popular as recently as seventeenth-century Europe. The term for this kind of philosophy is 'occasionalism'. As previously discussed, occasionalists see no other way for objects to interact than for God to become directly involved in even the tiniest causal events in the universe. Those who follow any of the major monotheistic religions are, of course, committed to the view that God created the world out of nothing. But few monotheists believe that God is directly involved in all everyday incidents of causation, even the most trivial. Yes, it is still common to hear people say that miracles occur frequently, or that everything happens for a reason. But occasionalism isn't merely the widespread belief that God intervenes to save a cancer patient, or prevents you from getting a desired job in order to save you for an even better one in the future. Instead, occasionalism is a truly radical view entailing that God is directly responsible for even the minutest swerve of an atom.

It's not just that God is indirectly responsible for everything because he created the laws of nature. It amounts to the extreme view that laws of nature are not even needed: direct divine intervention occurs even when a grain of dust drifts through some uninhabited piece of desert, or when the surface temperature of Jupiter rises from -110.03° C to -109.07° C. For occasionalist philosophers, even if two cars crash they do not affect each other directly; instead, their apparent physical contact is merely the 'occasion' for God to make it look like they have damaged each other through touching. But there is no real contact between the cars, or indeed between anything at all. Fire does not burn cotton; only God burns cotton. It's not surprising that most occasionalists have difficulty accepting the existence of free will: the power to make our own decisions and bear the consequences. What drives occasionalists to their extreme standpoint is the intuition that individual things are so self-contained, so cut off from one another, that they are incapable of engaging in relations unless God intervenes to make this possible. Pushed to the limit, this view entails the disappearance of all human freedom and responsibility; in many branches of monotheism, of course, this stance would count as a heresy.

In its most extreme form, occasionalism upholds two separate but related doctrines. Alongside the idea that no two entities can make direct contact, we sometimes find the additional view that even two moments of time cannot touch each other. In other words, time is not a continuum as Aristotle thought: it does not flow smoothly and continuously, and everything that exists must be recreated in every instant. The universe exists for only the flash of a moment before vanishing, so that God must recreate it in the following instant, and so on forever. In Early Modern Europe we find a number of philosophers who believe that such intervention is needed, due to the impossibility of any two things interacting directly. This includes such important figures as René Descartes, Nicolas Malebranche, Baruch Spinoza, George Berkeley and G. W. Leibniz. But the occasionalist way of thinking began in the Islamic world, where it had more explicitly religious roots.[2]

It first appeared in medieval Basra, Iraq, in the work of Abū al-Ḥasan al-Ashʿarī (873–936 CE). He was initially inclined towards a liberal rationalism, but at the age of forty turned in a more traditionalist direction. al-Ashʿarī's philosophy provides a fine example of maximum discontinuity or discreteness: only God touches anything; nothing else makes contact with anything else. These ideas lived on in the Ashʿarīte tradition that bears his name, above all in the influential conservative Persian thinker al-Ghazali (1058–1111).[3] Five centuries later – from the 1600s through the early 1700s – this current of thought finally entered Europe, where its hidden influence continues to this day.

During the twentieth century, there was a quiet resurgence of the occasionalist doctrine in the philosophies of the English metaphysician Alfred North Whitehead (1861–1947) and eventually the writings of Bruno Latour.[4] A mathematician by training, in middle age Whitehead received a surprise invitation to teach philosophy at Harvard; in that role he developed into one of the most important philosophers of the twentieth century. Whitehead says that we perceive reality only through the mediation of qualities ('eternal objects') contained in God: an idea foreshadowed by the French occasionalist Malebranche.[5] If I notice that someone across the room is wearing a blue shirt, this is possible only because I perceive it through the blueness contained in the deity. Latour's version of occasionalism is less orthodox, and consists in the claim that every causal interaction between two things has a local mediator rather than touching by way of God. For instance, Latour shows that politics and neutrons were first linked in France by the physicist Frédéric Joliot (the local mediator in this example), in his failed pre-war effort to convince the government to invest in atomic bomb research.[6] We will return to this example later. Despite his personal religiosity as a practising Catholic, Latour never invokes God as a causal agent in explaining everyday phenomena, which makes him probably the first secular occasionalist in the history of philosophy.

As for the occasionalist theory of time, it sometimes pops up in unexpected places. Even Bertrand Russell (1872–1970), one of the patron saints of twentieth-century rationalism, takes the occasionalist theory of time seriously in a version that does not involve God. For instance, he speaks of 'the illusion of persistence' and says that the continuous self is not a real entity, just as had been said two centuries earlier by David Hume. Russell writes as follows: 'The real man too, I believe, however the police may swear to his identity, is really a series of momentary men, each different one from the other, and bound together, not by a numerical identity, but by continuity and intrinsic causal laws.'[7] In other words, Russell seems to think that each of us does not remain the same person over the course of time, but is replaced in each instant by a similar but slightly different version of ourselves. His mathematical co-author Whitehead thought so as well, and even more intensely.

We saw that philosophers of extreme continuity, like Parmenides or the young Levinas, face the difficulty of explaining why individual entities at least *seem* to exist. Parmenides blames this on the delusions of the senses, while Levinas claims that the human mind carves a primordial world-lump into pieces. But making a philosophical argument isn't just about preaching to the choir, as the idiom goes: it is also about trying to convert opponents, and this works better when intelligible reasons are given rather than repeated solemn avowals of one's own views. This is where a philosophy of extreme discontinuity like that of al-Ashʿarī runs into a problem: namely, why should a single mighty exception (God) be permitted to account for all the relations and causal effects in the universe if nothing else is capable of doing so? A religious person already believes that God is omnipotent, and may not see any problem here. But if an occasionalist wants to convince non-believers that their philosophy is true, they will need to give some account of the mechanisms by which God is able to touch everything while nothing else can do so. One can always retreat into faith or prophecy as the anchor for a key idea, but in that case all dialogue with others has probably ceased.

Notice that both of these groups – the occasionalists and those we will call continuists (such as the young Levinas) – adopt extreme solutions to the problem of the continuous and the discrete. They either reduce the discrete entirely to the continuous, or they reduce the latter to the former instead. Against this we saw Aristotle's awareness that any adequate philosophy must make room for both. In other words, the continuous and the discrete are both genuine facets of reality that cannot be explained away by tightly embracing the opposite. Aristotle did this through a division of labour in which time, space, number and physical motion are treated as continua while substances and qualitative change are regarded as discrete. In the late nineteenth and early twentieth centuries, Bergson would attempt a new approach to this problem.

Bergson's Division of Labour

During an interview conducted in 1981, Levinas recalls his first encounter with Martin Heidegger's *Being and Time*, which he calls 'one of the finest books in the history of philosophy'.[8] Levinas goes on to provide us with his full list of great books: Plato's *Phaedrus*, Immanuel Kant's *Critique of Pure Reason*, Hegel's *Phenomenology of Spirit* and Bergson's *Time and Free Will*.[9] It is a remarkably varied collection. The *Critique of Pure Reason* is likely to be on every philosopher's list, given how foundational it has been to Western philosophy since its publication in 1781. Plato is also sure to appear on every list, though some readers would choose dialogues other than the *Phaedrus* as their favourite (my own preference is for the simple and straightforward *Meno*).[10] The *Phenomenology of Spirit* will rank high for some but not for others, given Hegel's deserved reputation for unusually demanding prose. And while many readers agree with Levinas that *Being and Time* is a classic, it is shunned by some due to Heidegger's shameful involvement with the Nazi Party. But the real surprise on Levinas's list of the finest works of philosophy is Bergson's debut book, *Time and Free Will*, whose original French title (*Essai sur les données immédiates de la conscience*) literally means 'Essay on the Immediate Data of Consciousness'. The surprise is partly because other works by Bergson are more famous, and partly because no work by Bergson would appear on most Top Five philosophy lists at all. But many readers, myself included, are repeatedly delighted by the wealth of fresh ideas in this early work.

Bergson was born in 1859 to Jewish parents, his father coming from France (though with Polish roots) and his mother from England. He is most famous for the books *Matter and Memory* (1896), *Creative Evolution* (1907) and *The Two Sources of Morality and Religion* (1932), whose topics are clearly indicated by their titles.[11] Bergson was internationally renowned during his lifetime. He is one of the few philosophers to have won the Nobel Prize for Literature, awarded in 1927 with the following citation: 'In recognition of his rich and vitalizing ideas and the brilliant skill with which they have been presented.'[12] He published *Time and Free Will* at the age of thirty, an astonishing performance in a field usually dominated by the middle-aged. He also conducted a famous debate on time with Albert Einstein, resulting in the 1922 publication of Bergson's *Duration and*

Simultaneity, though most observers now regard Bergson as having been the loser in that exchange. A good summary can be found in Jimena Canales's excellent book *The Physicist and the Philosopher*.[13] Often busy with political activities during his mature years, Bergson resigned in 1925 from his leadership position in the cultural organ of the League of Nations (the failed early predecessor of the United Nations), and withdrew from public life altogether due to poor health. For several decades before this retreat he dominated the French philosophical scene, especially among the educated upper classes; later, he seemed more like the relic of a bygone era. His death in 1941, after refusing to exempt himself from registering as a Jew during the Nazi occupation of Paris, came as a shock, primarily because many had thought him already long dead due to his extended silence. The recent revival of Bergson's reputation owes much to the influence of Gilles Deleuze, a tireless admirer of his work.[14]

Bergson's mesmerizing style is visible even in his debut book, where lucid explication is peppered with poetic images. As a philosopher, he is one of the proverbial champions – along with the pre-Socratic thinker Heraclitus – of a world in constant flux. He defends imaginative intuition over hidebound rationalist method and freedom over determinism. Nonetheless, Bergson begins this first book with what sounds like a rather familiar duality: 'within our ego, there is succession without mutual externality; outside the ego, in pure space, mutual externality without succession'.[15] That is to say, in our minds everything blends together continuously, while the external world is made up of discrete and mutually external beings. On the inside we have quality, intuition, heterogeneity and vague borders between distinct experiences; on the outside everything is quantity, rationality, homogeneity and clear-cut outlines separating one thing from another. This formulation is Bergson's contribution to the present book.

It was mentioned that Bergson is an emblematic philosopher of flux and flow, displaying a sharp personal preference for the continuity, heterogeneity and dynamism that he finds in human experience as studied by philosophy, psychology and literature. He opposes this sharply to what he sees as the discrete, homogeneous and static conditions of the outside world as studied by natural science. Yet he doesn't weigh these opposites as unequally as might be assumed: Bergson does not automatically grant special favour to the smooth and continuous character of our mental world over the chunky discreteness of nature. In fact, his philosophy can be viewed as a struggle to do justice to both sides of the duality between waves and stones. Reconciling the two sides of this story is not treated by Bergson as a brief and facile exercise: it is the central problem that animates his career. In effect, Bergson tries to chart a course midway between the extreme continuism of Parmenides and the young Levinas and the extreme discreteness of al-Ashʿarī and the later European occasionalists. In this respect his approach is strikingly similar to Aristotle's, though with different results.

For all his innovations, there is one important sense in which Bergson is a typical modern European philosopher: namely, he takes mental life to be fundamentally different in kind from everything else. This is hardly a surprise, given that he is writing in the wake of Descartes's dualism between thought and physical extension and Kant's analogous rift between the phenomenal that we experience and the noumenal thing-in-itself that we cannot. Bergson believes that what's inside us is like a molten river of lava, with no sensation or feeling ever repeating itself in exactly the same way; this corresponds to what he considers the true, inner self.[16] As soon as we 'dig below the surface and get down to the real self', he writes, what will happen is that our 'states of consciousness' will 'begin to permeate and melt into one another'.[17] This molten depth of the self is inexpressible, since 'there is no common measure between mind and language'.[18] Recall Aristotle's point that a thing can never really be defined, since things are concrete but definitions are made of universals.

Let's consider an example of how, for Bergson, the inner self differs from the quantifiable external world. Imagine that you have returned home hungry from work. You pull a frozen pasta meal from the freezer, follow the printed instructions to microwave it for four minutes, stir, then heat it again for an additional minute. You calculate that this process uses up around five minutes of your waking day, or roughly half of 1 per cent of your non-sleeping time. If you keep detailed records of such things in order to stream-line your activities, the information it yields may prove useful, though it might also be regarded as obsessive-compulsive behaviour. But even this type of detailed record tells us nothing at all about what those five minutes were really like: the qualitative character of the time that has passed. You wait for the pasta to cook, with a mixture of impatience and anticipation. Your mood fluctuates. The hum of the microwave provides a steady background drone as you sift through all the jubilant or embarrassing memories of the day. Perhaps you feel some resentment that there has been no time to eat anything since breakfast. Once the first four minutes of cooking are finished, the bell on the microwave rings. You carefully remove the plastic covering while trying to avoid a steam burn to your hand, and become annoyed as you feel the heat and struggle with the flimsy plastic; you stir your meal half-heartedly and put it back in the microwave, waiting for the final minute of cooking to elapse. Finally, the pasta is ready to eat. In a quantitative sense it is true that a mere five minutes have elapsed. But in another sense a million thoughts, sensations and memories have unfolded in the meantime. For Bergson, to quantify time simply by counting it means mistreat-ing time as a form of measurable, homogeneous space.

If we want to consider time on its own terms, we need to treat it instead as a continuous and unrepeatable experience (this implies that the discrete is always repeatable, and that is in fact what Berg-son thinks). Even if you come home late tomorrow night and prepare the same meal in an identical period of five minutes, the experience will not be the same as the one you had tonight. By then you will have gone through another day of your life, and even if it is filled with nothing but predictable, mediocre incidents, you will be slightly older and have additional memories that give the experience of microwaving a slightly different colouration. Fans of

French literature won't be shocked to learn that the entangled reminiscences filling the novels of Marcel Proust were partly inspired by Bergson's philosophy.[19] In fact, Bergson's wife, Louise, was a cousin of Proust; the novelist himself served as best man at their wedding.

To repeat, Bergson thinks it is wrong to say that we can speak of internal experience in the same way we do of the outside world. For instance, in daily speech we often say that an internal sensation 'can be twice, thrice, four times as intense as a sensation of the same kind'.[20] We might say that tonight's headache is 'twice as bad' as last night's, or claim that William Shakespeare is '100 times' the playwright Henrik Ibsen was. The foundational insight of Bergson's philosophy is that such claims are categorically false, even in the apparently quantitative case of physical effort. If we lift a fifty-pound weight and then a hundred-pound weight, we are tricked by the numerical doubling into thinking that the second experience of lifting was twice as hard as the first. But in fact, the two experiences are qualitatively rather than quantitatively different. Pay close attention, and you will find that lifting the heavier weight involves a different range of muscles and causes stress in parts of the body that were not involved at all when lifting the lighter one.

An even simpler example given by Bergson is that of clenching one's fist. Do it right now: clench your fist, and see what it feels like. Now try to clench it twice as hard. Although you may think there is a sheer quantitative doubling at work, what actually happens is that in the second case 'the sensation which was at first localized has affected your arm and ascended to the shoulder; finally, the other arm stiffens, both legs do the same, the respiration is checked; it is the whole body which is at work'.[21] The same happens if you press your lips together with increasing intensity until you reach the maximum possible pressure. While focusing all attention on your lips, it may seem like there is nothing but quantitative increase in the area of the lips themselves. But here just as with the double-heavy weights, each increase in pressure brings into play new muscles, sensations and emotional intensities.

This shows yet again that experience is primarily qualitative rather than quantitative in character, as seen from other examples of how intense concentration brings new areas of the body into play. Bergson reports that the radical German scientist Gustav Fechner (1801–87) discovered that when we try to remember something, our scalp muscles tighten.[22] In similar fashion, the French psychologist Théodule-Armand Ribot (1839–1916) noted that as we pay increased attention to anything our eyebrows rise, and sometimes our mouth even drops opens.[23] Overwhelming sounds or odours might seem to have 'triple' or 'quadruple' the impact of faint ones, but again, the real difference between the cases is the qualitative variety of each experience. Once a sound reaches a certain volume, it might give rise to a headache that was not previously there.[24] In an era when natural science was taking over numerous problems that were previously considered philosophical in character, Bergson tried to set limits to the quantification of everything brought about by the scientific method. While quantity might be sufficient for our knowledge of the discrete physical world, it fails when applied to the swirling continuous landscape of our intermingled thoughts and sensations.

As Bergson sees it, numbers and exact multiples do not tell us about our lived experience, but pertain only to the homogeneous external world. The exact quantitative relations between various entities do not shed much light on inner life. To use them amounts to an illegitimate translation of the 'unextended' (the non-spatial) into the 'extended' (the spatial). In other words, it is a doomed attempt to turn quality into quantity.[25] The experience of time in the inner, qualitative and continuous sense is what Bergson calls *durée*, a term often left in the original French. Although the word 'duration' is the usual English equivalent, the French word itself means 'the endured' or 'that which is endured'. In any case, *durée* is 'without precise outlines'.[26] It consists of unrepeatable experiences of varying intensities that cannot be precisely measured, in which things and moments blend together in such a way as not to be entirely distinct from each other. *Durée* is an ally of the continuous in its everlasting war with the discrete.

The Second Self

In another sense, Bergson is prepared to acknowledge that human experience is not just a turbulent, molten flux. We also have a stable sense of personal identity that ties together our actions and experiences across the years. Nonetheless, he treats this enduring identity as superficial: 'The self, insofar as it has to do with homogeneous space, develops on a kind of surface, and on this surface independent growths may form.'[27] This surface self functions as a kind of interface between the molten self of pure qualitative duration and the outer world of stable, petrified things. The way he puts it is to say that there is a 'second self' positioned between the more flowing, deep-seated persona and the outer world: one that engages in the familiar activities of language, abstract thought and social life. Bergson regards this second self as the bridge between the inner homogeneous or continuous world on one side, and the heterogenous or discrete outer world on the other.

From the time of infancy, we learn to distinguish the various experiences we undergo at any given moment: the barking of a dog, the ringing of church bells, the glare of sunlight on a window or the gnawing hunger in our stomachs. All of them have the air of discrete realities, clearly differentiated from each other. But if we take a closer look, we will find that in mental life all these separate entities blend together into something like a seamless whole, one not entirely different from the *apeiron*. Thanks to the work of language and reason, we wrongly think of experience as made up of discrete chunks, so that many of our ideas 'float on the surface, like dead leaves on the water of a pond'.[28] Articulate language 'overwhelms or at least covers over the delicate and fugitive impressions of our individual consciousness'.[29] In this way, the second self betrays our true self by deferring to the petrified rules of everyday life. Bergson criticizes this derivative self for its 'parasitic' character, and sadly concludes that most people are stuck on that level and never live as their true selves.[30] This idea may have influenced

Heidegger's later analysis of how most human life is imprisoned in the dismal everyday habits of the anonymous public he calls 'the they', to such a degree that authentic human existence is difficult to achieve.[31] Yet without this second self we would remain stuck in the flowing, inarticulate duration of feelings that runs beneath all language and society like a hidden river, and would be unable to account for any enduring things in the world.

As a rule, Bergson also identifies the flowing inner self with time, and the petrified outer world with space. This is a recurring theme throughout his works; whenever Bergson links anything with space, it is not meant as a compliment. In particular, when he speaks of homogeneous space what he means is the scientific notion of space as a system of three-dimensional co-ordinates through which the movement of physical objects can be measured. But that is not all that Bergson means by the term. As he sees it, scientific space is not just an empty container in which everything is located (as Isaac Newton believed), but a more general intellectual form that arises even in cases that we would usually take to be non-spatial, such as the quantitative analysis of music. In any case where we think of the world as made up of durable, measurable things, Bergson thinks we are making use of space in his sense of the term. Homogeneous space, he says, 'enables us to use clean-cut distinctions, to count, to abstract, and perhaps also to speak'.[32]

Obviously, this is not altogether bad. Bergson agrees that this scientific sense of space is a great cognitive achievement. For instance: 'the higher we rise in the scale of intelligent beings, the more clearly do we meet with the independent idea of a homogeneous space'.[33] In the realm of biology, he notes the 'surprising ease' with which vertebrates and some insects manage to 'return almost in a straight line to their old home, pursuing a path which was hitherto unknown to them over a distance which may amount to several hundreds of miles'.[34] In humans this capacity becomes extremely powerful: we do not just navigate space as a set of practical paths towards personal goals, but are also able to conceive of space in abstract geometrical terms. Indeed, this skill is often viewed as one of the crown jewels of the human intellect. In the recent past, it enabled our species to launch the satellite-powered GPS system that allows us to drive with ease to destinations previously unknown.

But despite Bergson's attempt to balance the two worlds, it's safe to say that his true passion is for the intuitive side of the human mind and its basically continuous experiences. Consider the fact that we understand the difference between left and right even though it is difficult to define in words. When pressed by a young child to explain the difference between the two, is there any way to do so other than simply pointing? The skill of distinguishing between left and right is crucial for everyday life, and is eventually mastered by almost everyone, but in a way that is hard to quantify or put into words.

Freedom

Time and Free Will also employs the central distinction between *durée* and space to make a novel argument that human action is free of causal determination. For Bergson, we are not just machines acting in accordance with inevitable laws of nature. With his typically modern idea that the self 'cannot be compared to that of any other force',[35] he declares that human thought is something utterly unique in the cosmos. Much like Descartes and Kant, he treats consciousness (in the form of heterogeneous *durée*) as something so important that it deserves to make up a full half of philosophy, though in another sense the human mind is also just a trivial speck in the vastness of the universe.[36]

Even so, Bergson's human exceptionalism is not without insights. He begins by describing most everyday decisions as arid and predictable: 'our daily actions are called forth not so much by our feelings themselves, which are constantly changing, as by the unchanging images with which these feelings are bound up'.[37] Stated differently, most of our actions are shaped by the idea of how someone *should* act in this or that situation; we are guided by that infamous dullard, the second self. When dressed up as our second selves we do not always rise to the challenge of a serious decision, either because we are distracted by 'sluggishness or indolence' or because we irresponsibly outsource our decision to the counsel of others.[38] I once had an office colleague who held a strange grudge against another colleague, simply because that person had advised her to marry someone who turned out to be a disastrous mate. Although people routinely seek marriage advice from family and friends, this is clearly such a serious life decision that we ourselves must bear ultimate responsibility for the choice.

There are other times when we find ourselves on the verge of making a formulaic or conventional decision, when suddenly, something in 'the deep-seated self' rebels and pushes us in a different direction.[39] An example from the history of science comes to mind. The Italian physicist Enrico Fermi (1901–54) tried many ways of speeding up neutrons to create atomic fission in experiments, but nothing worked.[40] Then, suddenly and inexplicably, he chose a paraffin screen for the neutrons, even though paraffin slows neutrons down rather than speeding them up. Eureka! Counterintuitively, it turned out that slow neutrons were exactly what was needed, since the long-sought fast neutrons move too quickly to interact with the uranium atom at all. With this discovery, which arose from somewhere in Fermi's deep, intuiting self, the rapid march towards the atomic bomb and atomic power had begun.[41] In Bergson's words: 'It is at the great and solemn crisis' that we 'choose in defiance of what is conventionally called a motive'.[42] Even more poetically, he describes the deep-seated self as one that 'lives and develops by means of its very hesitations, until the free action drops from it like an over-ripe fruit'.[43]

Bergson's Dualism

In his youth, Bergson was a promising mathematician. This is worth remembering when we read the following passage in *Time and Free Will*: 'Number may be defined in general as a collection of units, or speaking more exactly, as a synthesis of the one and the many.'[44] The number fourteen is a unified whole, yet it also consists of fourteen separate units. Any of these can be subdivided further into halves or fourths or tenths, though most of us tend to visualize numbers as built of indivisible number ones.[45] More to the point, while it is true that the fourteen separate units that make up the entire number can always be cut up further and further, when dealing with any specific number we tend to regard it as indissoluble. When multiplying sixty by thirty-six, we will probably not stop to think of all the ones of which these two larger numbers are composed. The lesson Bergson wants us to learn from this is that 'the formation or construction of a number implies discontinuity'.[46] In other words, 'it is always by jerks, by sudden jumps that we move forward or from one unit to the other'.[47] This cuts against the grain of Aristotle's continuous conception of number.

Bergson offers a vivid example. If we hear a church bell tolling somewhere in the distance, we usually encounter it in the manner of pure duration. Each stroke of the bell is a slightly different experience, with the final stroke of midnight leaving a very different impression from the first stroke in the series. But with many other sounds there is something different going on: namely, an unexpected connection with space. For instance, 'when we hear a noise of steps in the street, we have a confused vision of somebody walking along. Each of the successive sounds is then localized at a point in space where the passerby might tread: we count our sensations in the very space in which their tangible causes are ranged.'[48] While Bergson usually regards such quantification of inner experience as a mistake, he is right that in cases like this we imagine sounds as emanating from distinct positions in space rather than focusing solely on their qualitative difference.

As we have seen, he is adamant that the opposing poles of heterogeneous inside and homogeneous outside should not be confused with each other. It is intellectually mistaken to break the smooth flow of our lived experience into discrete segments measured by a stopwatch, and equally mistaken to join Einstein and his former Zurich professor Hermann Minkowski (1864–1909) in reducing lived qualitative time to a four-dimensional space-time.[49] Yet here as with every dualism, we should ask where the two opposite realms are able to interface; otherwise, we will have two separate parallel universes with nothing to link them. How can the continuous be translated into the discrete, and vice versa?

The problem is analogous to Descartes's need to find the place where mind and body are able to meet and interact, which we saw that he located quaintly in the pineal gland of the brain. Bergson turns to this problem of interface late in *Time and Free Will*. He begins with the seemingly clear assertion that space has independent existence outside the mind: 'There is a real space, without duration, in which phenomena appear and disappear simultaneously with our states of consciousness.'[50] The person walking in the street is not just a phenomenon in my mind, but actually moves through objective space. But how do we reconcile this person with the parallel person who haunts my inner temporal experience where everything blends together into something like a seamless whole?

Bergson thinks the way to bring the physical world into contact with the inner world of duration is to give *durée* 'the illusory form of a homogeneous medium'.[51] In other words, we need a way to translate the language of time into the language of space, and vice versa. If we do so, Bergson thinks we will find that 'the connecting link between these two terms, space and duration, is *simultaneity*, which might be defined as the intersection of space and time'.[52] If we look at time in the purely objective terms of the outer world, events can be assigned to their specific position in an unfolding linearity: fifty years ago I was a child, but I am now a middle-aged adult. Yet if we remember the narrator in Proust's *In Search of Lost Time*, he weaves different moments of his past life together in the present, with various odours and sounds serving to blend together objectively at different times.

Along with simultaneity and the second self, another term used by Bergson for the crossroads where time meets space is 'endosmosis', the scientific word for what happens when water permeates the wall of a cell. In his own words, between 'succession without externality' (time) and 'externality without succession' (space), we find that 'a kind of exchange takes place, very similar to what physicists call the phenomenon of endosmosis'.[53] In Bergson's view, this concept allows us to comprehend 'the contradictory idea of succession in simultaneity'.[54] By this he means that there is, indeed, a place where space and time can meet.

In a sense, Bergson's position is an upside-down version of the theory of motion of the medieval Islamic thinker an-Nazzam (777–845), a theory quite different from Aristotle's own.[55] Whereas Aristotle maintained that both space and motion were continua not made up of any definite number of points, for an-Nazzam (at least according to Peter Adamson) space is a smooth continuum but motion is not. That is to say, an-Nazzam thinks that I move towards the doorway through a series of infinitesimal jerks and jumps, even though space remains continuous as I do so. But whereas an-Nazzam conceived of jerky movement through smooth space, Bergson offers the inverted model of smooth motion through a series of punctuated spatial points. For the French thinker, motion is always smooth and continuous insofar as it is experienced in the flowing *durée* of inner conscious life, while space is viewed not just as an infinitely divisible continuum, but as an actual series of already subdivided points.

Rather than finding smooth flow everywhere in the universe, Bergson finds it only in conscious experience, while the outer world (with all its objects) is treated as a series of frozen or immobile poses in homogeneous space. For him the physical movement of entities outside the mind occurs in the manner of dancers viewed under strobe lights, or claymation monsters in low-budget adventure films. As he writes: 'each of the so-called successive states of the external world exists alone; their multiplicity is real only for a consciousness that can first retain them and then set them side by side by externalizing them in relation to one another'.[56] Here he is close to Kant, who also conceived of time as something synthesized by the human imagination rather than existing independently of our minds.

With such theories, Bergson effectively launches an early version of C. P. Snow's 'two cultures' battle between science and the humanities, referring to the increasing post-World War II gap that hampered their ability to communicate.[57] Bergson does so with his claim that mathematics cannot capture lived time, and that 'science cannot deal with time and motion except on condition of first eliminating the essential and qualitative element'.[58] In present-day philosophy there is an additional split between two cultures, one that closely mirrors the conflict bemoaned by Snow. Analytic philosophers generally defer to science on questions pertaining to the external world, whereas the so-called continental tradition tends to view science as merely 'correct' or 'accurate' but somehow not up to the task of addressing the deepest questions about reality. As the continental philosopher Heidegger once wrote: 'science does not think', a phrase that inspires contempt among analytic philosophers.[59]

A remaining ambiguity in Bergson's treatment of the difference between *durée* and space is the status of animals in his thinking, since it is hard to know where they would fit by comparison with the human experience of time. Is the time experienced by animals a lesser version of human *durée*, or simply a different one? Western philosophy across the centuries shows a range of differing standpoints about the status of animals, though generally it has treated them as lesser versions of humans. Aristotle's brief but challenging classic *De Anima* (*On the Soul*) names plants, animals and humans as the three basic kinds of creatures, depending on their possession (or lack thereof) of the increasingly sophisticated nutritive, perceptive and intellective faculties.[60]

From there, St Thomas Aquinas expanded the Aristotelian model by splitting animals into two types: (a) those like shellfish that can perceive but not move, and (b) those that can also move, such as cats, horses and humans.[61] Descartes launched modern philosophy by cruelly dividing all entities between those with full-blown human minds and those (including all types of animals) that are mere machines lacking in thoughts or even feelings.[62] Heidegger humanizes Descartes somewhat by contrasting humans (who are 'world-forming') with stones (which are 'worldless'), so that animals inhabit an ill-defined intermediate zone (called 'world-poor'), while no room is left for plants at all.[63] Heidegger's less famous contemporary Helmuth Plessner (1892–1985) makes things more interesting with his focus on the emergence of multicellular life and the central nervous system as key moments in the history of evolution.[64] More recently, the French philosopher Quentin Meillassoux (b. 1967) took a stab at classifying animals with a series of jumps between matter, life and thought, though this also seems to compress both plants and animals into a single category ('life'), and is generally too traditional and commonsensical in outlook.[65] Where does Bergson fall along this spectrum of possible attitudes?

We know that Bergson associates duration with consciousness. But even if an animal is conscious, does it automatically experience *durée*? He will have more to say about various forms of life in *Creative Evolution* eighteen years later, but one of the key remarks in his first book suggests a typically modern, privative view of animals. As he puts it: 'probably animals do not picture to themselves, besides their sensations, as we do, an external world quite distinct from themselves, a world which is the common property of all conscious beings'.[66] This leaves us with something of a paradox. For although Bergson usually champions the excellence of *durée* over the blandness of spatial homogeneity, we recall that he also measured the intelligence of various species by how far they are able to depart from duration and advance towards an abstract world of language, social customs and homogeneous three-dimensional space. More than once in *Time and Free Will*, he makes a vivid point that is emblematic of his philosophy as a whole. Namely: 'if all the motions of the universe took place twice or thrice as quickly', none of the equations of science would need to change, since they have no bearing on the *experience* of time.[67] By way of contrast, we can imagine a frantic internal experience, a *durée* in which things sped up to such an extent that our lives would flash by impossibly fast. Indeed, this is a common phenomenon for human beings as they age, or as they become more familiar with a new city that initially seemed to move in slow motion. At the beginning of my career I spent sixteen years teaching in Cairo, Egypt. During the first few months each day felt like a week, as I learned a new alphabet, numbers, vocabulary, customs and historic monuments. Yet by the time I returned to America, a day in Cairo felt no different from a day in Chicago or anywhere else. It wasn't just me: an older Canadian colleague in Egypt had already warned me this would happen.

Just as human *durée* can vary in this manner, there is evidence to suggest that different animal species have different rates of experienced time, though Bergson does not cover this topic. Think of the terrier with a ball in its mouth who expertly moves it just millimetres out of reach as our hand approaches; perhaps the dog sees humans as lumbering giants trapped in sloth-like motion. In similar fashion, mosquitoes often escape the intended death blows of even the most athletic human arm. Another example is given by the ecological philosopher Jakob Johann von Uexküll (1864–1944), just five years younger than Bergson. Uexküll reports that humans only notice that a light is flashing if it does so fewer than eighteen times per second; more than that, and it appears to us as a steady light. Slower than us are snails, who are able to notice only three to four flashes per second, while fighting fish can perceive somewhere between thirty and fifty.[68] Hence a specific *durée* seems to be built into the sensory apparatus of every animal species, perhaps with some room even for variance among individuals. This strikes a blow against any idea of homogeneous time, since it suggests that the specific *durée* of a creature results from its inborn biological equipment. But since Bergson remains silent on this matter, we can only guess what he might say about it.

The Problem of Endosmosis

Although separated by more than two millennia, Aristotle and Bergson confront the same basic philosophical problem: where is the interface between the continuous and the discrete? Where can they meet in order to interact? Discrete substances (Aristotle) and discrete points of space (Bergson) are not unaffected by what happens inside continuous space, time and motion (Aristotle) or continuous *durée* (Bergson). That being said, how can these opposites ever influence each other?

We saw that for Aristotle it happens in contact or touch, defined as something midway between the total fusion of things and their total separation. For Bergson, who is positioned squarely within our modern insistence on the special status of human thought, it is the mind that does the work of interfacing with the world through the endosmosis that follows from our 'second self', conceived in a spatial rather than temporal sense. This second self is a kind of inauthentic or diplomatic version of me, capable of converting my flowing river of continuous experience into the common currency of exchangeable quantities. While the term 'endosmosis' is perfectly adequate for denoting this process, for Bergson it is less an exchange than a one-way street. For although the external world cannot translate itself into terms of *durée* for our human convenience, our internal world can transform into a second self, that compromised but successful public character that manages to make its way in the world.[69] For modern philosophers like Bergson, the human mind is entrusted with the starring role in every play.

But there's a complication. At the close of the previous chapter we discussed the dialectic, in which two apparently opposite principles seem to reverse roles in ways that are often complex, as noted in the cases of Kublai and Timur. We now find another example at the heart of Bergson's philosophy. For we have seen, as a rule, that for Bergson 'there are two kinds of multiplicity: that of material objects, to which the conception of number is immediately applicable; and the multiplicity of states of consciousness, which cannot be regarded as numerical without the help of some symbolical representation'.[70] In other words, space represents the discrete and *durée* represents the continuous. But at the same time he treats space as the location of continuously varying causal forces (such as fifty-pound or hundred-pound weights) while viewing the internal life of *durée* as the realm of sudden jumps through qualitatively discrete sensations (such as the different muscular tensions triggered by varying weights).

It is a curious paradox in Bergson's vision of reality, one that he seems not to notice. This suggests that his own concept of interface – under the differing names of endosmosis, simultaneity or the second self – is more intricate than we imagined. Just as the situation requires, endosmosis serves as a place where the continuous meets the discrete. But unlike with Aristotle, there is no simple taxonomy where certain parts of reality do the work of the continuous and others the work of the discrete. In this respect Bergson's approach to the topic should be considered a step forward: in philosophy, simply classifying different kinds of entities can never be better than a partial success.

We have now considered the fate of the continuous and the discrete in the cases of two major philosophers and a number of violent historical warlords. Yet we have still said very little about science, the gold standard of knowledge in modern civilization. How do the continuous and the discrete play out in the scientific world? Let's take a first look.

5.

Paradigm Shift: On Scientific Revolutions

Is scientific knowledge built up gradually and continuously, or does it move in sudden, discontinuous leaps? For a long time it was believed that progress in science is cumulative, with each new discovery building slightly on the existing body of knowledge. This view was openly opposed by the American philosopher of science Thomas Kuhn (1922–96), who spent the latter portion of his career at the Massachusetts Institute of Technology. Even if you haven't heard of Kuhn, you've probably heard of his signature concept: the paradigm shift. For Kuhn, scientific advance consists primarily in jerky forward jumps known as scientific revolutions.

Kuhn introduced the concept of paradigm shifts to help explain the intermittent great leaps forward that have occurred in the past four centuries of science, such as the 1543 heliocentric astronomy of Copernicus, Newton's 1687 theory of gravity, and the advent of quantum theory and General relativity in the early twentieth century. Ever since Kuhn published his book *The Structure of Scientific Revolutions* in 1962, the phrase 'paradigm shift' has spread like wildfire, far beyond the walls of the academy and the lab. We now hear talk of paradigm shifts even with the introduction of new consumer products like the smartphone, or with the current transition from gas-powered to electric cars.

Over the past sixty years, Kuhn's book (I'll refer to it here as *Structure*) has gone through many printings. It is a landmark work in the philosophy of science, the subfield that studies the foundations and methods of scientific knowledge. *Structure* has attracted as many enemies as allies, though most general readers find it accessible and at least partly convincing. Kuhn's basic outlook on science is easy to summarize: major scientific advances happen more through discontinuous sudden leaps than through the patient and gradual solution of puzzles. In the first case he speaks of paradigm shifts or scientific revolutions. The second, more plodding approach is what Kuhn calls 'normal science', which he also treats as very important even if less glamorous than revolutions.

Revolutions and Normal Science

When *Structure* first appeared, the reigning image of science was one of cumulative discovery: over time, science always comes to know 'more' than it did in the past. Knowledge was treated as if it were a series of deposits in a locked savings account, always growing and never decreasing. By contrast, Kuhn's understanding of scientific progress is that the continuous development of science is punctuated by exceptional moments, or revolutions. By 'paradigm' Kuhn is referring to a basic model of how some aspect of the world works. It makes a big difference whether we think that the universe revolves around the earth (Ptolemy), around the sun (Copernicus), or that there is no centre around which everything revolves (today's astronomy). The science of light is very different depending on whether we think of light as made of particles (Newton), as consisting of waves (Young and Fresnel), or as displaying both wave and particle aspects (today's theory). The passage from an old paradigm to a new one is a scientific revolution, and by definition such revolutions are rare. By contrast, the goal of everyday normal science is not to shake science to its foundations, but to show diligence and ingenuity in solving previously resistant problems, explaining them within a broader framework of scientific consensus. For example, a scientist in the lab might notice a slight anomaly from expected experimental results, and write a peer-reviewed journal article leading to a small but respectable modification of existing science.

The young Kuhn, initially trained as a physicist, was unimpressed by the existing model of cumulative progress approaching the truth more and more closely. It amounted to the view that development would consist in 'the piecemeal process' by which facts, theories and methods are added, 'singly and in combination, to the ever growing stockpile that constitutes scientific technique and knowledge'.[1] If this incremental picture were accurate, it ought to be easy to answer such questions as these: 'When was oxygen discovered? Who first conceived of energy conservation?'[2] But as a practising historian of science, Kuhn often found it difficult to answer questions of this sort. That was his first clue that something was wrong with the mainstream account of scientific progress.

Another problem Kuhn noticed early on was that from our present-day standpoint, many past works of science look ridiculous, including some of the greatest classics. He recalls his first impression that many of the claims in Aristotle's *Physics* were obvious nonsense; this made him wonder how anyone could have foolishly believed such things for so many centuries.[3] But after further reading, the young Kuhn noticed that Aristotle simply had a different concept of motion from Newton and later figures. We have seen that motion for Aristotle refers to any kind of change at all: an acorn growing into an oak tree, a healthy person becoming sick, or a planet or stone moving from one place to another. By contrast, motion in modern physics refers only to physical movement in space. Aristotle's broader concept of motion also meant that his way of doing physics was not mathematical in the way we've come to expect since the works of Galileo and Descartes.[4] Stated differently, Aristotle's science was qualitative rather than quantitative. Suddenly, Kuhn understood Aristotle's greatness within his own ancient context, and grasped that his physics was not just a bundle of errors or myths. Instead, it was a system that made perfect internal sense, whatever limitations may have been exposed by later breakthroughs.

Historical factors of this sort finally led Kuhn to change careers from physics to the history and philosophy of science. The difficulty of knowing exactly when scientific discoveries were made, and sometimes even who made them, had already led him to doubt the cumulative picture of scientific history taught in the textbooks. These doubts were only increased when he grasped the vast intellectual distance separating Aristotle's ideas from those of modern physics. We have not changed our ideas from Aristotle's by way of a large number of tiny changes, he realized. Instead, there must have been one or more significant shifts in physical paradigm between ancient Greece and today. And beyond this, it is possible that other profound shifts will eventually separate today's physics from future versions of the science.

Physics in our time is governed by General relativity (for gravity) and quantum theory (for everything else). Although these two theories remain maddeningly incompatible, both paradigms are binding on all serious scientists, at least for now. Any scientific article that ignored relativistic and quantum principles would immediately be refused publication, unless it were a revolutionary piece of writing that argued persuasively for a profound change in our understanding of nature. Along with the paradigms that dominate any period in the history of science, there is also a deeper commitment by all scientists to the underlying principle that nature can be explained: that the order of things must ultimately be intelligible. Without this unspoken belief that nature can somehow be known, no one can be a scientist.

As already seen, Kuhn is a thinker who emphasizes sudden, discrete leaps rather than gradual and continuous change. At one point he even says that a new paradigm can emerge 'all at once, sometimes in the middle of the night' during periods of scientific crisis.[5] Kuhn notes further that the person who discovers a new paradigm is often either very young or very new to the field in which they are working, since these factors partly insulate them from the dogmatic force of the status quo. Kuhn finds little supporting evidence in the historical record for the view of science as a cumulative enterprise; instead, the history of science is really a history of heroic and revolutionary change.

Kuhn defines paradigm shifts or scientific revolutions as 'non-cumulative developmental episodes in which an older paradigm is replaced in whole or in part by an incompatible new one'.[6] Such shifts can be recognized by their suddenness as well as their incompatibility with earlier theories. Due to the presence of these two features, he frequently compares paradigm shifts to the 'gestalt shifts' known from optical illusions.[7] Some of you may have seen a picture of the 'duck-rabbit' that the philosopher Ludwig Wittgenstein (1889–1951) borrowed from psychologist Joseph Jastrow (1863–1944), who had borrowed it in turn from a German humour magazine.[8] It is a drawing that can be seen with equal justice as either a duck or a rabbit, but never as both at once. In Kuhn's view, however, such gestalt shifts differ from the paradigm shifts of science in a decisive way. If we switch from seeing a rabbit in the drawing to seeing a duck instead, we can move freely back and forth between the two standpoints as often as we wish. But scientific revolutions are asymmetrical. Once we switch from Newton's smooth and empty space and time to Einstein's bent and buckled space-time, there is no going back. In contemporary parlance, we cannot 'unsee' Einstein's world once we have seen it.

Kuhn tells us that a paradigm must have two basic features. First, its 'achievement is sufficiently unprecedented to attract an enduring group of adherents away from competing modes of activity'.[9] Second, a paradigm must be 'sufficiently open-ended to leave all sorts of problems for the redefined group of practitioners to resolve'.[10] In some ways this second feature is of even greater interest. What it shows is that while paradigmatic achievements in science present a powerful new model, they are never free of gaps or unexplained difficulties. In fact, the opposite is the case. Any established theory will already have been worked out in great detail by thousands of scientists, while a new paradigm might initially be applicable only to a small number of special cases. What attracts adherents to a new paradigm is the opportunity it presents to conduct novel research on its basis, and to fill in its remaining gaps through the work of refinement and improvement.

This piecemeal enterprise of filling the holes in a paradigm is, to repeat, what Kuhn calls normal science. His view is that the vast majority of working scientists at any given moment are not interested in making revolutionary discoveries, except perhaps as a vague professional fantasy. They work instead to make further discoveries within the established paradigm. Yet this is rarely a straightforward process. After all, the practitioners of a science often disagree as to what the most fundamental aspects of the current paradigm even are.[11] In an example provided by Kuhn himself, a chemist and a physicist could not agree as to whether a single atom of helium should count as a molecule: 'For the chemist the atom of helium was a molecule because it behaved like one with respect to the kinetic theory of gases. For the physicist, on the other hand, the helium atom was not a molecule because it displayed no molecular spectrum.'[12] There may be no way to decide which of these criteria is more important, and thus scientific debate may persist as to whether a helium atom counts as a molecule or not.

For Kuhn, this means that a reigning paradigm is something deeper than any particular 'concepts, laws, theories, and points of view that may be abstracted from it'.[13] Although any paradigm will have aspects where things are explicitly defined, a paradigm also works on science at a subverbal level in ways that may take decades or more to state clearly in words. Crucially, this multilayered nature of the paradigm undercuts any conception of science as a discipline made up solely of clearly formulated equations and prose statements. Kuhn elaborates: 'After about 1630, for example, and particularly after the appearance of Descartes's immensely influential scientific writings, most physical scientists assumed that the universe was composed of microscopic corpuscles and that all natural phenomena could be explained in terms of corpuscular shape, size, motion, and interaction.'[14] This presupposition made it difficult for scientists to keep their eyes open for evidence suggesting that nature might not be made up entirely of corpuscles: that it might also have wave-like aspects, for instance. In this way, alternative intuitions about the structure of the universe are often subtly denied a fair hearing.

When left unchallenged, underlying assumptions of this sort not only have harmful effects of the kind just mentioned. They also have the beneficial consequence that shared agreement on a paradigm 'forces scientists to investigate some part of nature in a detail and depth that would otherwise be unimaginable'.[15] For example, the reigning view that matter can only be changed into different forms, not created or destroyed (this is known as the conservation of mass), prevents scientists from following countless blind alleys. On the flipside, this means that a paradigm also functions as a dogma, one that obstructs the emergence of possibly exciting new theories.

In many sciences, Kuhn notes, we can identify the first paradigmatic figures: Aristotle for motion, Archimedes for statics, Joseph Black for heat, Robert Boyle and Herman Boerhaave for chemistry, and James Hutton for geology.[16] In genetics the first paradigm dates to the discovery of the double helix structure of DNA in the early 1950s.[17] Yet in any field, inquiry or questioning begins before there is any reigning paradigm at all, a situation Kuhn refers to as the 'pre-paradigm' stage of a science. It is his view that every science has a pre-paradigm history of this sort, even if in some cases – mathematics and astronomy come to mind – that early stage is now lost in prehistoric mist. Not only does pre-paradigm research often lead to nothing but repetitive debate over basic principles, it also has the disadvantage that at this stage, 'all of the facts that could possibly pertain to the development of a given science are likely to seem relevant', meaning there is no way to know which experiments might be the most important to conduct.[18]

Kuhn frequently mentions the early seventeenth-century pages on heat by Francis Bacon (1561–1626). These include motley cases ranging from volcanoes to the warm dung of animals.[19] Good luck trying to develop a science of heat just by listing examples of hot things! As great a thinker as Bacon was, his pre-paradigm theory of heat made no distinction between important and unimportant cases; hence, absolutely anything might prove to be relevant to his theory. In similar fashion, through the work of Bruno Latour we know about the pre-paradigm state of hygiene prior to Louis Pasteur (1822–95).[20] In the mid-nineteenth century, masses of data were mounting over what sorts of things seemed to be unhealthy (living near rubbish dumps, failing to wash one's hands, leaving food in the sun for too long) but there was not yet a unified theory of germs available to explain the root cause of illness. In such periods it can be difficult even to discover new facts, since it is hard to know where to look for them.

What about optics in the period before Newton? As Kuhn reports: 'there were a number of competing schools and subschools, most of them espousing one variant or another of Epicurean, Aristotelian or Platonic theory'.[21] Since many pre-paradigm figures make important discoveries, we have no choice but to call them scientists rather than mythmakers. But Kuhn notes that the result of activity prior to the first paradigm in any field amounts to something that can't quite be called science. He observes that even the most basic principles of optics remained under dispute in the pre-Newtonian era. In fact, the disagreements about light from this period read today more like general philosophical meditations (such as Bacon's wonderfully amusing remarks on heat), not the sort of focused research on precisely defined problems that we have come to expect from mature scientific fields.

A similar example stems from the state of research on electricity in the early eighteenth century. Some schools accepted electrical attraction and repulsion on equal footing; others dismissed repulsive electrical force as secondary, while still others viewed electricity as a fluid running through a conductor. It was the discoveries of American Founding Father Benjamin Franklin (1706–90) that first put an end to this pre-paradigm chaos. 'Only through the work of Franklin and his immediate successors,' Kuhn remarks, 'did a theory arise that could account with something like equal facility for very nearly all effects.' This new understanding had real-life implications. Kuhn recognizes that the growth in understanding of electricity that ballooned after Franklin 'could and did provide a subsequent generation of "electricians" with a common paradigm for its research'.[22] The clue was provided by researchers who believed that electricity was a fluid. This theory spurred them to construct the famous Leyden jar (invented in 1745), which clearly worked as a storage device for electricity, since it gave severe shocks to many early users. Unfortunately, the theorists of electric fluid were unable to shed any light on the varying attractive and repulsive forces of electricity. It was Franklin alone who finally explained the real workings of the Leyden jar. Kuhn thinks this was enough to make Franklin's theory a paradigm, since it provided some basic principles that allowed electrical researchers to make genuine progress for the first time.

However, Franklin's theory still couldn't account for all known instances of electrical repulsion. This leads Kuhn to make an important point. In order to count as a paradigm, a theory only needs to provide a more successful general model than its competitors. Never does it, or can it, provide solutions to all existing problems in a field. Consider that Newton's theory of gravity was at first unable to explain how three bodies attract each other rather than two; it was also not very accurate in predicting the motion of the moon. Yet these puzzles were solved some decades later by researchers working within the Newtonian paradigm, and their breakthroughs lent that paradigm additional strength. For Kuhn, the main benefits of a paradigm are that it ends 'the constant reiteration of fundamentals', allows for more focused effort on the most important issues, and inspires a sense of collective confidence that researchers are finally on the right track. Once a paradigm is established, 'both fact collection and theory articulation become highly directed activities'.[23] Recalling Latour's book on Pasteur, once it was shown that germs were the primary source of disease, hygienists were able to replace their lengthy lists of do's and don't's (such as avoiding slaughterhouses, or reading comic plays when depressed) with more focused methods for preventing the spread of germs: such as washing one's hands, or not spitting on the ground.

Paradigm Pressure

Not everyone immediately converts to a new paradigm. But once it becomes broadly convincing, those who do not jump on the band-wagon find themselves increasingly isolated. Even a towering figure like Einstein spent his later years in relative intellectual solitude due to his refusal to accept the basic findings of quantum mechanics, especially the challenge it posed to causal necessity in favour of statistical probability. As Einstein famously put it in a 1926 letter to the physicist Max Born (grandfather of the popular singer Olivia Newton-John), quantum theory 'produces a good deal but hardly brings us closer to the secret of the Old One. I am at all events convinced that *He* does not play dice.'[24] (By 'the Old One', of course, Einstein means God.) When a paradigm changes, many survivors of an older world are left behind.

Paradigms are often associated with classic works in the history of science: Aristotle's *Physics* for the study of motion, Ptolemy's *Almagest* for geocentric astronomy, Newton's *Principia* for physics, Antoine Lavoisier's *Elements of Chemistry* for his field; or James Clerk Maxwell's *A Treatise on Electricity and Magnetism*.[25] But these days, books in science usually take the form either of university textbooks or popular treatments for a general audience. As science has grown more technical and specialized, it has become rare that scientific principles are packaged to appeal to an intelligent general audience in the manner of the old scientific classics. Instead, it is articles or series of articles in professional journals that have come to play the paradigmatic role. Einstein's theories of Special and then General relativity first appeared in difficult professional articles aimed at his colleagues rather than a broad readership.[26] Most general readers are quickly lost when trying to follow his arguments, whereas a wide range of readers would have been able to follow Franklin's writings on electricity at their time of publication. This is an unavoidable side effect of the increasing professionalization of science.

We have seen that for Kuhn most scientific labour is a matter of what he calls 'puzzle-solving'.[27] One important characteristic of any puzzle is that it must have a generally anticipable solution. Crossword and jigsaw puzzles test the ingenuity of those who try to solve them, but they are not true puzzles unless a solution exists. No one would purchase a jigsaw puzzle without the implicit guarantee that the pieces in the box can be made to fit together. By contrast, Kuhn notes, 'the really pressing problems, e.g., a cure for cancer or a lasting peace, are often not puzzles at all, largely because they may not have any solution'.[28] While new scientific paradigms often involve mind-blowing philosophical implications (quantum theory comes to mind), problems of this magnitude usually lie beyond the scope of the paradigm itself. Hence they are not the ones that most scientists try to solve: they are 'rejected as metaphysical, as the concern of another discipline, or as just too problematic to be worth the time'.[29] A typical scientist rarely attacks such problems as the deep philosophical meaning of quantum theory, since there are always plenty of intriguing puzzles at hand to be solved, such as discrepancies in existing measurements. As Kuhn puts it: 'On most occasions any particular field of specialization offers nothing else to do than solve puzzles, a fact that makes it no less fascinating to the proper sort of addict.'[30]

It is Kuhn's view that the quasi-dogmatic commitment of scientists to a paradigm does not just provide a solid framework for normal science. The very rigidity of a paradigm also paves the way for eventual revolution, by pressing researchers up against the edge of the framework in which they operate. As a paradigm reaches greater completeness, certain anomalies appear, and 'discovery commences with awareness of anomaly'.[31] For example, today's physics cannot yet explain the astronomical discovery of 'dark matter', which refers to the observation that there seems to be too much gravity in galaxies compared to the amount of matter we can see. It also cannot account for 'dark energy', which refers to the fact that the universe seems to be expanding more quickly than astrophysics says it should. As Kuhn sees it, anomalies or new discoveries can only be detected against the background of a relatively rigid paradigm that leads researchers to expect one result but find another. An otherwise stable scientific environment allows a new discovery to stand out like a strawberry in the snow.[32] For instance, before there was any paradigmatic theory of gravity, no one could have noticed dark matter in the first place. In some cases, however, scientific anomalies can persist for long periods without causing scientific crisis. Consider the failure to find the predicted elementary particle dubbed the Higgs boson until 2012, four decades after it was first baked into the Standard Model of particle physics.[33] If the Higgs boson had never been found, the Standard Model might have begun to appear shaky, though a four-decade delay was not enough to cause widespread scientific panic. More gravely, the two pillars of the past century of physics (quantum theory and General relativity) are still incompatible, but physics forges onwards nonetheless.

But even in cases where counter-evidence begins to gather against a reigning paradigm, there are still at least two different things that can happen. One scenario is that most scientists continue to see mere puzzles where a smaller number sees full-blown anomalies. In such cases normal science continues as usual, aside from a fringe group of would-be innovators who feel a crisis at hand.[34] Departments of physics are dominated today by String Theory, which assumes that all matter is made of tiny, vibrating strings. Although a certain portion of physicists are alarmed by problems with the theory, others view them as eventually soluble, and continue to promote strings among their doctoral students. It is still too early to know what will happen in this case.

Another possible scenario is that the anomalies lead to a collapse of faith in the existing paradigm more generally, leading to what Kuhn describes as times of 'pronounced professional insecurity'.[35] For instance, in the period just before the heliocentric astronomy of Copernicus was introduced, there was widespread frustration with the state of Ptolemy's theory, which assumed the earth to be at the centre of the universe.[36] During this era, 'astronomy's complexity was increasing far more rapidly than its accuracy', always a bad sign for a scientific discipline.[37] It is discouraging when a scientific theory is held together with ad hoc adjustments and needlessly complex equations, just as it would be unnerving to drive a car held together by duct tape and super glue. Such crisis situations, which Kuhn says may last for 'a decade or two' before a new paradigm emerges, have certain recognizable characteristics.[38] For one thing, in crisis periods we often find a 'proliferation of different versions of a theory'. A good example would be the massive number of competing theories of combustion that existed prior to Lavoisier's theory, which finally demonstrated that oxygen is what enables fire to burn.[39] For another, scientific debate begins once more to resemble the pre-paradigm period, with researchers arguing over basic concepts and struggling with problems that are often more philosophical than scientific in character.[40] To some extent this is happening today with various competing theories of dark matter and dark energy. One example is the Modified Newtonian Dynamics proposed by Israeli physicist Mordehai Milgrom (b. 1946), which aims to eliminate the problem of dark matter by making a slight modification to Newton's Second Law of Motion without incorporating Einstein's General relativity.[41]

More often than not, the crisis of faith in an existing paradigm is resolved only when a new paradigm is ushered in. Barring a fresh alternative, few will abandon the old one simply because of its mounting difficulties.[42] Instead, scientists will tend to make a series of minor adjustments to keep the old theory afloat, like patches on a rusting ship.[43] A textbook example would be the numerous 'epicycles' added to the circular orbits of planets around the earth in Ptolemy's geocentric astronomy. Whenever the supposedly circular motions of Mars or Jupiter failed to match the theory with sufficient precision, astronomers would add circles onto circles, and sometimes further circles onto the new ones, making clumsy and confusing modifications to what was originally an elegant geometrical system.

Kuhn adds the intriguing point that the germ of a solution to any scientific crisis was often thought up by someone in the past, though it was not adopted at the time because no crisis had yet arisen.[44] As early as the third century BCE, the Greek astronomer Aristarchus of

Samos had already proposed a heliocentric theory of astronomy. Unfortunately for Aristarchus, there was no good reason for his theory to catch on at the time: the geocentric view that placed the earth at the centre of the cosmos was doing quite well, and seemed a better fit for the apparent movement of the sun through the sky in everyday life. It was only later that the general crisis in Ptolemaic astronomy led Copernicus back to the heliocentric option. Another, later example concerns the nature of space and time, as debated in the early 1700s between Newton's follower Samuel Clarke (1675–1729) and the philosopher G. W. Leibniz.[45] Whereas Newton claimed that space and time were empty containers unaffected by anything happening inside them, Leibniz countered that space and time must be generated by the relations between objects themselves. Since there was no observational evidence at the time to support Leibniz, Newton prevailed in that controversy. Yet from today's perspective Leibniz looks like a prescient forerunner of Einstein's notion of space-time as stretched, compressed and bent by mass.

The Two Faces of Paradigm Shifts

There is a further subtlety in Kuhn's view of paradigm change that is seldom if ever discussed, though it has decisive importance for our broader discussion of the continuous and the discrete. This comes when he distinguishes between two moments in the process of discovery: there is the discovery 'that' something is, but also the discovery of 'what' it is. At first glance this distinction may look boring or pedantic; in fact, it turns out to have deep philosophical roots. Kuhn's initial example comes from the aforementioned difficulty of knowing exactly who made any particular scientific discovery. Consider the case of oxygen. The first person to prepare a pure sample of the gas was one C. W. Scheele (1742–86) in Sweden, though his achievement is usually ignored since it was not published until after others had already written in detail on the topic. That leaves two major contenders for the prize: Joseph Priestley (1733–1804) in Britain and Antoine Lavoisier (1743–94) in France. Priestley had collected oxygen, but not in pure form. He also managed to misidentify it twice: thinking in 1774 that it was the already known nitrous oxide, and in 1775 that it was 'dephlogisticated' air ('phlogiston' was once thought to be an element released in combustion). As for Lavoisier, he announced in 1775 that the gas was 'the air itself entire' (which is untrue given that oxygen is just one component of the earth's atmosphere), but by 1777 he had concluded that it was something new altogether. For the remainder of his life, until he was guillotined during the French Revolution, Lavoisier wrongly believed that oxygen existed only in combination with 'caloric', a substance then believed to be the source of heat.

Kuhn's tentative conclusion from all this is that oxygen was not discovered before 1774, though it was surely discovered by 1777 'or shortly thereafter'.[46] Between these two dates we have a grey zone in which it is impossible to pinpoint the discovery of oxygen with precision, given the ambiguity between the detected features of the gas (the 'what') and the fact 'that' a new gas had been discovered. Kuhn later toys with the formulation that Priestley 'discovered' oxygen and Lavoisier 'invented' it, linking these familiar opposites with the distinction between 'that' and 'what', though I think the reverse formulation would make more sense. After all, Priestley was the first to determine its features ('what') while Lavoisier came closer to realizing that a novel substance was involved ('that'). In any case, this distinction adds an interesting feature to Kuhn's theory of paradigm shifts, one that is actually the chief lesson of the present chapter. Although Kuhn had previously said that paradigms can emerge suddenly, even 'in the middle of the night', it now turns out that this does not occur instantly, even if it can happen relatively quickly. Kuhn and others regard the discovery of oxygen as the spur to what historians today call the chemical revolution. But rather than occurring in a single Eureka moment, it was a discovery that 'involved an extended, though not necessarily long, process of conceptual assimilation'.[47]

Another of Kuhn's examples concerns the 1895 discovery of X-rays by Wilhelm Röntgen in Würzburg, Germany. This began when he noticed that a screen was glowing unexpectedly. It might have seemed to be a minor discrepancy, since it was already noticed by others in the past: William Morgan as early as 1785, Johann Hittorf in 1869, and Philipp Lenard in 1888.[48] But Röntgen soon concluded that the glow was caused by something new and important, though time was needed to work out its exact properties. As Kuhn remarks: 'We can only say that X-rays emerged in Würzburg between November 8 and December 28, 1895.'[49] There is the additional intriguing fact that X-rays did not violate any existing scientific paradigm, but were simply not included in the mainstream science of the time. This was a bit like discovering a new animal that did not contradict the existing principles of zoology. Nonetheless, the new rays 'were greeted not only with surprise but with shock'. The prominent scientist Lord Kelvin (1824–1907) mistakenly declared them an 'elaborate hoax'.[50]

There were good reasons for this resistance. Since the cathode ray tubes that led to the accidental discovery of X-rays had long been used in laboratories, much of the work that had been done with these tubes would have to be redone: after all, the previously undetected rays might have influenced the outcome of many experiments without anyone realizing it. This meant that X-rays presented a number of hassles for scientists that made the existence of these rays highly inconvenient. Here again we see the inner tension between the fact 'that' a thing is (as a discrete but still unknown entity) and the question of 'what' it is (which refers to a more continuous range of possible properties). Despite his earlier contention that paradigm shifts can happen in the middle of the night, Kuhn notes that major transitions do not occur in an instant, but unfold in an intermediate grey period. Yet the final step must occur with relative suddenness: 'because it is a transition between incommensurables, the transition between competing paradigms cannot be made a step at a time, forced by logic and neutral experience. Like the gestalt switch, it must occur all at once (though not necessarily in an instant) or not at all.'[51] Although the phrase 'though not necessarily in an instant' is enclosed in parentheses, as if it were almost an afterthought, the cases of oxygen and X-rays show that a genuine stretch of time is needed to nail down both the 'that' and the 'what' of a new discovery.

This ambiguity between two moments of a paradigm shift is made even clearer in Kuhn's later book on black-body radiation.[52] In physics, a black body is an imaginary case of an object that perfectly absorbs all radiation while reflecting none. Idealized models of this sort can be very useful for science even though they never occur in such perfect form in nature. Once a black body reaches thermal equilibrium with its environment, it will begin to emit radiation. There were a number of difficulties with calculating this radiation, and with ensuring its consistency with the values measured by experiment. The eminent German physicist Max Planck (1858–1947) eventually caused a revolution with a surprising result: the problems with black bodies can be solved if we assume that the radiation cannot be emitted in just any amount, but only as multiples of some smallest unit of radiation. In other words, black bodies can emit one unit of radiation, or two, or seventy-five, or 1 million, but not 3.5 units or 76.1435 units. This smallest unit of radiation is the famous 'quantum' after which quantum theory was named. But just as with oxygen and X-rays, there is a question as to when the quantum was first discovered. Although Planck's breakthrough work on the topic occurred in 1900 and 1901, Kuhn argues that in these early years Planck only viewed the quantum as a convenient mathematical device.[53] It was not until 1909, after digesting the criticism of such leading scientists as Einstein and Paul Ehrenfest (1880–1933), that Planck realized quanta must truly exist as the tiniest physical units in nature.[54] Here again, a paradigm shift needed two steps to be completed. We will return to the point in the concluding section of this chapter.

Incommensurability and Metaphor

One of Kuhn's major themes is incommensurability. This refers to the incompatibility between our respective views of the world before and after a paradigm shift occurs. Under the old view of science as a series of cumulative changes, one could say that Aristotle gave us his theory of physics, that Newton then got us a bit closer to the truth, and then Einstein got us even closer. Kuhn insists that this is not how scientific advancement works. Incommensurability means that 'the referents of some of the terms which occur in both theories are a function of the theory within which those terms appear'.[55] What he means is that even when the same word is used in two different theories, the meaning can be so drastically different in these cases that the terms cannot really be compared. Furthermore, 'there is no neutral language into which both of the theories as well as the relevant data may be translated for purposes of comparison'.[56]

Let's stick with the example of Newton and Einstein. It is common to hear that Newton's physics was not refuted by Einstein's, but simply expanded to cover a broader range of cases. This would mean that Newton's equations are still correct when describing objects that are not very massive or not moving very close to the speed of light, whereas Einstein's theory covers such outlier cases as huge black holes and travel at close to light speed. Kuhn disagrees emphatically with this view. The reason is that mass has an altogether different meaning in the cases of Newton and Einstein: the word is the same, but the differing ideas it signifies are incommensurable. 'Mass' for Newton is conserved, meaning that matter can only be changed into a different form of matter. But for Einstein mass is convertible with energy, as in $e=mc^2$, his best-known equation.[57] It only took forty-seven kilograms of uranium (the so-called 'critical mass') to make the first atomic bomb, showing that a relatively small amount of mass can be transformed into enough energy to destroy a city. Newton's physics was utterly incapable of predicting such a thing. In an important sense, Kuhn concludes, scientists today live in a completely different world from those who followed Newton, even if they are using some of the same terminology. Therefore it is not simply a matter of scientists interpreting the same reality differently, but of living in altogether incompatible worlds.[58]

Along the same lines, Kuhn notes the contrasting meanings of the word 'planet' for astronomers before and after Copernicus. In the pre-Copernican period the moon was treated as just another planet orbiting the earth. The earth itself was thought to be motionless and therefore not a planet at all (the word 'planet' comes from the ancient Greek word for 'wanderer'). But after Copernicus, earth was understood to be one of the planets orbiting the sun, while the moon was reclassified as something called a 'satellite', a previously unknown category. In Kuhn's words: 'Eliminating the moon and adding the earth to the list of individuals that could be juxtaposed as paradigms for the term 'planet' changed the list of features salient to determining the referents of that term.'[59] In short, the word 'planet' entails entirely different sets of qualities before and after Copernicus, something Kuhn thinks will happen following any paradigm shift. Nor did the process end with Copernicus. In 2006 I happened to be in Prague on the exact day that the International Astronomical Union (IAU) met there and voted to demote Pluto from 'planet' to 'dwarf planet'. There were several reasons for this, but they boiled down to the fact that if Pluto were still to be considered a planet, then many other dubious objects would have to be called planets as well: including the large asteroid Ceres, along with such exotic Kuiper Belt objects as Orcus, Makemake and Sedna. The definition of 'planet' will no doubt continue to evolve, especially as new planets continue to be discovered in orbit around stars distant from our own.

The question of how to deal with these variations in the meaning of key words was discussed in an important current in early 1970s analytic philosophy known as the theory of direct reference. Such thinkers as Saul Kripke (1940–2022), Hilary Putnam (1926–2016) and Keith Donnellan (1931–2015) argued that we can use a term to point to the same object over time even if we turn out to be completely wrong about its meaning.[60] For example, although statements such as 'Christopher Columbus discovered America' or 'Albert Einstein invented the atomic bomb' are wrong, even when corrected no one doubts that the statements still refer to the same person as before.

The name 'Columbus' always points to the same human being, even if we were wrong about most or all of his qualities. This differs from earlier theories which claimed that a name is a shorthand abbreviation for all the qualities possessed by the object it names. Significantly, the theory of direct reference aligns perfectly with the earlier distinction we made between 'that' and 'what', which we saw to be a central feature of Kuhn's theory of paradigm shifts. The reference side of a word refers to the 'that', and the meaning side to the 'what'. For instance, at this moment in time the *reference* of the words 'Roman Catholic Pope' is Leo XIV (born in Chicago as Robert Francis Prevost), while the *meaning* of the phrase is something more like 'supreme authority of the Church, and successor of St Peter, who resides in the Vatican and generally serves for life'.

Although Kuhn appreciates the benefits of the theory of direct reference, he cautions us not to push it too far. While he understands that there is some continuity in the denotation of words – 'earth' has always referred to the place where we live, despite considerable changes in our understanding of that place – the vast differences in what Priestley and Lavoisier mean by 'oxygen' or what Newton and Einstein mean by 'mass' cannot be erased by assuming that the same word always points to the same thing.

To develop this point, Kuhn turns to the theory of metaphor. In particular, Kuhn shared my admiration for the 'interactive' theory of metaphor introduced by the philosopher Max Black (1909–88).[61] While it might be assumed that a metaphor draws our attention to pre-existing similarities between two objects, Black insists that metaphor is what first *produces* many of the similarities in question. For Black as for Kuhn, 'man is a wolf' does not mean the same thing as any literal statement, such as 'humans and wolves are both violent carnivores with hierarchical systems of leadership'. The metaphor goes beyond all literal statement, just as a paradigm in science goes beyond the set of explicit rules and procedures to which it gives rise, and just as an Aristotelian substance outstrips any possible definition of it. Much like a paradigm shift, a metaphor creates a new world, one in which humans and wolves are united in a new hybrid entity. For Kuhn, metaphors, analogies and scientific models are an 'irreplaceable part of the linguistic machinery of a scientific theory'.[62] Notice that unlike literal statements, metaphors are irreversible, a point that is my own rather than Kuhn's.[63] For instance, we can utter the literal statement 'Amsterdam is like Istanbul', perhaps in the sense that both are historic cities where water is a prominent urban feature. If we reverse the statement and say instead that 'Istanbul is like Amsterdam', the same comparison holds, even though one can always dispute its accuracy.

But the same is not true of metaphor. We cannot reverse Black's 'man is a wolf' into 'a wolf is a man' without changing the metaphor completely. What this shows is that the two terms in a metaphor play different and asymmetrical roles. 'Man is a wolf' provokes us to imagine a thing called man with wolf-like qualities; 'a wolf is a man' does the reverse. Metaphor is therefore reliant on the distinction between 'reference', which points directly to a thing apart from its known features, and 'meaning', which describes the thing precisely by listing its features. As already seen, these terms are mirrored by Kuhn's own distinction between 'that' and 'what' in scientific discovery. Priestley and Lavoisier discovered certain features ('what') that they mistakenly ascribed to objects already known, and needed some time to find out that they belonged to a new and unexpected object ('that'). With Röntgen and X-rays, the same thing happened in reverse: he knew 'that' something new had affected his experiment, but needed some weeks to discover 'what' exact features these new rays possessed.

To name something in science does not mean that we understand all or even most of its features: instead, we always point vaguely at something whose exact character is still unknown. By way of example, Kuhn notes that when scientists apply terms such as 'mass', 'electricity', 'heat', 'mixture' or 'compound' to nature, it is not usually by acquiring a list of criteria necessary and sufficient to determine the referents of the corresponding terms.[64] This seems to favour Kripke and Putnam's theory that a name always points to the same thing even when its meaning (the qualities a thing has) changes. This gap between a thing and its qualities, between its reference and its meaning, also helps explain why we don't invent a new word whenever a meaning changes. Although 'planet' means something different before and after Copernicus, the word is retained. A certain core of meaning persists: the support of the movement that guides the etymology of the word.

Kuhn calls himself an 'unregenerate' realist, meaning he believes in a world that exists independently of the human mind.[65] Yet he is also frequently accused of being a relativist who holds that 'truth is constructed by society', a view he viscerally loathed. But in all fairness, Kuhn brought some of this on himself by comparing a scientist's allegiance to a new paradigm to a religious conversion.[66] How, then, can we reconcile the two faces of Kuhn? There is the realist face, which smiles upon an objective reality outside the human mind. Yet there is also the face which says that a change in meaning brings about an actual change in the world: that Newton and Einstein are not speaking of the same thing when they speak of mass.

Kuhn would say that what we find in the history of science are not gradually closer approaches to truth in the same direction, but the sudden reversals and counter-currents brought about by paradigm shifts. In important respects Einstein's General relativity is closer to Aristotle than to Newton, just as Picasso's cubism is closer to medieval Byzantine art in its flatness and abstraction than to the intervening paintings of the Renaissance. Realism, in its usual form, asks us to modify our language to match up better with reality. By contrast, Kuhn suggests that reality should match up with language instead.[67] This is because the meanings found in language change the very world in which we live. The pre-Copernican and the post-Copernican astronomer are not using the same words as part of different theories; instead, they inhabit different worlds altogether. If Kuhn were a full-blown relativist or social constructionist who thinks reality is produced by human thought, this view of language would come as no surprise. It is more jarring that Kuhn proclaims both realism *and* the apparently contradictory idea that the changing meaning of a term also affects its reference. This relates importantly to what Kuhn described elsewhere as 'a deep duality in the concept of meaning'.[68]

The Problem of Retroactivity

We have seen that for Kuhn revolutions occur retroactively, not in the moment they are later thought to have happened. This is why it is sometimes hard to know exactly when paradigm shifts take place, even long after the fact. As we have seen, Kuhn proclaims that oxygen was discovered sometime between 1774 and 1777, or was perhaps first 'discovered' and then 'invented' shortly thereafter. In his view this slippage happened again with Planck and quantum theory, which in one sense was launched in 1900–1901 but in another sense did not begin until 1909, when Planck was finally persuaded by critics that quanta must really exist in nature rather than merely amounting to a mathematical technique.

Based on this evidence, one might draw the conclusion that scientific revolutions or paradigm shifts are 'merely subjective', since it is only with hindsight that anyone is in a position to say when a revolution happened. We might address this problem by saying instead that a paradigm shift happens at some definite objective moment, but that historians have intellectual or psychological difficulty in figuring out exactly when. This is an improvement. When Kripke and Putnam argue that we can refer to objects rigidly without knowing exactly what qualities they have, they are not talking about a difference produced *ex nihilo* by the human mind, but one that exists in reality. In the terms of medieval philosophy, the difference between existence (that there really is such a thing as quanta) and essence (that quanta have certain specific features) is not found only in human understanding, but is there in the world itself.

Let's consider the important eighteenth-century Scottish philosopher David Hume (1711–76), who denied that objects exist as anything more than bundles of qualities.[69] As Hume has it, when certain sets of qualities occur together frequently – such as red, spherical, hard, cold, juicy and sweet – we give them a joint nickname: 'apple'. Yet for Hume and his devotees there is not really any such thing as an apple, because we cannot say what an apple consists in other than the aforementioned qualities themselves. From Kuhn's standpoint things do not look quite this way. Imagine that something unexpected is discovered. Initially it is unclear whether something new has appeared, or whether this is a mere glitch in experimental technique. There is no sense yet that we are hunting for any sort of new object. But eventually, through a modification of the thing's observed qualities past a certain critical point, it is clear that we are no longer dealing with a bundle of familiar qualities, such as 'red' and 'juicy' for an apple, but with an underlying object whose existence was previously unsuspected. The same can happen in reverse: perhaps we are absolutely sure that we have discovered an unprecedented object, but it takes a good deal of time to work out what qualities it has.

In any scientific discovery, there is a crucial human element at play, and that element is *belief*. In the late 1950s the Hungarian-born chemist and philosopher Michael Polanyi (1891–1976) published a book, much admired by Kuhn, about how knowledge entails a degree of personal commitment.[70] In similar fashion, Kuhn seems to suggest that a new object is finally established when Röntgen is convinced that X-rays are there, or when Planck is finally persuaded that quanta exist in nature and not just in calculations. The role of belief in science helps explain why Kuhn controversially likened new paradigms to religious conversions, a view that earned him a good deal of abuse from critics.[71] But once again, it is the theory of metaphor that will help explain why Kuhn saw a link between science and belief.

Earlier we noted that literal statements are reversible but metaphors are not. We can easily flip the order of 'Amsterdam is like Istanbul' into 'Istanbul is like Amsterdam', but cannot do this in the case of 'man is a wolf' without changing the nature of the metaphor. Yet there is an even more important difference between the two cases, one foreshadowed by the branch of philosophy called 'speech-act theory'.[72] This subfield, of which the British philosopher J. L. Austin (1911–60) was among the leading figures, begins with an important difference between what it calls 'constative' and 'performative' statements.[73] If I write 'the temperature in Los Angeles is currently 29 degrees Celsius', this at least seems to be a neutral, descriptive statement. But the situation seems different if I say the following: 'I promise that I will give you the accurate figure for the temperature in Los Angeles after I check it in just a moment.' The latter statement entails a clear commitment on my part, and has certain consequences. If it is later discovered that I lied about the temperature, or if I fail to send the reading quickly or in legible form, then my reputation for honesty and reliability will be damaged. Another way of putting it is that in the second case I am more deeply committed to my statement than in the first, since I myself have become an explicit ingredient of the proposition.

An analogous difference can be found in the cases of literal and metaphorical statement. When I say 'Amsterdam is like Istanbul', I am making a constative or literal statement about the resemblance between two cities, drawing attention to what I take to be their shared qualities. It is true that someone could disagree, pointing to many important differences between these two places, leading to a factual dispute of the sort that we find in some conversations. But with a metaphor, even in the relatively banal case of 'man is a wolf', the intimate participation of the reader is required. If we are not bored or distracted and actually try to grasp the metaphor, we will find that the literal equation of man and wolf does not work. We are unable to stay at the level of merely comparing qualities, since the resemblance of human and wolf is far-fetched in any literal sense.

In other words, with Black's 'man is a wolf' we have a mysterious object 'man' that somehow has wolf-qualities in a way that it normally would not. And while wolf-qualities are not hard to come by in our minds (ferocity, pack hierarchy), a wolf-like man is unfamiliar, Hollywood werewolf films notwithstanding; note that at any given moment a werewolf is either wolf or man, while in the metaphor it is both at the same time. As a result, we must step in ourselves and try to produce the new unified object that is said to fuse wolf-qualities with 'man'. We ourselves mentally become the man with the qualities of a wolf, and the same would be true even if the metaphor were 'a lion is a wolf': in that case, we would simply try to perform mentally the mysterious lion that is said to have wolf-qualities. In Austin's terminology, the metaphor is more a performative act on the part of the person creating or hearing it than a constative or literal one, just as so-called 'method actors' try to become mentally the person or thing they portray rather than simply imitating their salient features.[74]

This is why Kuhn's distinction between 'that' and 'what' entails that discoveries or paradigms can only be retroactive. This helps clarify one of the problems that stumped him early in his career: why is it often so difficult to date scientific discoveries with any exactitude? A deep involvement by the observer is needed to reconcile the two mutually resistant poles: one object fusing with the qualities of another. This point had not gone unnoticed by Kuhn's great predecessor in the philosophy of science, Karl Popper (1902–94). Popper famously proposed that science is less about 'verifying' our theories (since even cranks and pseudo-scientists claim to have verifying evidence) than about aggressive attempts to 'falsify' them.[75] In other words, scientists must actively test their current theories in a genuine effort to prove them wrong, not to prove them right or establish them as airtight. We can never say that any given scientific theory is true, only that it has not been falsified so far. Statements can be considered scientific only if there are possible events that would lead us to abandon them.

Less famously, Popper added that attempts to falsify scientific theories must be 'sincere', a word not often encountered in scientific writing. Stated differently, Popper claimed that scientists must be fully involved with or committed to the entities whose existence they defend. A scientist who tested certain theories 'ironically' would be a disturbing figure indeed, even if they discovered interesting results. We also remember the case of Planck, who according to Kuhn was not a real quantum theorist until 1909, nine years after he supposedly discovered the theory. The latter year was when his newfound commitment to real indivisible units in nature made him a scientific radical for the first time.

The French philosopher Alain Badiou (b. 1937) makes a similar point that is relevant here. In his view philosophy is always shadowed by an 'anti-philosophy' that emphasizes lived commitments over abstract theories.[76] Such anti-philosophy can be found in the works of Blaise Pascal (1623–62), Søren Kierkegaard (1813–55), or Nietzsche, all of whom despised academic discourse for its lack of connection with the basic decisions that structure human life. This is probably why Badiou, in somewhat Kuhnian fashion, holds that truth has an inherently retroactive character. For instance, no political event is a revolution except in the fidelity of later militants who commit themselves to it after the fact.[77] For the arch-Leftist Badiou, the French Revolution happened most fully not in its moment of occurrence, but in the attempts to re-enact it by later revolutionaries in Russia and China.[78]

Such considerations help explain Kuhn's doubt that science comes 'increasingly close' to the way nature really is. While Kuhn says he remains a realist who believes there is more to the world than minds, he is also aware that the mind itself is part of the world: human decisions are as much a part of reality as oxygen, X-rays or quanta. Many philosophers take pride in their ability to 'carve nature at the joints', as if the distinctions made by their theories were the same as the distinctions present in reality itself.[79] But we must remember that when we construct theories about the world, the mind itself is yet another joint to be carved. Objectivity comes not from pretending that humans stand outside nature as external observers, but from recognizing that human thought and decision are no less real (but also no more real) than apples, or than the smaller particles of matter that make up an apple. What Kuhn's theory of retroactivity suggests is that the mind is yet another causal force in the construction of nature itself.

But more importantly for us, Kuhn's recognition that scientific revolutions are linked with the gap between the fact 'that' a thing is and 'what' it is gives a new twist to the problem of the continuous and the discrete. For it has turned out that every scientific object incorporates both of these moments. A thing either exists discretely in the world as a real thing distinct from others, or it doesn't, yes or no. But whether it exists in reality or only in the mind, it possesses a certain number of qualities that can vary continuously in intensity without turning the thing into a different thing. On this note we turn our attention to evolutionary biology, one of the fields where the debate between the continuous and the discrete has echoed the loudest.

6.

Creeps and Jerks: Evolutionary Theory

The American palaeontologist Stephen Jay Gould (1941–2002) was one of the most prominent recent authors to have defended a theory of reality as consisting of discrete jumps rather than gradual and continuous change. Along with co-author Niles Eldredge (b. 1943), Gould was one of the founders of 'punctuated equilibrium', a theory claiming that the evolution of species happens with surprising suddenness. This opposes the more orthodox rival theory – Eldredge and Gould call it 'phyletic gradualism' – which maintains that species change happens in the continuous Darwinian manner of reproduction and mutation followed by the slow work of natural selection. The dispute between the two camps is often summarized in the following humorous terms. When Gould's enemies mockingly described punctuated equilibrium as 'evolution by jerks', he replied that the gradualist account should be called 'evolution by creeps'.[1]

We will find Gould in dispute with Richard Dawkins (b. 1941), born the same year and in many respects his chief intellectual enemy. Whereas Charles Darwin (1809–82) saw natural selection as unfolding primarily on the level of individual organisms, his admirer Dawkins shifted our focus to an even smaller level of selection: genes. By 'levels' I mean the different scales of life on which evolution might be thought to occur, from genes to entire groups of species (or 'clades'). Dawkins argues for genes as the primary level of evolution, while for Gould there are multiple levels of evolutionary selection rather than just one. In Gould's words: 'The hierarchical theory of selection recognizes many kinds of evolutionary individuals, banded together in a rising series of increasingly greater inclusion, one within the next – genes in cells, cells in organisms, organisms in demes, demes in species, species in clades.'[2] Given that we human beings are organisms, we tend to think of the organism as the level where everything important happens. This means we run the risk of assuming that all evolution works in the way most relevant to us as individuals: namely, adaptation to our current environment.[3]

By contrast, Dawkins's famous idea of the 'selfish gene' assumes that individual creatures are just vehicles or containers for genes, which play the primordial role in evolutionary struggle.[4] The chief aim of genes is to reproduce themselves even if it harms us as individuals. Dawkins's theory, put forward to great acclaim in his 1976 classic, *The Selfish Gene*, has both an appealing simplicity and a fascinating counterintuitive power. Domestic pigs, ducks and rabbits might seem foolish as they linger placidly in human captivity only to be slaughtered and eaten every time. Yet if we view this situation from Dawkins's standpoint, we find that the genes of these animals do rather well. Although every animal is eventually eaten, their genes reproduce more reliably in human captivity than they would in the wild. It's almost as if individual animals were tricked by their genes into delivering themselves to annihilation in a way that benefits the genes alone. Is it possible that our human genes play similar terrible tricks on us? The psychoanalyst C. G. Jung (1875–1961) once mused that if every human were to choose a mate based solely on instinctive sexual attraction, it would probably be healthier for the species as a whole (he seems to assume this would produce the strongest offspring) but psychologically miserable for individuals.[5] Although Jung is speaking hypothetically, he can hardly conceal the dark implication that human sexuality is a parade ground of choices disastrous for individuals, its decisions steered by the interests of a different level of selection than the individual organism.

Since genes are very small, and since they mix only intermittently through sexual reproduction, it can take many generations for specific genes either to establish their dominance in a gene pool, lose prominence or even disappear. This lengthy process requires

Dawkins to commit to a gradualist model of evolutionary change rather than the theory of sudden jumps preferred by Gould. Before getting into the long-standing debate between gradualists and saltationists (the latter view evolution as a series of discontinuous leaps), let's look more closely at punctuated equilibrium and why this theory was proposed in the first place.

Punctuated Equilibrium

In 1972 Eldredge and Gould published their path-breaking article 'Punctuated Equilibria: An Alternative to Phyletic Gradualism'.[6] Their title was inspired by the famous gaps in the fossil record long faced by palaeontology. In many or even most cases, we lack fossil evidence of the slow transition between species predicted by Darwin's work. Such gaps have generally been treated in one of two opposing ways. The first option, favoured by orthodox Darwinians and most other scientists, says that changes from one species to another are simply the long-term byproduct of the reproduction and selection of individuals over extremely long periods of time. In principle this means that we ought to find lengthy series of fossils displaying gradual changes from one to the next. But although there isn't much evidence to back this up, Darwin's theory seems so basically sound, and so powerful in explaining so many things, that his adherents understandably stand firm. With time, they say, the fossil record ought to become more complete and the problem of gaps will disappear.

It's worth mentioning that Darwin himself, known for his intellectual honesty, was personally troubled by this mismatch between his theory and the fossil record. In his great book on the origin of species he wrote as follows: 'Why then is not every geological formation and every stratum full of . . . intermediate links? Geology assuredly does not reveal any such finely graduated organic chain; and this, perhaps, is the most obvious and serious objection which can be urged against my theory.'[7] Strong commitment to

Darwin's ideas often leads his admirers to de-emphasize this very problem, though their hero admitted it clearly enough. There is no better example of how allegiance to a specific theoretical model can encourage researchers to downplay facts that speak directly against their own favoured views. Kuhn already noted this point when discussing the dogmatic force of reigning paradigms; over the past century and a half, Darwin's paradigm has been nothing if not dominant.

The second, opposite attitude is typically found among creationists. For them the gaps in the fossil record prove straightaway that Darwinian evolution cannot have occurred. God must have intervened to create all the new species throughout the history of the earth, assuming he did not just create all species at the start, as the Bible suggests. But contrary to popular belief, religiously minded creationists are not the only ones to have emphasized the gaps in the record. In previous centuries such reputable scientists as Georges Cuvier (1769–1832) and Louis Agassiz (1807–73) argued for 'catastrophism' in geology as a way of explaining the empty spaces in the available evidence.[8] That was because they saw both evolution and geological change as resulting from intermittent disasters rather than gradual change.

A different sort of non-gradual approach to evolution was proposed by the Dutch botanist Hugo de Vries (1848–1935). His 1886 discovery of strange primrose plants that produced offspring bizarrely different from their parents led him to formulate a 'mutation theory', according to which evolution occurred through sudden genetic modifications rather than gradual change.[9] His wide influence ended only in the 1930s at the hands of the so-called 'modern synthesis' (a term coined after the fact by humanist biologist Julian Huxley). The modern synthesis blended Darwinian evolution with the previously ignored plant-breeding experiments of the Austrian monk Gregor Mendel (1822–84).[10] Celebrated though he is today, Mendel's pioneering work had only been cited three times in footnotes from its appearance in 1865 till 1900, or approximately once per decade. Although de Vries was among those who rediscovered the work of this diligent monk, it was ironically the resurgence of Mendel's ideas that would put an end to de Vries's own influence, by giving a new lease of life to Darwinism.

The modern synthesis addressed perhaps the central problem faced by Darwin's theory: the lack of any firm explanation of how heredity worked. It was clear enough from experience that blond-haired and blue-eyed humans give birth to children with these same features, and that fast horses tend to have fast baby horses. As to why such things happened, Darwin could offer no explanation, but Mendel could. This is what made the combination of Darwin's and Mendel's ideas so powerful. With the modern synthesis of these two figures, the attention of biologists shifted to population genetics, the exploration of how the gene pools of a species shift over time. This helped preserve the Darwinian model of a gradual process governed by natural selection. Continuism lived to see another day.

But according to Eldredge and Gould, Darwin's vague hope that the fossil record would become more complete at some future date rendered his account of species change 'virtually unfalsifiable', meaning that it was accepted more or less as a matter of scientific faith.[11] This is a problem, since the gradualism of Darwinian theory implies a number of consequences not supported by the available evidence.[12] Eldredge and Gould note with befuddlement that their contemporary Gerrit Neef found evidence of 'saltation', an abrupt evolutionary change in a group of Plio-Pleistocene snails, but failed to draw the needed conclusion that Darwin was wrong in his gradualist views. Instead, after briefly considering that a new account of evolution might be needed, Neef drew the feeble conclusion that a non-Darwinian approach 'cannot be entirely excluded'.[13] This is the academic version of hedging one's bets rather than taking a justified risk.

More specifically, Eldredge and Gould complain that Neef remained trapped in the workings of what Kuhn called 'normal science'; in making this point they even cite Kuhn by name. As we saw, puzzle-solving normal science within the textbook framework of the existing paradigm is 'a strenuous and devoted attempt to force nature into the conceptual boxes supplied by professional education'.[14] Eldredge and Gould state their own very different intentions boldly: 'We wish to consider an alternate picture to phyletic gradualism; it is based on a theory of speciation that arises from the behavior, ecology, and distribution of modern biospecies.'[15] What they were announcing was no less than a theory of evolutionary change as discontinuous rather than continuous.

The authors state that in support of their theory, they will not be relying on the fossil record. To be clear, no particular theory of evolution can be confirmed by direct examination of this record, which to this day remains relatively sparse. Fully aware that this is the case, Gould and Eldredge note as follows: 'if discrepancies are found between paleontological data and the expected patterns, we may be able to identify those aspects of a general theory that need improvement. But we cannot formulate these improvements ourselves.'[16] This is to say, they are open to interpretations of the gaps in the record other than their own. Yet this cautious statement understates their findings considerably, since they are perfectly convinced that evolution must be sudden rather than gradual.

Even if Eldredge and Gould cannot seal their case with data from the fossil record, they are confident in their model because of a different theoretical breakthrough: allopatric speciation. The word 'allopatric' in Greek refers to something that comes from another country. When the term is used in conjunction with 'speciation', it means that species evolve primarily through geographic isolation. This theory was foreshadowed in the early 1950s by the colourful Venezuelan botanist Léon Croizat (1901–51), known for his flashy clothing, charismatic storytelling and various personal eccentricities.[17] It then took definitive form in the writings of Ernst Mayr (1904–2005), a pre-war German birdwatcher who ended his career as one of the most formidable biologists of the twentieth century.[18] Eldredge and Gould lean heavily on Mayr's theory, which they summarize as follows:

The central concept of allopatric speciation is that new species can arise only when a small local population becomes isolated at the margin of the geographic range of its parent species. Such local populations are termed *peripheral isolates*. A peripheral isolate develops into a new species if *isolating mechanisms* evolve that will prevent the re-initiation of gene flow if the new form re-encounters its ancestors at some future time.[19]

In other words, a small population (the peripheral isolate) is somehow cut off from the rest of its species. This population faces unusual conditions in its marginal location, and rapidly evolves under these new environmental pressures. Eventually, a sufficient number of changes occur that the peripheral isolate is no longer able to mate with its ancestral population, thereby marking the emergence of a new and distinct species.

This process has been documented in some specific cases. Perhaps the most famous involves the divergence of shrimp on the two sides of the Isthmus of Panama, as studied extensively by coral reef biologist Nancy Knowlton (b. 1949) and her colleagues.[20] It was only around 3 million years ago that North and South America joined together at the isthmus: a narrow strip of land that closed the previous ocean gap between the two continents. When that happened, the local shrimp population became geographically split between the two sides of Panama, with some located in what is now the Pacific Ocean and the others in the Gulf of Mexico. Although the two groups of shrimp do not appear visibly different, Knowlton found that when males and females from opposite sides of the isthmus were brought together they did not engage in the usual courtship rituals, but snapped at each other violently. Three million years of isolation was enough to turn the two groups of shrimp from potential mates into mortal enemies, and they are now classified as different species.

Another example is the spotted owl of the Western Hemisphere, which is known to have three separate subspecies. The most broadly diffused of them is the Mexican Spotted Owl, which ranges from Mexico itself up through Utah and Colorado. At the other geographic extreme, Northern Spotted Owls are found from British Columbia down to the north-western portion of California. Located between these two subspecies is the California Spotted Owl, a possible medium of hybridization and transition between the other two. Genetic differences between these subspecies were eventually discovered, leading one team of researchers to note that the 'region of northern California has a history of isolation that led to the evolution of this previously undocumented lineage'.[21] In this way, geographical factors help explain why three separate populations of the spotted owl became isolated from each other and evolved into different subspecies.

As the researchers conclude: 'the California Cascades area of northern California is in many respects a naturally isolated landscape region, with the Shasta Valley and the Sacramento River Valley providing wide divisions with unsuitable habitat between the areas that we investigated and other proximate areas with Northern Spotted Owl habitat in the Oregon and California Klamath provinces'.[22] In other words, the isolated area where the California Spotted Owl is now found, through its very isolation, was practically destined to produce new species. The island nation of Madagascar is another such case: separated by 400 kilometres from the neighbouring African continent, it hosts countless unusual plant and animal species that cannot be found anywhere else on earth.

Far from changing gradually as Darwin imagined, living species tend to have astonishing stability. For instance, sharks have dwelled on our planet in roughly the same form for hundreds of millions of years, although they are now in rapid decline due to human activity. One reason for this stability is the 'gene flow' that maintains a common genetic pool for individuals, barring those cases of isolation thought by Mayr to provide the spur for the appearance of new species. Sharks circulate freely through the world's oceans, and thus it is hard for them to become geographically isolated enough to generate new species or even especially variable gene pools.

Allopatric speciation entails that the evolutionary adjustments found in a new species should occur relatively soon after isolation, as a result of unique local selection pressures. In this spirit, Eldredge and Gould observe that 'we should not expect to find gradual divergence between two species in an ancestral-dependent relationship'.[23] This directly contradicts Darwin's vision of slow evolutionary change. Moreover, given that populations tend to form an equilibrium with their environments, a species will prove to be relatively stable once it comes into existence. With this in mind, rather than gradual changes in the fossil record of any species, what we should see is a pattern 'of oscillation in mean values'.[24] For example, we will not see any given subspecies of spotted owl become gradually smaller or larger over time, but should find it fluctuating between different average sizes, perhaps due to shifts in food supply as environmental factors waver between extremes. If the Northern Spotted Owl ever managed to spread far beyond its current home and gain a foothold in Mexico, the fossil record in that country would show it appearing suddenly in geological history, whether or not the native Mexican Spotted Owl were to go extinct as a result.

Eldredge and Gould caution that they cannot say definitively that gradualism is incorrect, only that an '*a priori* picture of phyletic gradualism has imposed itself upon limited data'.[25] In other words, people are so drunk on the overall conceptual power of the Darwinian paradigm that they fail to emphasize those cases in which it falls short. While Eldredge and Gould acknowledge that there is a relative paucity of fossil remains, they do provide fossil evidence on their own behalf. One case involves Gould's personal study of snails in Bermuda; during his lifetime he was a world-renowned expert on this creature. The evidence led him to conclude that any gradualist interpretation of species variation in these snails is insufficient.[26] Another example concerns the now extinct life form known as trilobites, ancient creatures of the ocean with a somewhat insect-like appearance. Trilobites appear suddenly in the fossil record, then become widespread in the seas for 270 million years before disappearing entirely during the late Permian mass extinction: the one just before the more famous extinction that killed off the dinosaurs. The late Permian event is named after the city of Perm, near the Ural Mountains in Russia, and it is now thought that this mass extinction occurred due to widespread volcanic eruptions in Siberia.[27] Since this occurred a mere 252 million years ago, trilobites actually survived on this planet for longer than they have now been extinct, making them one of the most successful living species ever documented. By comparison, our own species, *Homo sapiens*, has existed for perhaps 300,000 years, or less than one eight-hundredth the longevity of trilobites, and even less if we only count the 160,000 or so years of the modern version of our species.

After reviewing the record of different trilobite variants during a specific period in North America, Eldredge and Gould conclude that a gradualist explanation fails once again. The way these species developed, they hold, only makes sense if the allopatric model of speciation is true. They note that although the examples of snails and trilobites are widely separated in geological time, they 'have much in common as exemplars of allopatric speciation. Both required an attention to *geographic* distribution for their elucidation . . . Both are characterized by *rapid* evolutionary events punctuating a history of stasis.'[28] Echoing Mayr, they remind us that 'speciation occurs in *peripheral* isolates because only geographic separation from the parental species can reduce gene flow sufficiently to allow local differentiation to proceed to full speciation'.[29] That is to say, species can be viewed as homeostatic systems in a stable state. One example of homeostasis is a thermostat, which tends to maintain a relatively constant temperature in a room unless windows or doors are left open.[30] Another example is that our bodies tend to remain at a stable temperature. Human blood sugar levels also remain naturally within a stable and healthy range, except for those who suffer from diabetes. To view species themselves as basically homeostatic means that something unusual must happen for new species to emerge from old ones.

While there does seem to be evidence for certain gradualist trends in the fossil record, Eldredge and Gould turn these apparent counter-examples to their own advantage. Such trends, they argue, are less likely to be the result of gradual genetic change than of what happens when a previously isolated species spreads over a greater range and comes under selective pressure to adapt slightly to its new environment. For example, if a species of bird spreads from its ancestral homeland into a desert region, it is likely to take on a gradually less colourful look for the purpose of protective camouflage, as in the case of the desert cardinal. Although Darwinian gradualism continues to dominate the popular imagination, Eldredge and Gould are adamant that this is not the key to the evolutionary story.

Evolutionary Pluralism

Gould makes the same basic objection to both Dawkins's selfish gene and the orthodox Darwinian focus on the individual organism. Namely, he retorts that evolution occurs on multiple levels rather than just one. To this end he considers all possible scales, not just the smallest level of the gene. It should be mentioned that Gould uses 'gene' in a sense loose enough to cover chromosomes, organelles, bacterial plasmids or any other entity smaller than the cellular level.[31] In doing so he warns against passing judgement on genes solely according to whether they are 'good' or 'bad' for the organism in which they are housed.[32] Gould's model of selection is hierarchical (by which he means 'multi-levelled'), and assumes a vertical evolutionary structure from genes at the bottom through ever-larger units at the top. If we view genes on their own terms, he argues, we will find that many are purely neutral for the organism, at least at any given moment in evolutionary history. It is well known that a large amount of DNA has no effect on the outward form (or 'phenotype') of a living creature, and is therefore not subject to natural selection because it provides no evolutionary advantage or disadvantage.[33] Because of this, we must also consider the role of randomness at the genic level, as in the phenomenon of 'genetic drift' studied by the American geneticist Sewall Wright

(1889–1988).[34] The purely random inheritance of alleles (either the dominant or recessive parts of a gene) in any population can lead rapidly to the pre-eminence of some traits and elimination of others. Simply put, natural selection cannot explain everything about species change.

In the late 1960s, the Japanese biologist Motoo Kimura (1924–94) worked along the same lines in developing his 'neutral theory' of molecular evolution, which Gould joins others in calling 'the most interesting revision of evolutionary theory since Darwin'.[35] Although Kimura fully accepted the importance of Darwinian mechanisms at the level of organisms, he took note of the vast amount of molecular change that occurs randomly in living species. Here Gould adds an intriguing point. As we rise from the level of the gene to that of the organism, randomness fades in importance: a person is more likely to die from health problems related to their genetic composition than a freak accident due to being in the wrong place at the wrong time. Even so, randomness regains importance at the level of the species and the clade. The clade is the highest level considered by Gould, and refers to a group of multiple, related species. For example, humans are joined with chimpanzees, gorillas and orangutans in the Hominoid clade. Just add Old World and New World monkeys to this group, and we have a broader clade called the Anthropoid. A clade's survival depends heavily on environmental factors beyond its direct control, and in this way it becomes vulnerable to random change once more.[36]

One of the recurring themes of Gould's account of evolutionary hierarchy is the way that each level tends to suppress variation at the one just below it, for the same reason that an army or sports team has greater success if it forces its members to behave in uniform fashion. Nowhere is this more evident than in the relation between an organism and its component cells. In Gould's words: 'multicellular organisms are effective in suppressing the differential propagation of subparts as a necessary strategy for maintaining functional integrity, the definitive property of individuality at the organismal level'.[37] In other words, it's important for an organism that its smaller parts behave predictably and reliably.

One prominent case where this fails is cancer, in which some cells run amok and proliferate wildly in a way that endangers the survival of the entire organism.[38] This packs an especial punch given that long-time cancer patient Gould was dying of the disease as he wrote his huge final book, perhaps due to asbestos exposure at Harvard University's Museum of Comparative Zoology.[39] Fortunately for us living organisms, most differential cellular proliferation within our bodies is successfully repressed in the name of functional integrity, or the good of the organism as a whole.[40] This is so much the case that even potentially beneficial mutations (super-speed or super-strength) tend to be blocked in the interests of organic caution, as cells lose their presumed former ability 'to construct major evolutionary innovations'.[41] If each human individual had bizarre genetic mutations, some of us might become super-human, but the vast majority would die quickly after birth: not a sustainable path for any species to take. It is Gould's view that cellular variation has been so effectively suppressed by individual organisms that it is no longer a significant factor in evolution.

Although Darwin was focused on the level of the organism, Gould views it as the most overstudied topic in the history of evolutionary theory. It is true that the functionally cohesive organism must often adapt to challenging circumstances as it seeks its fortune in the world; this is the level of 'organized adaptive complexity' so beloved by Darwinians. The philosopher Daniel Dennett (1942–2024) often writes as if adaptations were the sole fact of evolutionary history, as in his book *Darwin's Dangerous Idea*, which Gould describes as 'narrowly focused'.[42] When it comes to the usual complaint that species-level selection cannot be important because it is incapable of generating complex organisms, Gould jokes that this reminds him 'of the cook who didn't like opera because singing couldn't boil water'.[43] His point is that each level of the evolutionary hierarchy does different things well, acting under widely differing pressures, and should not be assessed according to the missions of the others.

For Darwin, unlike Gould, all evolution happens at the level of individual organisms. Species emerge and disappear only as the bulk result of what occurs when countless individuals are born, reproduce and die. According to this view, a species itself is nothing more than a collective nickname for the actions of vast numbers of individual beings. As Gould jokingly puts it, there is something odd about Darwin's book title *The Origin of Species*, given that he does not fully accept the existence of species in the first place.[44] The world of Darwin is a world of organisms, not species.

Gould died of cancer in May 2002 at sixty years old. Despite the human tragedy of his early death, he was able to see his mature work into print during his lifetime. With just two months to live, Gould published *The Structure of Evolutionary Theory*, a compulsively readable summary of his key ideas, weighing in at 1,343 pages even without the bibliography and index. We have seen that punctuated equilibrium theory covers the theme of intermittent species change *diachronically*, or over the course of evolutionary time. This is one way of challenging the gradualist or continuist picture of biological change made popular by Darwin's legacy. But there's another way to look at the picture of evolution: *synchronically*, meaning within a single instant of historical time. To do this we can ask about the primary level of natural selection that exists on our planet at any given moment. Alongside Gould's thought on the punctuated equilibrium of species appearing and disappearing in geological time and emerging through geographical isolation, there exists another kind of saltation or sudden jump. This refers to the divisions within any given moment between the levels of gene, cell, organism, deme, species and clade, which we might call 'structural' saltation in opposition to the temporal kind. These six hierarchical levels, although they now exist simultaneously, emerged one by one in the course of geological time. Gould notes that prior to the emergence of multicellular life and sexual reproduction, there were only four levels where evolution happened: gene, cell, clone and clade.[45] As mentioned, Gould uses 'gene' in a loose sense to refer to anything smaller than a cell. As for cells, they are the basic units of life, consisting (in animals, plants, fungi and protists) of a nucleus and various organelles or organ-like units. By 'clone' Gould is referring to the product of binary fission without sexual exchange of genes, which is how bacteria reproduce. And clade, as mentioned, refers to a group of closely related species. Since we have already mentioned the gene and cell, since the clone lost its universal level once sexual reproduction began, and since Darwinians have already given the individual organism all the attention it deserves, let's continue this discussion with the fourth of Gould's levels: the deme.

The deme refers to a particular interbreeding subpopulation of a given species. For example, all the deer in the Rocky Mountains might constitute a deme rather than a unique species. Gould knows that objections to the evolutionary importance of the deme are numerous, and therefore he spends little time discussing it. But he does refer to evidence that selection among demes seems to favour large female populations. A deme with more females will tend to have more offspring than demes with more males, but since organisms themselves flourish best when there is a more balanced sexual population, this partially counteracts the strong female preference at the deme level. Taken together, these two opposing forces may average out to the limited female prevalence found in many animal species.[46]

Darwin's own attitude towards demes was mixed. On the one hand, since altruism is hard to explain as any sort of advantage to the individual organism, he opted to treat it as a collective social advantage for those demes that practise it.[47] On the other, he viewed demes as something like transient dust storms too fragile to have any stable identity, since they quickly rise and fall due to changing environmental factors.[48] Gould's strongest evidence for the importance of demes comes from Mayr's theory of allopatric speciation, as with the spotted owls encountered earlier. After all, any isolated deme is already a potential new species in waiting.

The level of selection that Gould cares about most is species, since it fits best with his theory of punctuated equilibrium. While in the Darwinian tradition organisms dominate 'microevolution', meaning changes in individuals, Gould contends that species must dominate 'macroevolution', referring to the broad-scale evolution-ary changes studied by palaeontology. He goes so far as to say that macroevolution makes up a separate subdiscipline with its own rules.[49] Gould acknowledges that the fossil record at present is hardly rife with instances of species selection, though he contends that this is due to more than a century of Darwinian bias in favour of individual organisms. Yet there are still some good examples of species selection, such as the work on Cretaceous era molluscs by geophysical scientist David Jablonski (b. 1953) at the University of Chicago.[50] It appears that some of these molluscs were able to feed while floating, and thus spent more time aloft in the water; others did not feed this way, and hence spent more time on the ocean floor. An interesting consequence is that while the 'floaters' obviously showed wider geographic range, the 'non-floaters' showed higher rates of speciation due to their greater isolation from each other, leading to more species diversity on their side of the divide. Impor-tantly for Gould's vision of palaeontology, which is so reliant on the notion of sudden species change, Jablonski 'also showed that selec-tive forces can change rapidly during episodes of mass extinction'.[51] This may also point to a mechanism at work in the next higher level of selection, the clade.

The clade is the broadest evolutionary level considered by Gould. Technically, 'clade' can refer to any group of organisms (of whatever size) that share a common ancestor. But in the context of Gould's theory, a clade refers to a group of multiple, closely related species. Gould seems strangely unconvinced by the importance of selection at this level. In his own words: 'I am not sure that clade selection plays a major role in evolution.'[52] Despite this hesitation, he still thinks clades have some evolutionary role. Consider the dramatic forces at work in mass extinctions, some of which proved capable of wiping out entire clades in a single stroke. It is now believed that the dinosaurs became extinct after an asteroid smashed into the Yucatan Peninsula some 66 million years ago. Despite the dark fate of the dinosaurs, other creatures such as mammals and birds survived and developed. Referring to this event, Gould asks: 'if mammals survived in part by virtue of small body sizes, and dinosaurs died for a set of consequences related to invariably . . . larger body size, couldn't we say that mammals, as a clade, possessed genetic determinants . . . that all dinosaurs lacked as a result of their own evolved cladal distinctions?'[53] This mechanism of clade extinction looks similar enough to that of species extinction that the real problem may be different: perhaps the clade lacks sufficiently distinct evolutionary mechanisms to count as anything more than a larger type of species. In other words, perhaps the clade is only a name, rather than a genuine emergent level of evolution. In that case the evolutionary hierarchy would come to an end at the level of species, though the effect of this on Gould's theory would be relatively minimal.

Levels of Punctuation

We have already discussed some of Gould's objections to the theories of gene-based selection defended by Dawkins and his one-time ally George C. Williams (1926–2010).[54] But what, exactly, is wrong with treating genes as the heroes of the evolutionary process? Philosopher of biology David Hull (1935–2010) argued that in evolution we must distinguish between replication and interaction while respecting the importance of both.[55] Replication refers to an entity making a copy of itself, while interaction designates its ability to survive and flourish in a given environment. Gould's interpretation of this distinction is that, in opposition to Dawkins and Williams, Hull makes interactors rather than replicators the place where natural selection happens. Genes make more or less accurate copies of themselves, but since they are confined to the interior of an organism, they have limited opportunity to engage directly in the drama of natural selection.

One could argue, like Dawkins, that genes are still the hidden puppet masters in the interactions between an individual organism and its environment. But this would amount to denying the existence of what are called emergent properties. It would be like saying that the workings of the Los Angeles Police Department can be understood as resulting from the movements of billions of subatomic particles: a truly fruitless method for analysing law enforcement, given that particles are not directly involved in police activity. To extend this thought, if we want to keep track of the evolutionary movement of a population or species, then genes (when available) are unparalleled as a measuring instrument for examining this process. But Gould's concern is that precisely because genes serve as such good evolutionary bookkeeping devices, they have been elevated to a causal pre-eminence in evolution that they do not deserve. For him there are more important things at play, even if we can't track them as easily.[56]

Simply put, an organism is not merely an additive sum of genes. Much of what happens in a living individual is shaped by the interaction among its higher-level components: for instance, genes play no direct ongoing role in the reciprocal workings of bodily organs. Nor do genes produce organisms directly, as seen from increased attention in recent decades to what happens at the embryonic stage of development.[57] Dawkins himself tries to dodge what happens in the uterus due to its sheer difficulty, saying that 'embryonic development is controlled by an interlocking web of relationships so complex that we had best not contemplate it'.[58] One can almost hear Gould laughing in the background. As we've seen, he argues that selection occurs at multiple different levels. For him this means that a 'hierarchical model almost automatically prevails once we accept this model of causality'.[59] By 'hierarchical' Gould is not suggesting that one level of selection is more important than the others. Quite the opposite: he's saying that evolution takes place simultaneously at multiple levels with no intrinsic favouritism, whether it involves genes, individuals or species as a whole. As he memorably puts it: 'There is hierarchy in the world of natural selection, but no aristocracy.'[60]

Gould wonders aloud why thinkers as intelligent as Dawkins and Williams came to such different conclusions from his own, and takes a sympathetic guess at their motivations. Darwin sparked a revolution by descending to the smallest level of evolutionary analysis possible in his time: the individual organism. On this basis, Gould supposes, 'the more thoughtful gene selectionists' (such as Dawkins and Williams) 'worked by analogy, reasoning that if they could break causality down even further, below the level of organisms, similarly interesting, and perhaps revolutionary, consequences might follow'.[61] Gould doesn't fault this way of thinking; he just doesn't agree that the smallest level of analysis is always the best one. Reductionist thinkers in every field will always look for the tiniest unit and try to make it responsible for everything more complex. For instance, reductionists in physics try to explain everything in terms of subatomic particles, whereas chemists are concerned with larger units such as molecules. Much like the chemists, Gould calls for a focus on emergent higher layers containing properties found nowhere in their constituent elements.

While Gould concedes that evolutionary units replicate themselves, he denies that such replication is always faithful. He sees this point as yet another nail in the coffin for the selfish gene. Individuals, he holds, 'must contribute to the next generation by hereditary passage, and they must plurify their contributions relative to those of other individuals . . . But the contributions can be wholes or parts, faithful replicates or disaggregated bits of functional heredity. Selection demands plurifaction, not faithful replication.'[62] What he means is that the level of organisms, where interaction with the environment occurs so spectacularly, has no vested interest in the survival of any individual gene. Once we abandon the idea that replication must always be faithful (mutations do exist, after all) the door swings open to the idea of evolution happening at multiple levels.

We recall the case of demes, referring to particular subpopulations of a species. Here the exact proportion of red-haired humans or white-furred wallabies will always flicker a bit from one generation to the next, and certain 'unfaithful' genetic changes will inevitably occur due to radioactive damage and sheer randomness.

Despite these fluctuations in specific properties, such as the number of people with strawberry-blond hair, or variations in the distance between eyes found in individual humans, we can still meaningfully consider a population to be one and the same over long stretches of time. In this connection Gould happily announces that the idea of group selection (meaning anything larger than the individual) 'has risen from the ashes to receive a vigorous rehearing'.[63] He supports this claim by reminding us that our ultimate competition is not with other organisms, but with the environment itself. In other words, the competition between organisms tends to be indirect more often than direct. If the earthly dominance of humans should come to an end, this will not be because we are eaten en masse by sharks, wolves and lions, but because the climate heats up to a level dangerous for our species, dependent as we are on environmental and agricultural stability.

Hierarchies of Selection

We have seen that orthodox Darwinians treat the level of individual organisms as the sole site of natural selection, and that Dawkins thinks selection happens at the level of genes instead. What both approaches share is the reductionist bias that privileges the purportedly smallest level of nature, bestowing an evolutionary monopoly upon it. Reductionism is a form of continuism that envisions a cosmos without distinct layers each having their own autonomous mechanisms. For instance, a radical atomist might hold that even complex entities such as UNICEF or the Vietnam War could be explained entirely in terms of the motion of atoms, leaving no room for emergent complexity at higher levels. But stated bluntly, Richard Nixon's expansion of the Vietnam War by the bombing of trails used by Viet Cong guerrillas in neighbouring countries had severe ramifications that unfolded at a level much larger than atoms. To call protons and neutrons relevant factors in Nixonian geopolitics would border on absurdity.

Let's take a moment to explore a related controversy about evolution: specifically, whether it concerns the 'emergent character' of an organism or its 'emergent fitness'. The philosopher of biology Todd Grantham once dubbed this 'The Lloyd–Vrba Debate', since the two positions are defended respectively by Elisabeth Lloyd (b. 1956) of Indiana University and Elisabeth Vrba (1942–2025) of Yale University.[64] Gould co-authored papers with both of them at various times, though he eventually settled on Lloyd's side of the dispute.[65] Vrba wished to contest the standard Darwinian reduction of species either to individual organisms or to genes. For this reason she focused on isolating those selective traits that do not appear until we reach the species level. By way of analogy, we might consider the importance of personal reputation in human society. A bad reputation usually means greater difficulty in forming friendships as well as business and mating relationships. Obviously, individual humans would have no reputations if each of us simply wandered the wastes in solitude; reputation comes only with coexistence. Human reputation thus corresponds to what Vrba calls 'emergent character', since it takes an entire deme of humans for an individual to have either a good or bad reputation.

By contrast, Lloyd (joined later by Gould) believes in 'emergent fitness'. This means that a biological species as a whole has traits that cannot be found in individuals at all. Examples would include such species-level features as population density and geographical range.[66] If we define evolutionary success at all levels, including higher ones (as Gould does), we can see how certain properties of an entire species might be either helpful or harmful to its survival. A species whose range was limited to a single volcanic island would face extinction if there were an eruption on that island.[67] Moreover, a densely populated species might spread diseases among its members too easily, while a species spread too thinly might have difficulty in mating. Unlike the reputation of an individual human, which can exist as soon as there is a deme, the density or thinness of population spread occurs only at the level of the species as a whole. For this reason Gould eventually concluded that Vrba's model excluded too many important factors in the survival or non-survival of a species, and he shifted his allegiance to Lloyd.

Here Gould stresses the example of genetic variability within a species.[68] We can imagine cases in which all members of a species share a single trait that becomes a horrible liability in the wake of a major environmental change, with the species going extinct as a result. Take the recent COVID-19 pandemic, in which entire families were wiped out in a single stroke while others suffered nothing more than minor cold symptoms. Although some of this can be explained by environmental and income-related factors, it is likely that genetics also played a role. This points to a great deal of variation in the human populace as regards our respective abilities to resist the novel coronavirus. But now imagine another scenario, one involving a far more severe pandemic or less variable human population, leading to the total annihilation of our species. The importance of this for Gould and Lloyd is that variability belongs only to populations, not to individuals. If 10 per cent of the human populace has enhanced genetic ability to resist the imagined killer pandemic, it is far better for us than if it were only 1 per cent. As Gould asks: 'don't we also want to say that a species . . . survived by virtue of greater variability – a trait that does not exist at the organismal level, but that surely interacted with the new environment to preserve the species?'[69] Restated differently, there are evolutionary forces that affect the fate of entire species and not just each individual one by one.

To explain this point further, Gould offers the helpful fable of a pond that is home to two different species of fish: an optimally successful fish shaped perfectly by millennia of natural section, and a mediocre fish that barely scrapes by, but which has a good deal of genetic variability.[70] Imagine that at some point in time the pond becomes muddy and stagnant. Most of the middling fish die off right away, while the optimal fish survives with a high population for a considerable time. But eventually the pond becomes so muddy and stagnant that the optimal fish species goes extinct; it was accustomed to swimming in clear waters and thus perishes under the new, subpar conditions. Yet the genetic variability of the middling fish ensures that a few of them have what it takes to survive in the remaining handful of mud-clogged pools. Eventually the rains return and restore the pond to its previous size. Given the extinction of their formerly superior rival species, the mediocre fish (small, slow, stupid, ugly and weak) have now become lords of the pond. This is a case where species-level selection favoured fish that were clearly 'inferior' on an individual Darwinian level.

A relevant piece of terminology here, which Gould coined with his former ally Vrba, is 'exaptation'.[71] Whereas adaptation refers to beneficial traits evolved by creatures in direct response to the environment, exaptation concerns features that prove useful despite having been produced by natural selection for some other purpose, or perhaps no purpose at all. There is a non-biological analogy involving the dime, the American ten-cent coin that displays the profile of Franklin D. Roosevelt. Ten cents can no longer buy much of anything. But, as Gould notes, 'American dimes . . . also happen to fit snugly into the operative groove on the head of a standard screw – and dimes therefore work very well as adventitious screwdrivers.'[72] Most residents of the United States have found themselves making use of dimes in this way at least once. In a different case, I remember the moment in 1990s Chicago when some of my fellow graduate students discovered that the essentially worthless German one-Pfenning coin (no longer in use following the 2002 introduction of the Euro) was the same size and weight as the token for riding Chicago's public transit trains. This insight led to many illicit free rides for students: another clear case of exaptation, this time in a minor criminal sense. As for biological examples, Gould suggests that 'feathers function for enhanced thermoregulation on the arm of a small running dinosaur', but that the aerodynamic benefits of feathers did not become relevant until some of these dinosaurs evolved into birds.[73]

In the same spirit, Gould and Richard Lewontin (1929–2021) introduced the architectural term 'spandrel' into biology. Spandrels are residual spaces left on the sides or corners of an arch that serve no structural purpose. Yet over time these spandrels became a place for often spectacular painted decoration, to the point that we now almost expect to see a filigree, floral or pastoral design carved into the wood or stone around arches in cathedrals and elsewhere.[74] In similar fashion, rear bumpers were placed on cars to protect them from rear-end collisions. But this accidentally created a place for drivers to add bumper stickers proclaiming their support of various political parties or other causes. Likewise, non-Americans were long intrigued by the way Americans like to cover their refrigerator doors with magnets and written messages, a practice that has begun to spread worldwide. Such examples are everyday cases of spandrels.

It is no surprise that Dawkins dismisses exaptation and spandrels as belonging to a 'boring by-product theory', or that Gould is adamant in their defence.[75] The concept of exaptation aligns closely with Gould's idea that evolution can happen at levels higher than the adaptation of individuals. In fact, Gould believes that both adaptations and exaptations occur, with variations in frequency split largely along the lines of organisms and species. To be more specific, Gould surmises that most adaptations will occur at the level of individuals, given that individuals are more functionally integrated than higher evolutionary levels: it is individual cats who adapt to jumping over human-built walls, not the cat species as a unit, even if they all eventually do so. Likewise, exapatations are probably more frequent at the species level, where the encounter with largely random environmental factors is more prevalent. Another way of putting it would be to say that whereas 'survival of the fittest' rules at the individual level, at the species level there is also a degree of 'survival of the luckiest'. Birds and mammals were not 'well-adapted' to the random apocalypse of a giant rock hitting the earth, but dinosaurs were even less so: they became extinct just like the optimal fish in Gould's imagined muddy pond. Instead, birds and mammals probably survived the meteor through 'exaptation', such as their smaller body sizes being turned to advantage in a post-apocalyptic era with less available food.

This is our first encounter with Gould's important idea that the various levels of selection are not 'fractal', meaning that each level does not internally resemble the others.[76] Instead, each level is different, with 'varying strengths and modalities'.[77] Whereas individuals tend to survive when they personally adapt to changing circumstances, species are more dependent on broader environmental changes to which it is often impossible to adapt beforehand. Yes, an individual may die as the result of physical or mental deficiencies by contrast with other members of its species. But an entire species is more likely to go extinct through environmental change that challenges the basic physical make-up of the species. For instance, given the rate at which global warming is increasing the acidity of the oceans, it is too much to expect that all aqueous creatures will be able to evolve into acid-tolerant organisms. Some may succeed in doing so, while for others it will prove impossible due to built-in limitations. Similarly, one group of humans may be stronger and more fertile than other members of our species, but that will make little difference if a collapsing agricultural system leads their entire deme to starvation.

One factor largely excluded from Gould's hierarchical model of evolution is possible cases of lateral evolutionary forces. These operate when separate species exchange biological material, as often happens through the work of viruses. Gould is perfectly aware that such cases exist. He even cites molecular biologist W. Ford Doolittle's argument that 'the history of life cannot properly be represented as a tree', due to the wide extent of lateral gene transfer between species.[78] Gould himself refers to 'plants with extensive hybridization', and to 'prokaryotes evolving with frequent lateral transfer'.[79] Let's briefly discuss these prokaryotes.

Without referencing it directly, Gould's point about prokaryotes brushes against the Serial Endosymbiosis Theory (SET) of Lynn Margulis (1938–2011), an alternative model of evolutionary punctuation to that of Eldredge and Gould.[80] Margulis is famous for her theory that eukaryotic cells (which have nuclei and other organelles) emerged from prokaryotic cells (which do not) through symbiotic relationships between originally separate species. One life form initially entered another as a parasite, but they eventually proved mutually beneficial to the point that they became inseparable, so that they now reproduce in unison. In short, a new species has formed from a combination of two or more pre-existent species. I would say that this theory amounts to a more radical challenge to the status quo than Gould's refreshing but essentially reformist Darwinism, since it allows for the combination of organisms that were once thought to exist on completely different branches of the evolutionary tree.[81] Species often compete and kill each other off, but they also combine into new 'Centaur species' that appear as suddenly as those that emerge from geographical isolation, or which arise in the wake of catastrophes such as the one that destroyed the dinosaurs. Certain bacteria live in the roots of plants and help them by fixing atmospheric nitrogen, while in turn the plant provides the bacteria with a welcoming home.

Despite his odd silence about Margulis, even in his gigantic bibliography, Gould does defer somewhat to David Sloan Wilson and Elliott Sober's intriguing discussion of 'wingless insects as they move among resource patches' who carry 'various mites, nematodes, fungi, and microorganisms'. Although these parasites often harm the insects who carry them, they also seem to enable the insects to colonize new patches, thereby extending their geographic range.[82] In such cases, Gould says, 'the entire association may be evolving as a "superorganism"'.[83] Other examples come easily to mind, as with the well-known associations of pilot fish and sharks, ants and aphids, or humans and their domesticated livestock and pets.

But in the wake of James Lovelock's ambitious Gaia theory, which treats the earth's climate as a system created and stabilized by life itself, we might press further and ask whether the discussion of evolution needs to be limited to biology at all.[84] Consider the case of human politics. Although modern political theory is largely obsessed with the question of whether human nature is good or evil, and draws numerous consequences from the answer, there is increasing awareness that human political space is largely structured and stabilized by inanimate objects such as coins, walls and identification cards, not to mention environmental conditions.[85] Any ongoing physical evolution of the human species is now vastly eclipsed by our symbiotic relationships with everything from hammers and printing presses to websites and smartphones. Such devices change swiftly, while the basic human physique has remained relatively constant for over 100,000 years. Given that new technological devices appear suddenly and intermittently rather than continuously and gradually, the case for punctuated equilibrium seems even stronger in the socio-political sphere than in the fossil record.[86]

The Problem of Cascades

Let's return in closing to Gould's own model of saltation: the idea of sudden evolutionary leaps. While he expresses annoyance at those who find it implausible that a new species could appear in an instant of time, what Gould means by 'instant' is not a single generation of animals. Instead, what we call sudden evolutionary saltation 'may span several thousand years'. We may never be able to reach a more precise answer than that, given that layers of rock containing fossils can't be dated to a specific year, decade or even century – at least not yet. But there is something dissatisfying about this manoeuvre, despite its rhetorical effectiveness against gradualist objections. What Gould tells us is that by a moment in time he means a 'single geological bedding plane', a single layer of rock uncovered when digging for fossils. This is effectively the smallest unit of time available to a working palaeontologist, even if such a layer may cover thousands of years of evolutionary history.

And yet, the inherent vagueness of geological layers cannot be used as an excuse to dodge the important question of when exactly a species change occurs. Let there be no misunderstanding: I am not demanding that palaeontologists point to a specific week or year when trilobites first began to exist; we will surely never have the technical tools to answer that question. If we are forced to content ourselves with such results as 'the ankylosaurus disappeared between 68 and 66 million years ago', then so be it. The palaeontology of the future may be able to narrow that range somewhat, though it is doubtful how precise the figure could ever be. But the fact we can only *know* the date vaguely does not mean that it could have *happened* vaguely, and this is where Gould is inconsistent. A saltationist like him cannot avoid the conclusion that in one generation there was an ankylosaurus and in the next generation there wasn't. Our *knowledge* of the appearance of the ankylosaurus may forever be limited to a vague layer of rock spanning millennia, but this cannot mean (at least not for Gould) that the ankylosaurus emerged gradually over a long period of time. It is only the Darwinian gradualist, not the saltationist, who can get away with saying that a species began vaguely during a period of several thousand or million years: for them evolution is primarily a matter of individuals, and a species is effectively just a nickname for a large number of similar individuals. Species-level vagueness is therefore perfectly fine for the gradualist; it does not contradict their theory at all.

But for Gould, restated differently, species must be cleanly separated from each other. And even if we accept Gould's idea of species-level selection, he faces a problem with the mechanics of how a species comes to be in the first place. After all, he joins Ernst Mayr in proclaiming reproductive isolation to be the mark of a genuine species. That is to say, if a new species requires a reproductively isolated deme, then every individual organism is either on the bus or it isn't. Either a given bird can mate with a California Spotted Owl or it can't. In other words, the punctuation at the level of species must cascade down to the level of two individual owls, even if there is no hope of knowing *exactly* where and when this happened.

This becomes important when Gould critiques Dawkins's 1982

follow-up book to *The Selfish Gene*, entitled *The Extended Phenotype*. In one sense the new book continues the reductionist argument of the earlier one: Dawkins still argues that birds' nests and beaver dams are direct expressions not of the visible birds and beavers who made them, but of the underlying genes that exploit these bodily creatures as vehicles. So far, so good for the selfish gene theory. Nonetheless, the Dawkins of *The Extended Phenotype* leans hard on a form of conventionalism, even relativism. He does this in connection with the famous optical illusion known as the Necker cube, in which a given face of the cube first seems to be in front, but then suddenly appears to be at the back. In a surprising move, Dawkins's 1982 book compares his own work to this optical illusion: 'the vision of life that I advocate, and label with the name of the extended phenotype, is not probably more correct than the orthodox view. It is a different view and I suspect that, at least in some respects, it provides a deeper understanding. But I doubt that there is any experiment that could be done to prove my claim.'[87] What he means is that we can look at it one way and say that individual organisms alone are the primary unit of natural selection, or look at it another way and say instead that the things an animal builds for itself are also parts of those animals for evolutionary purposes. He turns the whole question into a matter of perspective, and concedes that one perspective is not necessarily better than another.

This is not only a wishy-washy statement from the normally fire-breathing Dawkins; it is also demonstrably false. Gould tells us why: 'Necker cubing will not apply to genuine cases of irreducible species selection because the nature of our world (not the conventions of our language) regulates the locus of causality.'[88] Although Gould speaks here only of 'language', he means more broadly that the place where any causal event occurs has a real location in the world, and is not something posited arbitrarily by the mind. In other words, causation takes place at discrete locations, not along a continuum. And when it comes to species change, the 'locus of causality' has to be a number of individual organisms no longer able to mate with their living cousins. Returning to the example of the ankylosaurus, it follows that this process must be far more sudden than Gould claims when he speaks of a species as disappearing only in the vague stretch of time represented by a single layer of rock. No saltationist can avoid biting this bullet. While our knowledge of reality may be vague, reality itself is not.[89] Causation unfolds at discrete points in the world, not at some indefinite point in a loosely defined plane of fossils.

7.

Fractures and Folds:
Architectural Theory, 1988–93

Some people care more about architecture than others. There are those with strong opinions about every building in sight, and others who barely notice the details of the city that surrounds them. At times, the battle of 'starchitects' in the news might look like an empty struggle over vain subtleties. But there is a simple reason why everyone should pay attention to architecture: as David Ruy notes, architecture provides us with our primary sense of reality.[1] Almost never do we find ourselves amidst some mythical, virginal nature. Even on hikes and camping trips, we are assisted by paths, signs, maps and GPS devices. Decisions about architecture and the built environment are nothing less than decisions about the reality we face collectively in daily life. Although architecture as a discipline can sometimes feel remote from our immediate needs, or as nothing more than a service industry for mansion-owning and tower-climbing plutocrats, it is hard to imagine anything more important than the basic features of our lived environment.

During my years as an academic visitor in their world, I have generally found architects to be well read, intellectually omnivorous and blessed with a knack for lively writing. Philosophy has often provided especial stimulus for this field, even though a certain number of architects have come to resent philosophers, as if we were pompously giving orders about how they should conduct their business. In large part this is a phantom worry, since those philosophers who have had the greatest impact on the profession have known little about architecture themselves, and would never have dreamed of offering commands to those who actually know how to build.

The dialogue between philosophy and architecture is certainly not new. Plato had a general atmospheric influence on the Italian Renaissance, and traces of his idealized and mathematized view of reality are easy to detect in the designs of that period. During the nineteenth century, the German philosophers Immanuel Kant and G. W. F. Hegel were widely studied throughout the Western world, and architects read them as much as anyone else. Yet it seems to me that the golden age of philosophical influence on architecture began in the middle to late 1960s, and continues perhaps to this day. The most likely reason is the slow-burn crisis of architecture brought about by the collapse of high modernism in the mid-1960s. Between the emergence of Le Corbusier and the Bauhaus in the early twentieth century and the erupting postmodernist doubts about austere glass and steel geometries in the mid-1960s, it is safe to say that modernist architecture hoped to disentangle itself from the history of the field. For this reason, the calling card of early postmodernism was the desire to salvage something from the vast fund of historical signs and symbols that modernism had exiled from the profession. As we will see, this moment in architectural history is closely linked with such architects as the American Robert Venturi (1925–2018) and the Italians Aldo Rossi (1931–97) and Manfredo Tafuri (1935–94). Of these figures we will pay closest attention to Venturi, who makes an especially clear case for architectural conflict and discreteness.

As for the philosophers most relevant to the postmodern period in architecture, interest passed in turn from Martin Heidegger (1970s), to Jacques Derrida (1980s), to Gilles Deleuze (1990s). Each of these thinkers was beloved by a particular group of architects: Heidegger was the guiding star of so-called architectural phenomenology, Derrida the half-suppressed idol of deconstructivism, and Deleuze the inspiration for a turn to architectural smoothness in the 1990s. In what follows, discussion is centered on the transition from the Derridean to the Deleuzean period in order to focus on two especially important years in the debate: 1988 and 1993. It was during these years that the clearest calls were made, respectively, for the discrete and the continuous as the basic principle of architecture.

Architecture Disjointed or Smooth

There are two extreme positions concerning the discrete and the continuous in architecture. The first maintains that architecture should maximize sudden transitions and ruptures, with the creation of jarring gaps between a building and its environment or even within a building itself. Partisans of this view often declare that this sort of architecture prods us to think, by making our otherwise bland surroundings less familiar. Generally speaking, this was the style preferred by the architects assembled in the famous 1988 Deconstructivist Architecture show at New York's Museum of Modern Art (MoMA), co-curated by Philip Johnson and Mark Wigley. It can be seen as a distant echo of another MoMA event co-curated by Johnson, fifty-six years earlier, entitled 'Modern Architecture: International Exhibition'. The 1988 show was influenced by both the philosopher Jacques Derrida (1930–2004) and the Russian Constructivist movement of the early twentieth century. This approach leads to structures that are intentionally disturbing or disorienting, often clashing with their urban context. A good example is Peter Eisenman's ominous Memorial to the Murdered Jews of Europe in Berlin, which no visitor will soon forget: a stark field of 2,711 concrete slabs in the heart of the cosmopolitan city. Indeed, the New Yorker Eisenman is one of the architects most associated with Derrida's thinking, along with Bernard Tschumi from Switzerland.

The opposite approach emphasizes continuous flows and folds in close connection with a building's physical surroundings, a style that became popular beginning in the early 1990s. This new architectural strategy, in part a response to the jarring elements of deconstructivism, found inspiration in the writings of a different French philosopher: Gilles Deleuze. Among the leading figures in this later movement were the architectural critic Sanford Kwinter

(b. 1955) and the prominent designer Greg Lynn (b. 1964). In 1993, the twenty-nine-year-old Lynn guest-edited a special issue of the British journal *Architectural Design*. His own contribution, entitled 'Architectural Curvilinearity', led off with a survey of recent architectural history: 'For the last two decades, beginning with Robert Venturi's *Complexity and Contradiction in Architecture* . . . and continuing through Mark Wigley and Philip Johnson's *Deconstructivist Architecture*, architects have primarily been concerned with the production of heterogeneous, fragmented and conflicting formal systems.'[2] This is the style that Lynn wants to leave in the past. He continues: 'Both Venturi and Wigley argue for the deployment of discontinuous, fragmented, heterogeneous and diagonal formal strategies based on the incongruities, juxtapositions and oppositions within specific sites and programs.'[3] By opposing this style, Lynn was making an audacious call for the end of deconstructivism, just five years after the MoMA show enshrined it in architectural history. He declares his own preference for an 'alternative smoothness' that involves the 'intensive integration of differences within a continuous yet heterogeneous system'. And further: 'Smooth mixtures are made up of disparate elements which maintain their integrity while being blended within a continuous field of other free elements.'[4] Rather than disturbing us with disconnectedness and ominous ruptures, architecture should become smooth and flowing in the manner of a continuum.

'Continuous yet heterogeneous', one of the phrases used by Lynn in the previous passage, is not as innocent as it sounds. It is one of the trademark phrases of our old friend Henri Bergson, as mediated by the eccentric and imaginative Deleuze. Another way to write 'continuous yet heterogeneous' would be 'continuous yet discrete', a clear amalgam of opposites. As we have seen in this book, to reconcile these two is often difficult, and certainly not something that can be stipulated into existence with words. 'Continuous yet heterogeneous' is, in fact, a phrase somewhat abused by the followers of Deleuze in order to conceal the one-sidedness of their focus on continuity.

In any case, Lynn was one of the leaders of the new architectural movement, and he was far from alone in his enthusiasm for Deleuze, who 'describes smoothness as "the continuous variation" and "the continuous development" of form'.[5] But what does Lynn really mean by architectural continuity, and what did his deconstructivist elders mean by discontinuous architecture? The path to understanding will take us on a whistlestop tour of some of the major landmarks mentioned by Lynn above. The tour begins in 1966, when Venturi's influential book *Complexity and Contradiction in Architecture* (hereafter, *Complexity and Contradiction*) was published.[6] From there we will jump forward to the 1988 MoMA show, before making a shorter leap to 1993 to witness the emerging alliance between Lynn and others in escorting Deleuzean smoothness to the architectural throne.[7]

Complexity and Contradiction

The rise and fall of modernism in visual art is a story that's been told and retold. Many historians date the origin of modern painting to Édouard Manet in the 1860s. This includes the influential American critics Clement Greenberg and Michael Fried, although they disagree importantly as to why.[8] Greenberg argues that from the Italian Renaissance onwards, painters had been trying to increase the realistic illusion of three-dimensional space, and that Manet reversed this trend by beginning to 'flatten' painting so that all objects seemed to inhabit the same visual plane. This led to a historic reversal of illusionist space into the sort of flatness we see in medieval icons, but this time in the cutting-edge cubism of Pablo Picasso and Georges Braque.[9]

Fried sees it differently. For him, Manet's new flatness was simply a byproduct of the 'facingness' of his paintings, referring to the way Manet's works tend to stage a direct confrontation with their beholders: this includes such techniques as having a central figure in the painting – often enough a naked woman – stare directly at us.[10] Other historians take impressionism to be the starting gate of modern art, while the philosopher Arthur Danto aims slightly later with his choice of the post-impressionists Vincent Van Gogh and Paul Gauguin.[11] But all would agree that with the friendly rivalry between the twentieth-century masters Picasso and Henri Matisse, we are squarely in the midst of modernism.[12]

A succession of great, mostly European painters kept modernism vibrant up through World War II. The scene shifted from Paris to New York in 1948 with the professional arrival of Jackson Pollock and his fellow abstract expressionists, who dominated the field until the early 1960s. That was when Andy Warhol's commercial pop art first entered the spotlight. Op art and minimalism soon followed, as did performance art, explicitly feminist art, conceptual art, land art and other movements that cut against the grain of modernist painting and sculpture by way of a multifaceted approach across multiple media. Many of these new genres were inspired by an earlier figure initially overshadowed by the great modernists of his own generation: the prankish Frenchman Marcel Duchamp (1887–1968), with his famous 'readymades' including urinals, bicycle wheels and wine bottle racks introduced into galleries as artworks.[13] Most of the genres of art that arose from the 1960s onwards had at least one point in common: a postmodern suspicion of any dominant style or master narrative, a willingness to use media other than painting and sculpture, and (in many cases) the increasing influence of Left-leaning political discourse. As the art world became ever more entangled with the wealth of the investor and money-laundering classes, it found a new topic in the ironic denunciation of this very entanglement.

The history of architectural modernism may be less well known. The timetable is similar to that of visual art, though perhaps with a slightly later start date. The accelerating progress of industry and engineering in the nineteenth century made a sharp contrast, at least superficially, with the historicist flamboyance of Parisian Beaux-Arts architecture during that period, a style ridiculed (unfairly enough) in Ayn Rand's architecturally themed novel, *The Fountainhead*.[14] In the second half of the nineteenth century the Chicago School made breakthroughs in the design and construction of high-rise buildings, and Louis Sullivan (the father of skyscrapers) struck at least a verbal blow at historicism and its fascination with Graeco-Roman and Gothic flourishes with his maxim that 'form ever follows function'.[15] This declaration asserted that the appearance of a building should be in keeping with its purpose; it should not imitate historical styles simply for history's sake.

While its best-known designers had very different ways of building, modernist architecture became recognizable to the public through certain recurring features: the use of unadorned glass and concrete, a generally streamlined aesthetic, the tendency to surround buildings with abundant empty space and an inclination towards austere functionalism rather than historically based ornament. No one is shocked any more by such buildings, which are typical and even stereotypical elements of contemporary Western urbanism. Modernist architecture also entails the view, essential to modernisms in every field, that it is better to tear down existing reality and replace it with a rational plan than to accept the incremental and often accidental accretions of historical meaning. Bruno Latour and Émilie Hermant remind us that the crooked street called Rue St André des Arts (on the Left Bank in Paris) wonderfully preserves a Stone Age path used by early humans living in the area.[16] But any good modernist would no doubt prefer to level the whole neighbourhood and rebuild it as a right-angled grid.

Such European figures as Otto Wagner (1841–1918) in Vienna and Hermann Muthesius (1861–1927) in Berlin took important steps in a modernist direction. In America, Sullivan's former employee Frank Lloyd Wright (1867–1959) came to be known as the dazzling star of early modern architecture. But the heart of modernism can be found in a slightly later group: the foundational Bauhaus émigrés Walter Gropius (1883–1969) and Ludwig Mies van der Rohe (1886–1969) from Germany, and a remarkable Francophone Swiss named Charles-Édouard Jeanneret, better known as Le Corbusier (1887–1965). The latter's manifesto *Towards a New Architecture* is one of the foundational books of the modernist architectural movement, which came to be known as the International Style.[17] A towering figure of the following generation was Louis Kahn (1901–74), a childhood immigrant from the Russian Empire to the United States, and an unquestioned master viewed as the precursor of contrary trends that would become important at a later date. This can be seen in his concerns about the unique properties of specific building materials: in one instance he imagined himself in dialogue with a brick about whether or not to build an arch over a doorway.[18] Kahn also paid close attention to the interplay of light and shadow, and generally stressed the human experience of architecture rather than treating a building as a purely self-contained unit.

Just like modern visual art, modernist architecture was coming under heavy fire by the early 1960s, whether for its perceived dehumanizing vastness (brutalist public structures were often compared with Soviet buildings) or its failed aspirations to reshape civilization by architectural means. Examples of the latter include Le Corbusier's Unité d'Habitation in Marseille, a 330-unit residential block with shopping, dining and sport facilities, and other structures intended to rationalize urban living. A famous moment in the downfall of modernism occurred with the demolition of Minoru Yamasaki's colossal Pruitt-Igoe housing project in St Louis beginning in 1972, just eighteen years after its construction. This incident was famously described by the bestselling author Charles Jencks: 'Modern architecture died in St. Louis, Missouri on July 15, 1972, when the infamous Pruitt-Igoe scheme . . . was given the final *coup de grâce* by dynamite.'[19] In later years, other such experiments in modernist urbanism would come crashing down elsewhere, widely condemned for worsening rather than improving the social problems they were meant to address.

Three books in particular accelerated the move away from dogmatic modernist architecture. Alongside Venturi's 1966 *Complexity and Contradiction* (1966), there were Aldo Rossi's *The Architecture of the City* (also 1966) and Manfredo Tafuri's more openly political *Theories and History of Architecture* (1968).[20] As mentioned, the common link between these otherwise very different authors can be found in their renewed respect for the role of history. In their wake came postmodern architecture, which in one of its forms had an emphatically historical flavour. Here we'll focus on Venturi's breakthrough book, and to a lesser extent on his provocative 1972 follow-up with Denise Scott Brown and Steven Izenour: the amusingly titled *Learning from Las Vegas*.[21]

Venturi is opposed to the generally sparse and streamlined look of modernist architecture: think of concrete boxes and giant plate-glass windows. 'I like complexity and contradiction in architecture.'[22] So goes the opening sentence of his best-known book, in a riff on its title. Venturi thus loses no time in resuming a dispute among architects that goes back centuries. Through all the changes in architectural style since the Renaissance, there has generally been one camp committed to mathematical proportion and the triumph of symmetry, and another inclined towards the intricate, the ornate, the picturesque or the sublime. The second group is where Venturi belongs. By way of contrast, the ancient Roman architect Vitruvius, an employee of the first two Caesars and the earliest classic author in the discipline, is translated as saying that the three principles of architecture are 'commodity, firmness, and delight' (or usefulness, stability and beauty).[23] While the modernist hero Gropius argued that if commodity and firmness are present the delight will take care of itself, Venturi insists on a more nuanced relationship between the three, with their very triplicity ensuring a complex and contradictory architecture.[24]

Venturi's stance was not based solely on personal aesthetic taste, but also on his view that modernism cannot do what it claims: namely, remove all historical symbolism so that buildings look the way they do only because their function demands it. 'Orthodox Modern architects . . . in their attempt to break with tradition and start all over again . . . idealized the primitive and elementary at the expense of the diverse and the sophisticated. As participants in a revolutionary movement, they acclaimed the newness of modern functions, ignoring their complications.'[25] That is to say, the rejection by modernists of the past complexities of architecture often led them to downplay the intricacy of actual present-day needs. For instance, the visual symmetry preferred by modernists often clashes with the asymmetrical requirements of plumbing, heating or elevator infrastructure.

Such 'purist' architecture, as Venturi terms it, was in part a reaction to the excessive mixing and matching of the previous century: 'Gothic churches, Renaissance banks, and Jacobean manors were frankly picturesque.'[26] In breaking with this hotchpotch of historical references, modernist architecture aspired to purity of medium: a return to functional transparency and the supremacy of basic geometrical form, as seen in the most famous buildings of Le Corbusier. Against this trend, Venturi advocates a 'both-and' approach to oppositions rather than a forced 'either-or'.[27] He finds support for this strategy in such champions of irony and ambiguity as the literary critics Cleanth Brooks (1906–94) and William Empson (1906–84).[28] Architecturally speaking, he favours both 'crowded intimacy' and the 'double-functioning element' (such as a staircase having storage space underneath), both of them anathema to the cool openness and transparency of modernism: imagine a Swedish hotel lobby from the early 1960s.[29]

Venturi is deliberately provocative when he speaks of the relation between the inside and outside of a building: 'Contrast between the inside and the outside can be a major manifestation of contradiction in architecture,' he says, again embracing contradiction as a creative force. In his view this has not been sufficiently realized by recent designers: 'one of the most powerful twentieth century orthodoxies has been the necessity for continuity between them: the outside should be expressed on the inside'.[30] He firmly rejects such forced continuity. Modernist architecture sacrificed meaning on the altar of expression, in the sense that a stock exchange or a government building supposedly referred to nothing but itself instead of quoting earlier historical styles. Whereas a modernist would treat windows as having an inherent, self-contained form that should be designed for maximum efficiency and simplicity, Venturi reminds us that no one is ever seeing a window for the first time. Instead, there is a historical or memorial layer to a human's encounter with a pane of glass. To see a window is to see 'the image of the window – of all the windows you know plus the ones you find out about'.[31]

As Venturi has it, modernist architecture was guilty of saying one thing while doing another. For even while claiming to strip all symbolism from buildings, the modernists introduced a new set of symbols: 'despite their protestations', they 'derived a formal vocabulary of their own, mainly from current art movements and the industrial vernacular'.[32] For instance, the famous streamlined look of modern buildings is not always functionally necessary or even useful. Modernism is a style that Venturi's co-author Denise Scott Brown denounces as 'striving and bombastic'.[33] And further: 'modern architects have substituted one set of symbols (Cubist-industrial-process) for another (Romantic-historical-eclecticism) but without being aware of it'.[34] Whatever modernists may claim, they are not just building sleek pure functions. Instead, they are relying on a varied fund of symbols that *suggest* modernity, especially after the initial generation of modernist masters had established a stable repertoire of new symbols.

In the eyes of Venturi and his co-authors, the modernist effort at self-referentiality has backfired: 'architecture of today, while rejecting explicit symbolism and frivolous appliqué ornament, has distorted the whole building into one big ornament'.[35] The modernists also have a tendency to suppress whatever they dislike in their forerunners. For example, they ignore the abundant use of decoration in the nineteenth-century industrial architecture they otherwise adore.[36] The authors also lambast the modernists for their false claim of 'building outside a formal language', which is strictly speaking impossible. Instead, what we really need is a 'formal language suited to our times'.[37] As Venturi's ally Vincent Scully puts it: 'There is no way to separate form from meaning; one cannot exist without the other.'[38] Furthermore, meaning is always contextual, so that buildings can never be self-contained in the way that many modernists demand. A building cannot exist in a utopian space, separate from inherited formal principles or the daily machinations of human life.

What Venturi generally finds missing in modernist architecture are 'complex and contrapuntal rhythms' of the sort one would expect to find in complex classical music. He sings the praises of the hybrid, the compromising, the distorted, the ambiguous and the perverse, not to mention the boring, the conventional, the accommodating, the redundant, the vestigial, the inconsistent and the equivocal. 'I am for messy vitality,' he concludes, 'over obvious unity.'[39] He expands on this point in a memorable passage: 'the eye does not want to be too easily or too quickly satisfied in its search for unity within a whole'.[40] Consider the enduring appeal of Gothic architecture and pre-modern city plans, however 'illogical' they seem when contrasted with the grids of modernism. The easy symmetries of mathematics and geometrical solids do not satisfy us in the way that modern theoreticians assume. Instead, the human soul demands a bit of mystery and imaginative play.

Although Venturi's preference for history over unfettered modernism helped inspire the rise of historical postmodernist architecture in the 1970s, there are two key features that set him apart from others in that movement. One is his hearty enthusiasm for pop culture, as seen from his co-authorship of a book on the architectural merits of Las Vegas: hardly a favoured location for design snobbery. Another is his conviction that history cannot be imitated without the often difficult work of adaptation. He approvingly cites T. S. Eliot's words that tradition 'cannot be inherited, and if you want it you must obtain it by great labour'.[41] He also quotes Eliot's citation of Doctor Johnson as to how in the great Elizabethan writers 'the most heterogeneous ideas are yoked together *by violence*'.[42] However, he adds that aesthetic attractiveness of this sort is a difficult achievement, unattainable by the casual eclectic.

That brings us to one of Venturi's signature techniques: the pasting of one thing onto another, which is the literal meaning of the French word *collage*; or, as Venturi would have it, 'the appliqué of one order of symbols onto another'.[43] One of his favourite tricks is to graft literal meanings onto the expressive connotations favoured by the modernists. For instance, he is fascinated by the dominant role of road signs on the Las Vegas strip, seeing them as an obvious way to bring meaning back into architecture. He takes this to a logical conclusion in his co-designed Guild House for the elderly in Philadelphia, whose entrance is topped by a sign saying – what else? – 'Guild House'.[44]

In some passages Venturi knocks early on the door of the debate that would rage a quarter-century later between the jarring effects of deconstructivism and the undulating lines of Deleuzean smoothness. While these two trends are usually seen as polar opposites, Venturi seems to think he can integrate elements of both into a unified vision. An early hint of later deconstructivism comes when Venturi favours adjusting 'the scale or context of familiar and conventional elements to produce unusual meanings', before adding that 'the familiar that is a little off has a strange and revealing power'.[45]

As for smoothness, Venturi sees this occurring as a result of architectural 'inflection', in which one thing is defined by its relation to something else. In this way, 'the parts contain their own linkage: inflected parts are more integral with the whole than are *un*inflected parts'.[46] This means that 'extreme inflection literally becomes continuity'.[47] Put another way, the more explicit contrast there is between the various aspects of a building, room or other architectural unit, the more potential there is for a paradoxical sense of unity. The distinct parts bring attention to the spaces between them, creating an intricate total picture or sense of continuity, instead of the instant surface-level holism produced by the regulated features of modernism.

What does this tell us about the continuous and the discrete in Venturi's architectural programme? On the one hand we have a call for the familiar to be redone in a way that makes it slightly 'off': imagine a familiar architectural form with an unfamiliar colour, such as a pink skyscraper. On the other, there is the inflection of parts that combine towards the whole, whose logical outcome would be total continuity. But in that case nothing could really be 'off', since everything would blend together with everything else. This creates a tension in Venturi's writings between complexity and totality, one that would soon be echoed in the debates about architectural theory that peaked in 1988 and 1993.

Deconstructivism and Discontinuity

As mentioned, the 1988 Deconstructivist Architecture show at MoMA was co-curated by Philip Johnson (1906–2005) and Mark Wigley (b. 1956). A brief word about these two figures will help clarify the unique position of the Deconstructivist show in architectural history. Johnson led one of the most colourful lives of any architect in the twentieth century.[48] Born in Cleveland, Johnson spent his undergraduate years at Harvard, where he focused his studies on ancient Greek philosophy. While there he befriended his older fellow student Henry-Russell Hitchcock (1903–87), joining him in the late 1920s on an architectural tour of Europe, where Johnson first became acquainted with the masters of modernist architecture. It was Hitchcock and Johnson, along with their comrade Alfred Barr (1902–81), who would organize the landmark 1932 exhibition at MoMA. This show had such a distinctly Europhile flavour that the American superstar Wright, long a modernist icon, was included only with reluctance.

During the global political turmoil of the ensuing years, Johnson strolled an alarming path. Adopting a radical Right-wing populism, he returned to Europe as a journalist, where he reacted with initial enthusiasm to Adolf Hitler's Nazi movement. Yet the most surprising turn in Johnson's life was yet to come. On the eve of America's entry into World War II, he enrolled as an architecture student at Harvard's prestigious Graduate School of Design (GSD). Although still a novice in the field at age thirty-five – if an exceptionally well-informed one – Johnson would go on to become one of the most celebrated American architects of the century. Collaborating with such luminaries as Mies van der Rohe, in 1979 Johnson won the first-ever Pritzker Prize: the professional equivalent of a Nobel in architecture.

Yet the twists in Johnson's life story were seemingly endless, and the one that concerns us here occurred in old age. In 1988 came the second landmark show at MoMA. Johnson served as co-curator of the fabled Deconstructivist Architecture exhibition an astonishing fifty-six years after his first show on the International Style. The 1988 show featured eight architects, of whom only two – Peter Eisenman and Frank Gehry – had even been born at the time of the 1932 MoMA exhibition. The other architects and firms included in 1988 were Coop Himmelblau (Wolf Prix and Helmut Swiczinsky), Rem Koolhaas, Bernard Tschumi, Daniel Libeskind and Zaha Hadid. This curatorial selection showed remarkable foresight: while some of these figures were more established than others, all would achieve a high degree of eminence in the field.

Johnson's co-curator for the show was the fresh-faced New Zealander Wigley, then teaching at Princeton University, whose work provided further impetus to architecture's involvement with recent European philosophy. Just two years earlier Wigley had completed his doctoral dissertation on Derrida, the founder of deconstruction. Thus it comes as something of a surprise that Wigley's catalogue essay for the 1988 show does not mention Derrida by name. This was possibly due to pressure from Johnson, who may simply have been too old to have much patience for a new Parisian philosophical trend.[49]

Instead, Wigley's essay presents deconstructivism as a development internal to architecture itself, with a lineage traceable to the constructivism of early twentieth-century Russia. This tactical disavowal by an architect of a favourite philosopher is not surprising. With each of the various turns of post-1960s architecture to a new philosopher (Heidegger, then Derrida, then Deleuze) there has been lingering sensitivity about the professional autonomy of the architect, with the implied fear that it might look as if architects were taking their marching orders from some philosophical master.

This is partly due to an unfortunate asymmetry between the fields, with architects frequently reading philosophy for inspiration, while philosophers (through their own limitations) have remained unable to borrow much from architecture in return: other than occasional pieces of terminology, including 'postmodern' itself.

The Russian avant-garde of World War I and the early Bolshevik period (from the onset of war in 1914 to Lenin's death in 1924) was extremely fertile. There was the suprematism of Kazimir Malevich and his circle, devoted to an art of simple geometrical abstraction. There was the influential school of literary criticism known as Russian formalism, featuring Viktor Shklovsky and his idea of 'defamiliarization'.[50] And then there was constructivism, a movement in art and architecture founded in 1915 by Vladimir Tatlin and Alexander Rodchenko. The influence of these movements on later European styles is profound – even if relatively hidden – and it is hard to imagine what the twentieth century might have looked like without them.

Wigley begins by observing that the solid institutional edifices and reassuring residential districts provided by architecture have always offered a sense of stability and order. But once upon a time, he recollects, the Russian constructivists made a brief but glorious challenge to this ordered stability through 'skewed, geometric compositions', as can be seen in the various disjointed designs produced by this movement.[51] For the earlier constructivists, as the Russian Revolution drew near, their 'geometry became increasingly irregular'.[52] Wigley speaks of constructivist buildings in which 'pure geometric forms become trapped in a twisted frame', or where 'pure forms have broken through the structural frame', or a building's 'frame has completely disintegrated'.[53] Dark sayings. Yet Wigley believes that the more the constructivists turned towards architecture after the Revolution, the more their radical ideas grew tame, as though yielding to classical assumptions about the need for stability. On his reading this happened because they were never very interested in destabilizing the status quo in the first place. As a result, the constructivist experiment in architecture failed, tapering off into stylistic excess rather than creating powerful underlying structural change. Ultimately, the constructivist 'wound in the tradition soon closed, leaving but a faint scar'.[54]

But all was not lost. Wigley proclaims the 1988 MoMA show to be a reopening of the same wound, with the internal failings of modernist architecture inviting a renewed destabilization. Although modernism called for the abolition of superfluous ornament in the name of exposing 'the naked purity of the functional structure beneath', it ended up with nothing more than a right-angled aesthetic of steel, glass and reinforced concrete. (We saw that Venturi made a similar critique of modernism as early as 1966.) This amounted to 'replacing the classical skin with a modern skin', without 'transforming the fundamental condition of the architectural object'.[55] Wigley asserts that the architects of the 1988 show try to 'irritate modernism from within',[56] operating in a familiar modernist idiom while distorting its syntax in often unsettling ways. With deconstructivism, 'the dream of pure form has been disturbed'.[57] This passes beyond mere external disturbance, which would yield nothing but 'a decorative effect, an aesthetic of danger, an almost picturesque representation of peril'.[58] In more positive terms, the distortion 'seems to belong to the form, to be part of it . . . The form is distorting itself.'[59] Structure is not 'destroyed', but 'displaced'. Imagine a crack deliberately placed down the middle of a building, and you will gain some idea of displacement without destruction.

But this tension between a form and its internal unrest is not the only relevant one. Also under fire is the interaction between forms and their contexts. It was once a liberating idea to consider everything in terms of how it was shaped by its surroundings. Yet this technique has become so dominant in so many fields that the literary theorist Rita Felski opposed it with a primal scream: 'Context stinks!'[60] Wigley is inclined in the same direction as Felski, complaining that 'contextualism has been used as an excuse for mediocrity, for a dumb servility to the familiar'.[61] But rather than completely ignoring context, deconstructivism *defamiliarizes* it. This means that certain things are left as they are, while others are made strange and disorienting: 'towers are turned over on their sides . . . bridges are tilted up to become towers, underground elements erupt from the earth and float above the surface, or commonplace materials become suddenly exotic'.[62] Not only does the building exist out of joint with its context; it lacks even a smooth relation with its own internal components.

When Wigley warned about the risk of excessive contextualism, he might also have been thinking of the pre-Derridean current of thought known as structuralism, which dominated French intellectual life during the 1950s.[63] For the structuralist, any individual element of a system is defined by its place in the system as a whole. Language is a total structure rather than a sum of individual elements. The original structuralist is usually said to be the Swiss linguist Ferdinand de Saussure (1857–1913), whose ideas on language were published posthumously by his students as the classic *Course in General Linguistics*.[64] One of his signature ideas was that since linguistic terms are chosen arbitrarily, they gain meaning only through their mutual differentiation. A powerful analogous trend developed in anthropology under the leadership of Claude Lévi-Strauss (1908–2009), who argued on structural grounds that there is no inherent difference between 'civilized' and 'uncivilized' humans.[65] This idea differed greatly from colonialist views of the time, and opened up the possibility of treating all cultures on an equal footing. Structuralist leanings could in fact be found across most of the humanities during the 1950s. This includes the post-Freudian works of the French psychoanalyst Jacques Lacan (1901–81), who was deeply influenced by Saussure and Lévi-Strauss, as seen especially in Lacan's view that the unconscious is structured like a language.

Then came 'post-structuralism', whose very name suggests suspicion of the structuralist idea that everything can be explained by its place in a total system of meaning. In 1966 – the same year Venturi's major book was published – the young Derrida attended a conference on structuralism at Johns Hopkins University in Baltimore.

His unexpected and contrarian lecture, entitled 'Structure, Sign and Play in the Discourse of the Human Sciences', was interpreted by some as an act of sabotage against the conference as a whole, and caused considerable annoyance among many attendees.[66] Among the points Derrida made was that structuralists organize their work around certain central terms that are never questioned: such as the opposition between 'nature' and 'culture' in the works of Lévi-Strauss.

By Wigley's account the entrance of Derrida into architecture occurred in 1985, when Tschumi (one of the architects included in the later Deconstructivist show) invited him to collaborate on his design for the Parc de la Villette in Paris. The growing friendship between architect and philosopher resulted in Derrida publishing an article on architecture the following year.[67] It displays one of the features that has long made Derrida a disputed figure, with some philosophers dismissing him as a charlatan or fraud. I'm referring to his unusual writing style, which often resembles the difficult prose stylings of James Joyce. In 1992, a group of prominent analytic philosophers even published an open letter in *The Times*, urging Cambridge University (unsuccessfully) not to award Derrida an honorary degree. Their stated reason was that he was merely a producer of frivolous puns, using Dadaist tricks and gimmicks to hide thoughts that are either outright incoherent or – at best – false or trivial.[68] While I am not on board with the view that Derrida was a charlatan, I can certainly be counted among those who find his style more exasperating than brilliant. For readers unfamiliar with his writings, the opening paragraph of his 1986 architecture article should speak volumes. Here Derrida gives us a tortured meditation on the word *maintenant*, which simply means 'now' in everyday French:

Maintenant: this French word will not be translated. Why? For reasons, a whole series of reasons, which may appear along the way, or even at the end of the road. For here I am undertaking one road or, rather, one course among other possible and concurrent ones: a series of cursive notations through the *Folies* of Bernard Tschumi, from point to point, and hazardous, discontinuous, aleatory.[69]

Nonetheless, Tschumi admired Derrida greatly like so many others. On their first meeting Derrida asked why architects would be interested in deconstruction, since this philosophy is famously 'antiform, antistructure, antihierarchy'; Tschumi responded that this was precisely why he was interested.[70] When Tschumi eventually became Dean of the Columbia University School of Architecture he invited Derrida to lecture there, and found that the philosopher responded patiently to numerous hostile questions from the audience. This gets at the core of what Tschumi loved about him: 'Working in the margins, in the precarious interstices of culture,' he would 'inevitably put people off-balance, especially those who believed in the primary object of Derrida's assault – the permanent or ineradicable truth'.[71]

But perhaps the best place to look for Derrida's influence is in the work of the prominent American architect Eisenman (b. 1932), who grappled with the philosopher's ideas at length. While the young Eisenman of the early 1960s saw much of value in an architecture of basic geometrical forms as an aid to memory, by the late 1980s he was speaking out against unimaginative Euclidean constructions.[72] In Eisenman's view, modernist architecture simply revived the outdated humanism of the Renaissance, which placed human beings at the centre of all things: 'While the tabula rasa of modernism was not thought to be man-centered, it was, ironically, willed by man; his mythic shadow loomed more importantly than had been realized.'[73] Eisenman argues that this humanism ended with the explosion of the atomic bomb in 1945, putting the very existence of the human species at the mercy of an enormously powerful non-human object. It was impossible to continue believing in the supremacy of human beings when all human things could be incinerated in less than ninety minutes.

What Eisenman desires as an antidote to the purported humanism of modernist architecture is a 'formalist' architectural object:

one that is autonomous, cut off from all interactions with the human viewer and everything else, rather than being constituted by its relations with various other things. This is not exactly a Derridean position, since Derrida tends not to see anything as cleanly cut off from anything else, but Eisenman's work was always somewhat in tension with Derrida's despite his admiration for the philosopher. It is Eisenman's view that an object should not even have a negative relation to its surroundings, as with the stereotypical art gallery placed in stark contrast with the industrial wasteland where it is located. After all, discord creates a dialectical interaction between one thing and another, when the point is that the object must be self-contained rather than dialectically related to what it is not. Eisenman prefers discontinuity, and one could hardly imagine a greater degree of architectural discontinuity than a structure being built to exist solely in itself, with no relationship to its surroundings.

Despite this nod to self-containment, Eisenman is also on board with Derrida's central claim, which I myself reject: no object can have a stable identity, since every object is merely a cluster of differences from everything else. Following Derrida, Eisenman calls for the 'removal of the identity and significance from objects', meaning that every object exists out of phase with itself and is impossible to pin down as having certain qualities and excluding others. There are moments when Eisenman pushes these ideas in a remarkably literal direction. Most famously, he often engages in deliberate subversion of the expected convenience of human dwellings, as seen in a series of houses to which he gave numbers rather than names, in order to strip any human connotations from these projects. Speaking of his designs for House III and House VI, Eisenman seems pleased that 'several columns "intrude on" and "disrupt" the living and dining areas'.[74] These houses are just like they sound: we might find a column sticking through the middle of a dining table, or even through the chief bed of the residence. While most prospective homebuyers would find such features to be outrageous turn-offs, Eisenman boasts that 'these dislocations, these "inappropriate forms", have, according to the occupants of the house, changed the dining experience in a real and, more importantly, unpredictable fashion'.[75] There can be little doubt about that. What might be doubted is whether an architect *ought* to replace convenient dwelling with provocative conversation.

We have seen that deconstructivist architecture is a design trend laying great stress on gaps, breaks, fissures and discontinuities in the midst of what used to be reliably solid forms. It is an architecture of maximal discreteness. In the case of Eisenman's radical house projects, there is even an increased distance between buildings and the humans who use them, given that functional convenience is not one of his priorities. Elsewhere I have criticized this variant of anti-humanism as excessive, and as not really striking at the heart of the matter: it is not necessary to get rid of human usefulness in any literal sense to earn architecture its deserved autonomy.[76] Yet Eisenman's career has shown an unusually rigorous commitment to gaps and discontinuities of every sort. When Lynn – one of his most prominent students – helped launch the new trend of architectural continuity in 1993, he knew very well what he was fighting against.

Deleuze and Continuity

When I entered graduate school in 1990, the continental European philosophical tradition in which I work – as opposed to the analytic tradition that dominates the Anglo-American world – was still ruled by Derrida and Michel Foucault (1926–84). Neither of these two was among my heroes, but luckily there was still room for other influences to leak through here and there. One of these peripheral influences was Deleuze, whose bizarre imagination and irreverent sense of humour immediately won me over, though I never became a true devotee of his philosophy. It was only a few years later that Deleuzeans began to take the lead in continental philosophy. Later I learned that architecture had reached a similar point a few years earlier, with Derrida's influence already waning and Deleuze's star on the rise. In the early 1990s, Deleuze became the inspiration of choice for a group of architects newly preoccupied with continuity over the discontinuous ruptures of deconstructivism.

Deleuze began his authorial career in the 1950s with an unorthodox book on the philosophy of David Hume, famous for his scepticism.[77] There followed a series of equally offbeat works on other philosophers – Bergson, Nietzsche, Spinoza – and another on the novelist Marcel Proust.[78] The reaction to these works was enthusiastic, and established Deleuze in France as a guru among students intrigued by his unique approach to the history of philosophy. Indeed, Deleuze took almost perverse delight in offering strange interpretations of past figures. As he colourfully put it: 'I saw myself as taking an author from behind and giving him a child that would be his own, yet monstrous.'[79]

He reached what many consider his peak as a philosopher in the late 1960s, with the publication of the systematic book *Difference and Repetition* and the amusing *The Logic of Sense*, a series of meditations on Lewis Carroll.[80] Then came a period of co-authorship with the radical anti-psychiatrist Félix Guattari (1930–92): most importantly, their influential two-book series, *Capitalism and Schizophrenia*.[81] Deleuze's later years saw the publication of *The Fold*, on the philosopher G. W. Leibniz's relation to the Baroque, along with an intriguing two-volume work on cinema.[82] In November 1995 the world was shocked by the news that Deleuze had committed suicide in Paris, leaping from his apartment window after a long period of breathing difficulties. While this tragic event was somewhat buried in the press by the assassination the same day of Israeli Prime Minister Yitzhak Rabin, it left a deep impression on philosophers, architects, and others around the globe.

One of Deleuze's teachings that is especially important to architecture concerns the 'univocity of being', referring to the notion that everything that exists must exist in the same way. This thesis is drawn from the medieval philosopher Duns Scotus.[83] Aristotle had held that 'being' can be said in many ways: we saw that substance is one of them, along with essence, accident, actuality and possibility. But Deleuze is closer to Duns Scotus in the sense that Deleuze recognizes a single 'plane of immanence' where everything exists, although he asserts that the plane consists of a multitude of different intensities rather than being utterly unified. The French philosopher Alain Badiou is one critic who thinks that Deleuze never quite pulls it off: that Deleuze is really a philosopher of Oneness in spite of himself.[84] By contrast, Levi R. Bryant (b. 1974) has made an interesting defence of Deleuze centred in the practice of origami, the Japanese art of paper-folding.[85] If we think of how an intricate origami animal such as a crab or unicorn can be created from a single piece of paper, this is a useful analogy for how the One of being might be turned into the many of appearance.

Deleuze's interpretation of the philosopher G. W. Leibniz is a fine example of his iconoclastic approach. Leibniz was the co-inventor of calculus, and can certainly be interpreted as a mathematician of continuity. But Deleuze takes a significant risk by also positioning Leibniz as a *philosopher* of the continuous in his book *The Fold*, which became a bible of sorts among young architects in the early 1990s.[86] We've already touched on Leibniz's philosophy, which is centred in the idea of 'monads', or self-contained individual substances, including all humans and non-human things that are one by nature rather than by artifice. Since the monads 'have no windows', they cannot communicate directly with other monads.[87] Indeed, they cannot interact at all without a pre-established harmony between things ordained by God. In other words, since substances cannot influence each other, God had to ensure that they remained coordinated, which he did just once at the dawn of time so as to arrange all future events. It is hard to find a more committed philosophy of discreteness than this.

In Deleuzean circles, discrete individual objects are sometimes replaced with the term 'objectile', which the architectural historian Mario Carpo describes as 'a function that virtually contains an infinite number of objects'.[88] Whereas an object is conceived as something fixed and stable, the objectile refers to the object reconceived as a mutable process unfolding over time. In other words, what is primary is not individuals as Aristotle thought, but the process through which objects are produced. As Carpo remarks: 'Leibniz's differential calculus is for the most part the language still underlying the families of continuous forms that computers could now so easily visualize and manipulate'. This was an important concern for architects in 1993, at the dawn of the paperless studio as initiated by Tschumi as Dean at Columbia University.[89]

The American architect and former Princeton Dean Stan Allen (b. 1956) was another important figure in the shift of architectural discourse from discontinuity to smoothness. In a widely read article of 1997, he reported on his new theoretical standpoint, which he said was inspired by 'an intuition of a shift from *object* to *field* in recent theoretical and visual practices'.[90] What he (like physicists) means by field is a certain portion of space governed by a force with variable values at different spatial points. For instance, the gravitational field of our sun extends immeasurably far in all directions, though only bodies within the solar system experience gravitational values high enough to be visibly affected by it. Or think of a magnet: the closer we get to it, the stronger its magnetic field will be. Although Deleuze is never mentioned in Allen's article, the shift of interest to fields highlights an ongoing professional departure from the self-contained – if disturbingly edgy – objects of deconstructivism to more of a focus on smooth or gradual changes between one place and another. Allen's own version of the principle runs as follows: 'a field condition would be any formal or spatial matrix capable of unifying diverse elements while respecting the identity of each'.[91] In this spirit, he makes much of the structure of columns in the historic former mosque in Córdoba, Spain (fields), which he contrasts favourably with St Peter's in Rome, that Western monument of separate individual components (objects).[92]

At one point Allen recommends that 'we think of the figure not as a demarcated object but as an effect emerging from the field itself – as moments of intensity, as peaks or valleys within a continuous field . . .'[93] Yet moments later he takes the opposite tack, speaking of how flocks of birds are emergent patterns that arise from the simple rule-following behaviour of individual birds, meaning that fields are apparently generated by objects rather than the reverse.[94] I interpret this not as a stray inconsistency on Allen's part, but as an impasse that arises inevitably for those who try to ground the whole of reality in a continuum, reducing objects to nothing but derivative products generated by a deeper field. At some point the continuum theorist will always have recourse to the explanatory power of individuals, since to claim that everything is really just a field runs counter to our basic intuitions about reality. We can certainly speak of a field of birds acting in unison, but Allen is forced to admit that we cannot really understand such a field without considering the rules of behaviour followed by individual birds. Specific entities seem to be very much a part of the furniture of the universe, and few are likely to accept the theory of Parmenides that individuals are delusional byproducts of the senses. The Deleuze-inspired phil-osopher Jane Bennett (b. 1957), of Johns Hopkins University, runs a similar risk: 'One should understand "objects" to be those swirls of matter, energy, and incipience that hold themselves together long enough to vie with the strivings of other objects, including the inde-terminate momentum of the throbbing whole.'[95] In other words, Bennett sees fields as the primary stuff of reality, leading her to reduce objects to temporary stabilities amidst a wider cosmic flux.

On that note we return to Lynn's advocacy for smoothness in architecture, directed against the deconstructivist commitment to clashing and colliding forms. Lynn advances multiple arguments to support his case. The most amusing is surely his analogy from the kitchen: 'A folded mixture is neither homogeneous, like whipped cream, nor fragmented, like chopped nuts, but both smooth and heterogeneous.'[96] The same holds for his call for 'multiplicity', in the Deleuzean sense of something that is both one and many at the same time.[97] The problem here is that to describe something as one and many simultaneously is a major philosophical challenge, not something that can be stipulated into existence with words solely because it erases inconvenient difficulties.

Lynn also emphasizes a building's link with its environment. Indeed, he is convinced that one of the advantages of his smooth aesthetic over its deconstructivist predecessor lies in its superior ability to account for contextual relations. This is precisely the sort of thing that Wigley deplored as 'mediocre', given that it requires a degree of conformity between different buildings. Lynn takes the opposite view, asserting that the deconstructivist manner of compelling 'differences in conflicting forms often precludes many of the more complex possible connections of the forms of architecture to larger cultural fields'.[98] Stated differently, Lynn claims that deconstructivist buildings were too self-contained and did not interact sufficiently with their urban surroundings. This claim would be countered by Ruy in 2012, when he warned the profession about its excessive emphasis on connections between building and environment. [99] Ruy's primary worry was the risk of architecture turning into nothing more than a branch of ecology, with buildings reduced to their carbon footprints and other environmental statistics.

Lynn is seemingly unworried by this danger. He endorses the term 'affiliation', referring to means of architectural organization that invite external influences inside the building.[100] An example is found in his claim (with which I disagree) that Gehry's Guggenheim Museum in Bilbao, Spain, excels in its interrelations with the rest of the city. Lynn's 'affiliation' is meant to be a looser connection than Venturi's 'inflection'; we recall that inflection runs the risk of total unification, despite Venturi's interest in complexity. Lynn holds that affiliations (unlike inflections) can remain in the service of smooth topologies and the style of 'bending, twisting, and folding' that he prefers.[101] Nonetheless, he is concerned about architects taking this too literally and simply putting ornamental folded shapes on their buildings, which at times is exactly what happened. What Lynn proposes instead is a 'cunning submissiveness' in building, one that is capable of 'bending rather than breaking'.[102]

In 1999 Lynn went on to publish another book, the fascinating *Animate Form*. [103] In this work he pushes beyond the continuity of smooth curvilinear shapes and advocates a continuous notion of architectural time as well. Although it remains technically impossible to construct buildings that literally move across the earth, Lynn approaches movement through the theme of animation rather than outright physical motion. He begins with a critique of the usual manner of representing movement: 'The dominant mode for discussing motion in architecture has been the cinematic model, where the multiplication and sequencing of static snap-shots simulates movement.'[104] This is a clear echo of Bergson's critique of cinematic frames of time, in which the philosopher noted that time is continuous rather than made up of a transient series of individual instants. Instead, Lynn contends, 'animate design is defined by the co-presence of motion and force at the moment of formal conception'.[105] In other words, architecture should explicitly depict the dynamic forces at work beneath the surface of any apparently static building. The failure to do so leads to a falsely stable architecture that conceals its underlying dynamism.

Lynn reminds us, for instance, that a building's gravitational load on the surface of the earth is not the only force at play. Quite the contrary: there are 'multiple interacting structural pressures exerted on buildings from many directions, including lateral wind loads, uplift, shear, and earthquakes . . . Any one of these *live* loads could easily exceed the relative weight of the building and its vertical *dead* loads.'[106] Even a motionless building embodies a hidden drama, and Lynn seems to be looking for a way to reflect this in a building's appearance. He asks us to view stable forms in terms of a 'rhythmic motion', despite their necessary structural stability.[107] An architect should freely use gradients (gradual changes), flexible building envelopes and various sorts of smoothness, with the aim of allowing forces to flow into each other like rivers rather than sitting around in mutual isolation like wooden blocks.[108]

The Blob

The best-known result of Lynn's reflections on continuity is his defence of the 'blob' as a new architectural form, harking back to the formless *apeiron* of pre-Socratic philosophy.[109] Despite its humorous name with overtones of low-budget 1950s horror films, Lynn introduces the blob as a serious theoretical proposal. It is certainly compatible with the trend towards maximum smoothness found in Deleuze and his fellow French philosopher Gilbert Simondon (1924–89).[110] At first the blob might sound like the Deleuzean concept of affiliation pushed to an extreme in which everything is everything else, leaving us with nothing but a homogeneous lump; we have already seen that the *apeiron* ran this very risk. But Lynn views the situation differently. Consider a basic solid known even to children: a sphere. While the sphere appears to be a precise geometrical form, Lynn makes the unusual argument that this is only true because it is artificially cut off from the various pressures of its surroundings. The blob, by contrast, is to be understood primarily in terms of affiliations. In Lynn's own words: 'The sphere and its provisional symmetries are merely the index of a rather low level of interactions, while the blob is an index of a high degree of information, where information is equated with difference.'[111] The amorphous character of the blob has nothing to do with vagueness or indecision, but emerges naturally from its ultra-relational character. As Lynn puts it, the blob 'is capable of fluid and continuous differentiation based on interactions with neighbouring forces with which it can either be inflected or fused'.[112] The point seems to be that what the blob loses in detailed articulation, it regains in terms of a heightened flexibility and relational alertness to other nearby entities.

Although we are not told specifically how the blob engages in relations with its environment, the point seems to be that amorphous shapes are capable of relations with pretty much any surrounding structure, whereas more traditional architectural forms are too crisply self-contained, thereby hampering possible affiliations. Yet the difficulty with Lynn's blob, just as with the *apeiron* in ancient Greece, is that it cannot really be pushed all the way to the limit. At some point discrete entities will inevitably slip in through the side door, just as Allen's flocks or fields of birds turned out to be generated by individual birds rather than the reverse. Ironically, Lynn finds himself using terminology from Leibniz, one of the great champions of discontinuous individual substances. In Lynn's own words: 'These blob assemblages are neither multiple nor single, neither internally contradictory nor unified. With "blob" models,' he continues, 'geometric objects are defined as monadlike primitives.'[113] Here it seems to me that Lynn has followed Deleuze down a dead-end street. In no way can Leibnizian monads be described in terms of inflection or affiliation, except in the sense that God programmed them from the origin of the universe to *seem* to interact or affiliate. But divine intervention is not a plausible foundation for contemporary architectural style, to say the least.

Earlier, we encountered Wigley's claim that attention to context too often inspires mediocrity: a building must not slavishly devote itself to the pre-existent style of its surroundings. His aim in saying so was to point the way to a more self-contained architectural object. This aspiration does not mean that architectural relations with the external world are unimportant or non-existent. Obviously, such relations exist. It would literally be impossible for a contemporary building to lie outside any context at all, without even the sidewalks and roads that allow us to approach it. But to focus exclusively on relations, contexts and affiliations is to deny a building any reality of its own. This is why the Dutch architect Koolhaas emphasizes that buildings are not simply produced by a context, but shape their context in turn.[114]

The Problem of Inscription

We have seen that a debate exists in architecture between smoothness and discontinuity: between buildings that look curved and flowing and those that look angular, interrupted or disjointed. Although both styles have representative designers and theoretical champions, the opposition between them leads to what I will call the problem of inscription. What I mean by this term is that each of these architectural styles also needs to make use of its opposite.

For example, it is possible to design a building in basically blob-like fashion, but discreteness will inevitably leak through in any number of ways. One part of the blob's interior will have features making it more suitable for sleep or for dining, and one portion of the exterior is likely to be more approachable as an entryway than the others. There are sure to be abrupt contrasts of light and shadow at certain points, not just gently sloping shifts of illumination. In other words, it is impossible to have a completely unarticulated building that pays unlimited homage to smoothness.

The converse also holds: an architecture of total discontinuity is equally impossible. A truly 'punctuated' building would require something like tiny monk's cells – indeed, infinitely smaller rooms than that – with no doors or windows at all. Yet the basic human need to move around inside a building requires some means of passage between its various sections via corridors and other paths. Disruptions work most effectively if interspersed with stretches of continuity. Even if we were to build a house with sixty or seventy provocative interior columns that obstruct normal human activity – blocking our path to the oven or placed at various angles in front of doorways, so that we would need to climb over them to get into and out of rooms – such a house would still need at least minimal stretches of continuous hallway and floor space. Otherwise, no one could navigate between one of Eisenman's intriguing disruptions and the next.

The point is not that basically discontinuous and basically continuous architecture are impossible. It's easy for newcomers to the field to see that Libeskind generally works with the first type (see his unusual and interesting Jewish Museum Berlin), and 'the Queen of Curves' Hadid with the second (consider her beautifully flowing Heydar Aliyev Building in Baku, Azerbaijan). The point is that architecture must always consist of both discontinuous and continuous elements inhabiting the same space. No matter which of these one adopts as the primary style, whether it be smoothness or discontinuity, the problem is how the lesser style can be inscribed within the dominant one. This adds greater emphasis to a point we considered in connection with Aristotle: that the continuous and the discrete are both irreducible ingredients of reality, such that neither can be fully subordinated to its opposite.

This is an important point that goes far beyond architecture. From Kuhn we learned that any good theory in any field (whether it be gastronomy or astrophysics) needs to be guided by a dogmatic paradigm at its core, one that distinguishes it from sandbagging or mediocre statements that try to include everything but end up leaving room for nothing in particular. Recall the disappointment felt by Niles Eldredge and Stephen Jay Gould when Gerrit Neef offered a merely lukewarm challenge to Darwin when discussing abrupt changes in Plio-Pleistocene snails. Yet some room must be left for what any given theory excludes, since all theories (especially the best ones) are bound to be exaggerations. Something is sure to be left out, overstated or insufficiently articulated. For example, Karl Marx is generally at his best when strictly declaring that literature, art, religion, war and marital choice are all dominated by economic questions, as opposed to those blurry disciples who maintain that such claims about their master are exaggerations.[115] But there is also a hidden breadth in Marx's famous discussion of 'commodity fetishism', referring to the way that consumers ascribe value to commodities themselves rather than to all the human labour needed to produce, ship and sell them. For even here, Marx is careful to offer exceptions: natural air and water, or tithes paid to feudal lords, or goods traded in barter. These cases escape the logic of commodities, and are therefore not subject to fetishism.[116]

Sigmund Freud (1856–1939) shows similar awareness of what escaped his early theory of dreams. Initially he explained dreams as wish-fulfilments disguised by symbols.[117] But what about nightmares, which seem to be anything but the fulfilment of wishes? Freud's early theory treated nightmares as the fulfilment of wishes so forbidden that dreamers would deliberately frighten themselves, as a sort of alibi 'proving' that they did not really desire the objects secretly granted in bad dreams. But in the aftermath of the horrors of World War I, Freud treated many veterans suffering from recurring war trauma nightmares that could not possibly have represented wishes. What he eventually concluded is that the nightmares of war veterans were evidence of a previously unknown 'death drive' in the human psyche, a drive that counters the 'pleasure principle' that seems to guide everyday life.[118] Kuhn would say that it was precisely the rigid dogma of Freud's early theory of dreams that eventually allowed him to notice that war trauma nightmares did not fit that theory, which led him in turn to revamp it. To do so took intellectual courage. Not every theorist is willing to admit they got it wrong and refine or even discard key concepts.

Returning to architecture, Hadid's office colleague Patrik Schumacher (b. 1961) says that finding a proper balance between the two architectural extremes (discontinuity and smoothness) requires ingenuity.[119] Zaha Hadid Architects tends to favour an overall effect of smoothness, which presents the challenge of how to incorporate discontinuous phenomena into their style: doors and windows, for instance. In fact, one good method for assessing the skill of architects is to ask how well they assimilate features that run counter to their dearest design strategies. How well does Lynn, a strong adherent of continuous flow, articulate the interior and exterior of a blob?[120] How adeptly does Eisenman, who champions obtrusive disruptions and legible surfaces, still allow for serviceable movement through a building? These questions are worth keeping in mind for anyone examining the professional achievements of these figures. Whatever one takes to be secondary needs to be carefully subordinated, while still permitted to do the job that it alone can do. So it is with the continuous and the discrete in architecture, and wherever else this pair of opposites appears. It is reminiscent of the interplay of figure and ground in Gestalt psychology: these opposite terms distinguish between whatever in the visual field commands our attention and interest (figure) and the amorphous remainder that usually escapes our notice (ground).

8.

The Pope and the Horseman: How Many Magisteria?

Writing in *Foreign Affairs* magazine in 2020, political scientist Ronald F. Inglehart reports that he and colleague Pippa Norris observed a statistical decline in religious belief beginning in 2007, in 'the over-whelming majority of the countries we studied – 43 out of 49', including both wealthy and impoverished nations.[1] This downward trend was celebrated by the ardent secularist Robyn E. Blumner, president of the Richard Dawkins Foundation, who links the decline (at least in Western countries) with the publication around that time of a number of influential books.[2] Among them was Ayaan Hirsi Ali's controversial *Infidel*, which offered a severe critique of the status of women in Islam.[3] At roughly the same time, a cluster of volumes was published by the Anglophone authors known collect-ively as the Four Horsemen of New Atheism. This term was coined by one of the members of the Horsemen: California native Sam Harris, who rose to prominence with his books *The End of Faith* and *Letter to a Christian Nation*.[4] The American philosopher Daniel Dennett added his own flavour to the drink with the explicitly anti-religious book *Breaking the Spell*.[5] Meanwhile, the British journalist and one-time professional Leftist Christopher Hitchens (1949–2011), appalled by religious influence amidst the ruins of the 11 September attacks, weighed in with the provocatively titled *God Is Not Great*.[6] There was also the runaway bestseller *The God Delusion* by Richard Dawkins himself, the self-styled intellectual heir of Charles Darwin.[7] Although vehement atheism has been an available public option in most Western countries since at least the nineteenth century, the combined force of these authors – and their aggressive pursuit of publicity – made them one of the most prominent groups of public intellectuals in the early twenty-first century.

The reason I've brought up the Four Horsemen is not for the purpose of discussing religion. Instead, it's because their joint atheistic project takes a definite stand on yet another aspect of the debate between the continuous and the discrete. Two of the Horsemen, Dawkins and Harris, held public discussions at the Alex Theater in the Los Angeles area in 2016. Early on the first night they spent several minutes critiquing the prolific Stephen Jay Gould, who by then had been dead for a generation.[8] What bothered them most about Gould was his formulation of NOMA, an acronym for 'Non-overlapping Magisteria', which he introduced in a 1997 article and developed further in his book *Rocks of Ages*.[9] Harris called NOMA a 'dogma' and calmly dismissed it as 'wrong-headed and destructive', while Dawkins expressed regret that the idea had caught on among some working scientists.

Nonoverlapping Magisteria (NOMA)

The gist of Gould's NOMA principle is that religious belief is not incompatible with science. As he put it: the 'lack of conflict between science and evolution arises from a lack of overlap between their respective domains of professional expertise – science in the empirical constitution of the universe, and religion in the search for proper ethical values and the spiritual meaning of our lives. The attainment of wisdom in a full life requires extensive attention to both domains.'[10] Gould certainly wasn't motivated by any pious devotion; in the same article he describes himself as a Jewish agnostic.[11] Instead, his commitment was to a robust pluralism in intellectual life. Gould admits that the domains of science and religion are inextricably intertwined. The resolution to let science do its work and religion do its own 'might remain all neat and clean if the nonoverlapping magisteria (NOMA) of science and religion were separated by an extensive no man's land. But, in fact, the two magisteria bump right up against each other, interdigitating in wondrously complex ways among their joint border.'[12]

The Four Horsemen actively combat religion as a threat to both scientific understanding and the values of liberal civilization. Gould, eminent scientist though he was, claimed that science is out of its depth in the quest for ultimate meaning. Religion need not be viewed with contempt, since it adds something to the picture of reality offered by science. Referring to a 1950 letter to the Catholic community by Pope Pius XII entitled *Humani Generis*, Gould summarizes the Pope's view that Catholics should be free to believe scientific claims about evolution of the human body, as long as they also accept that God 'had infused the soul into such a creature'. Gould continues:

> I also knew that I had no problem with such a statement, for whatever my private beliefs about souls, science cannot touch such a subject and therefore cannot be threatened by any theological position on such a legitimately and intrinsically religious issue.[13]

The theory of evolution teaches a continuous family relationship between humans and other life forms, ranging from apes to pine trees to fungi. Yet Pius XII did not view such evolutionary gradualism as incompatible with human exceptionalism: the idea that we humans stand at the top of the ladder of being. Somewhere along the line, God must have injected primates with the immortal soul that serves for many as the bedrock of religious faith. It should be noted that Gould's phrase 'Nonoverlapping Magisteria' is a direct reference to the Catholic Church, where *Magisterium* is the term for the Church's teaching authority. It is adapted by Gould to argue that there may be numerous magisteria rather than just one. There can be more than one authority to which we turn for guidance and wisdom; multiple magisteria are needed for intellectual life to thrive.

Despite the Pope's favourable view of the coexistence of knowledge and belief, at the height of its power in the Middle Ages the

Church found itself firmly on the other side of the fence when it came to intellectual pluralism. In the thirteenth century a French bishop, Étienne Tempier, denounced a number of scholars at the University of Paris for defending more than 200 supposedly dangerous ideas that were henceforth forbidden. This so-called Condemnation of 1277 criticized the view upheld by certain scholars that 'there were two contrary truths . . . as if against Sacred Scripture, there is also truth in the sayings of the condemned pagans'.[14] The 'pagans' to whom he referred were Aristotle and his admirers, who at that time were still found primarily in the Islamic world. Much like Gould in our own time, some medieval thinkers claimed to be able to pursue philosophy that openly contradicted Church doctrine without actually renouncing God. Although no specific scholars were named, two of the primary targets are believed to have been the prominent thinkers Siger of Brabant and Boethius of Dacia (not to be confused with the earlier Boethius who was famously executed in Italy in 524).

Siger and Boethius were two of the most visible Latin Averroists, a term referring to those Europeans who were deeply influenced by the great Muslim thinker Averroës (Ibn Rushd), whom we met earlier. Born in Córdoba in what is now southern Spain (but was then the Islamic territory of al-Andalus) Averroës also lived at various times in Morocco. He is known to this day as one of the outstanding thinkers in the tradition of Islamic philosophy.[15] Though originally a judge, from the medieval period onwards Averroës has been considered the greatest commentator in any language on the major works of Aristotle. It was in large part through his influence that Aristotle re-entered European philosophy over the following century, as reflected in such major thinkers as Albertus Magnus (*c.* 1200–80) and St Thomas Aquinas (1225–74). From that point forward Europe also showed renewed openness to Aristotle's scientific works.[16] Without the return of Aristotle to Europe, it is unlikely that the modern scientific revolution would have occurred as soon as it did, or that it would have occurred in Europe rather than elsewhere.

Among other things, Averroës tried to thread the needle in a conflict between the ancient Greek view that the world had always existed and the monotheistic teaching that God created the world out of nothing. The startling compromise he offered was that the material world had always existed, but that God had given form to this matter at some definite point in time. This was one of the claims adopted by the Latin Averroists, who in effect were championing an early version of Nonoverlapping Magisteria: Averroës gives us one kind of truth, and the Bible gives us another. Trouble started to brew when it became clear that the Latin Averroists had taken the liberty of borrowing ideas from a 'pagan' Greek beloved by Muslim thinkers: ideas that were strictly incompatible with those of the Catholic Church. This Averroist deviation from the literal creation story of the Bible was something that Bishop Tempier, strict enforcer of Church dogma, could not tolerate. In a strange reversal, centuries later the agnostic Jewish scientist Gould found himself on the same side as Pope Pius XII, while the cry for a single magisterium (a sole authority for thought) was taken up in the name of science rather than the Church. Thus the baton was ironically passed from the thirteenth-century Bishop of Paris to the Four Horsemen of the twenty-first century.

Dawkins has said time and again that he has no patience for Gould's version of pluralism, in which scientific teaching on evolution may benefit from religious guidance when it comes to the meaning of life. Writing shortly after Gould's NOMA article appeared, Dawkins offered his own Condemnation of 1998, so to speak. In words clearly targeting Gould, Dawkins laments a certain 'cowardly flabbiness of the intellect that afflicts otherwise rational people confronted with long-established religions', despite their willingness to dismiss 'younger traditions such as Scientology or the Moonies'.[17] He observes that 'traditions too, however anciently followed, may be good or bad, and we use our secular judgment of decency and natural justice, to decide which ones to follow, which to give up'.[18]

The point he makes against Gould is that our secular judgement should guide us in moral matters as well as scientific ones. It's not hard to find examples from Scripture that offend the modern sense of human rights and decent behaviour, and Dawkins eagerly lists them: 'the obligation to stone adulteresses, execute apostates, and punish the grandchildren of offenders'. He calls out the biblical God for his 'pitilessly vengeful jealousy, his racism, sexism, and terrifying bloodlust'.[19] If someone objects that we cannot judge an ancient culture like the biblical Hebrews with present-day standards of ethical behaviour, Dawkins retorts that this is precisely the point: 'Evidently, we have some alternative source of ultimate moral conviction that overrides Scripture when it suits us.'[20]

Consider a later remark by Pope John Paul II, made in the same spirit as the earlier words of Pius XII, to the effect that 'with man . . . we find ourselves in the presence of an ontological difference, an ontological leap, one could say'.[21] What he means, in keeping with Church doctrine, is that humans are something utterly different in kind from other animals. In response, Dawkins asks sarcastically when the infusion of souls into humans supposedly occurred: 'A million years ago? Two million years ago? Between *Homo erectus* and *Homo sapiens*? Between archaic *Homo sapiens* and *H. sapiens sapiens*?'[22] For Dawkins, this belief in a massive divide between humans and everything that came before 'is an anti-evolutionary intrusion into the domain of science'.[23] It's worth noting that the idea of a great gulf between humans and all other living species is not just a matter of Church doctrine, but is found at the heart of modern Western philosophy as well. Beginning in the 1600s with René Descartes, modern thinkers have generally been committed to the special status of human thought by contrast with everything else in the universe.[24] Although dissenting voices are on the rise, human exceptionalism generally still holds among philosophers.

Dawkins tries to land heavy blows with his argument that science and religion already overlap, since the Church makes a number of claims that might well be tested (and possibly refuted) by the methods of science. He offers the following example: imagine that DNA evidence was discovered in some cave in the Middle East, proving definitively that Jesus had no earthly father.[25] Would we expect the current pope to wash his hands of the news, insisting that the discovery spoke neither for nor against the Immaculate Conception of our Saviour because it was a scientific discovery made outside the province of religion? Dawkins is probably right to suppose that, far from ignoring this finding, the Church would trumpet such a result as evidence of biblical truth. But cases like this are not just hypothetical. For Dawkins, religion already intrudes so far on the findings of science that the world is not big enough for both religion and science: between them, one must choose.

Introducing Scientism

Another of the famed Horsemen was the Boston-born philosopher Dennett, long a jolly champion of a scientific approach to nearly any intellectual dispute, to such an extent that some of his fellow philosophers thought he went too far. Dennett's 1991 book, *Consciousness Explained*, for instance, attacks the idea that consciousness is irreducible to scientific explanation.[26] According to Dennett, we need not ask about consciousness in a spirit of pious wonder; science will eventually be able to tell us everything we need to know about the mind, and to a large extent it already has. For this reason, some critics have joked that his book should instead have been called *Consciousness Explained Away*.

By way of contrast, Australian philosopher David Chalmers (b. 1966) argues that consciousness is a genuine philosophical problem: how could collections of atoms and molecules give rise to something as immaterial and immediate as the feeling of conscious awareness?[27] For Chalmers there is a genuine leap between the material and mental spheres, and explaining this jump is what he calls the 'hard problem' of consciousness. To prove how difficult it is to define what consciousness actually is, Chalmers turns to the undead. He greatly popularized Robert Kirk and Roger Squires's concept or thought experiment of philosophical 'zombies', sentient

creatures who possess both the neural subcomponents of consciousness at the bottom level of reality and observable human
behaviour at the top level, but without having conscious experience
of any sort.[28] For Chalmers, it is absurd to say that the zombie is
an adequate picture of what we are: for along with our physical
brains and our public behaviour, we also have conscious experience.
Surely there must be a difference between (1) a 'zombie' version of
Vladimir Putin who looks and acts like Putin but has no conscious
experience, and (2) the real Putin who can be presumed to have
thoughts and sensations?

Dennett counters this argument about zombies, which he calls
'preposterous', by claiming that if objects such as gold and silver
can be scientifically explained without losing their metallic sparkle,
then surely the same can be done for consciousness without erasing
the subjective 'feel' we have of it.[29] In principle, scientific explanations should be able to give exhaustive accounts of both metals
and people. But Chalmers maintains that consciousness is special in
this respect, and cannot easily be reduced to its physical base. For
my own part, I think that while the physical and the mental might
well be explained in similar terms, both kinds of explanations fail to
exhaust their topics.[30] We saw already in our discussion of Thomas
Kuhn that poems and metaphors cannot be paraphrased in literal
language; I would push the claim further and say that literal scientific statements are unable to give an adequate account even of
their own subject matter. While Chalmers thinks this is true only
of consciousness, I contend that we are equally unable to do justice to inanimate beings. We have encountered Aristotle's point that
things are always specific and concrete, while language and thought
work in the medium of universals. Since a thing is not built out
of universals, there will always be a significant degree of incommensurability between thought and what it thinks about. What
this means is that consciousness is not the only hard problem. No
matter the topic, there is a guaranteed mismatch between one thing
and another. Recall our earlier discussion of undermining, overmining and duomining. It can be helpful to reduce a thing downward by

focusing on its pieces, or upward by focusing on its effects, but this will never be enough to create a perfect model of that thing.

Dennett's science-heavy approach to the solution of philosophical problems is often tagged by his enemies with the name 'scientism', referring to the view that natural science is an all-purpose utensil uniquely qualified to solve any problem under the sun. Those targeted as champions of scientism often dismiss the charge as a childish insult. Yet the term is useful for its clarity: it perfectly captures the view that natural science is the sole magisterium, the court of final appeal in all intellectual disputes. James Ladyman and Don Ross strike the same note as Dennett in the opening chapter of their book, *Every Thing Must Go.*[31] There they claim that science should have the final word on all of the problems usually taken up by metaphysics, the branch of philosophy that speculates on the ultimate structure of reality. This means that quantum theory, the theory of relativity and evolutionary biology would simply replace existing philosophical speculation on space, time, causation and individuals, ignoring that philosophy's ways of studying the world do not fully overlap with those of science. Indeed, Ladyman and Ross proudly refer to their view as scientism, and they are not alone. A revealing dialogue between the philosopher Thomas Metzinger and the neurophysiologist Wolf Singer leads them to agree that neuroscience will eventually solve all major philosophical problems.[32] Philosopher Ray Brassier demands that science be granted 'maximal authority', since its findings are susceptible to correction only by the improved future findings of science itself.[33] One can only speculate as to why a philosopher would be so keen to relinquish their own sort of authority to science.

But despite scientism's association with remorseless modern enlightenment, there is a sense in which it is just the latest face of Bishop Tempier's authoritarian demand for a single magisterium. For the scientistic thinker, there is a unified human capacity called 'reason' that should be our guide in all aspects of life, while everything else – religion, art, emotion, poetic language – must be strictly subordinated to an all-powerful rationality. On this view, most of the problems in human society arise from the unwillingness of the flabby cowards dismissed by Dawkins to submit themselves to the rigours of cold, hard reason. It is telling how often reason is described by its champions as cold and hard, as if coldness and hardness were the very point of the exercise. Yet this pious reverence for the strict and the rigid is no doubt just as emotionally motivated as the opposing doctrines it condemns.

The other side of the debate includes the likes of Gould and Chalmers, who are open to the possibility that some parts of life are less exact than physics tries to be. This pluralist conception recognizes multiple magisteria, each capable of delivering independent findings even when they overlap or come into conflict. Another member of this more tolerant camp was Bruno Latour, who unlike Gould was a practising Catholic.[34] Latour's lengthy book *An Enquiry Into Modes of Existence* can be read primarily as an effort to challenge the force of the scientistic (and economic) dominance of modern intellectual life.[35] In this work Latour argues that the modern world consists of fifteen independent 'modes', each of them working as if in a different musical key.[36] As he sees it, science is important, but in its scientistic variant it is wrong to claim a monopoly on truth. Science is not always the best tool in our toolbox.

Latour gives the example of law.[37] Even the most scientistic thinker would have to admit that while law has some connection with what we usually call truth, it is not primarily concerned with proving scientifically obtained findings. For instance, key pieces of factual evidence are sometimes excluded by a judge because they were obtained illegally, or were not submitted by the required deadline, or even because the judge thinks the evidence would unfairly prejudice the jury against the defendant. Instead, science is given an important but subordinate role in criminal trials, with defence lawyers permitted to challenge the carefulness and motivations of lab technicians, or to investigate the police who handled evidence prior to testing. Simply put, scientism is not allowed to overrule the judge and demand that an entire trial focus exclusively on scientific fact. Some decades earlier, the philosopher Michael Polanyi had noted the legal rule of thumb that if a husband and wife drown together in an accident, it will be assumed (at least in some legal systems) that the older partner died first, although there is no possible way of knowing whether this is true.[38] For the purposes of life insurance and inheritance it could be important to establish which person died first, and in this case the law is required to fabricate a fact that cannot otherwise be determined. The law cannot function without such makeshift precedents, unrelated to scientific fact.

There can even be plurality within the mode of law, one that seemingly overrides evidence or even logic. Many foreigners were baffled by the fact that although the American celebrity athlete O. J. Simpson was found not guilty on all charges in his 1995 double-murder trial, he was later found liable for wrongful death *in the same case*, and ordered to pay $33.5 million to the families of the two people he had already been found not guilty of murdering. I know of one prominent European philosopher (not Latour) who was confused by this apparent contradiction to the point of anger. The reason for the difference hinges on the varying standards of certainty required in American criminal and civil trials. Whereas someone can be convicted of a crime only if the jury finds them guilty 'beyond a reasonable doubt', they can be liable for financial damages based solely on 'the preponderance of the evidence'. Conceivably – though this seems not to have been the case – all twelve jurors in the double-murder trial might have thought it highly likely that Simpson was the murderer, but 'highly likely' would not have been enough to convict him on criminal charges.

Alongside law, another example from Latour's book is politics. The case here is more controversial, since there are numerous science-minded people who do think the world's political problems are caused primarily by 'irrational' conflicts. In 2016 the astrophysicist and television host Neil deGrasse Tyson posted this tweet: 'Earth needs a virtual country: #Rationalia, with a one-line Constitution: All policy shall be based on the weight of evidence.'[39] The idea seems to be that the beastly power struggles of the earth would cease if everyone would stop being petty and selfish and calmly use reason to settle their quarrels. Some might say that deGrasse Tyson is naive about human nature; Latour's point would be that he is naive about the nature of truth. Where does 'the weight of evidence' fall, for instance, in the dispute between Egypt and Ethiopia as to what percentage of Nile River water should belong to each country? By digging into the details of the case you might be able to reach an informed conclusion, but that conclusion will never be enough to convince perfectly rational people on the other side of the debate. Even if a binding international court were to rule, say, against Egypt, this legal finding would be insufficient to sway an Egyptian President who felt that the very existence of the country was at stake. It is quite possible that they would simply make a political decision to ignore the court's legal authority. The Right-wing political philosopher Carl Schmitt (1888–1985) even saw in such cases the very essence of the political: the sovereign's sole authority to decide in matters posing an existential threat to their country.[40]

In similar fashion, it would be hard to determine a simple 'weight of evidence' when it comes to the advisability of nuclear power, an issue that divides even the green-minded Left between the horrors of carbon dioxide and the horrors of Chernobyl and Fukushima. What does 'the evidence' tell us about the expansion of NATO into Eastern Europe, except that one side claims the right of self-determination for all countries while the other complains of a grave threat to national security? Here as with law, political questions can benefit from the input of expert knowledge and scientific evidence, but politics is fundamentally a domain where the question of truth is settled very differently from how this happens in other disciplines, including science and law. Sometimes this involves armed conflict, which would be ridiculous in the case of science or law, though it is merely regrettable rather than ridiculous in the political sphere.

Returning to our central question, let's look at what these opposing views tell us about the continuous and the discrete. Regardless of what you might think about science or religion and their various claims to supremacy in one area or another, if there is rightfully a single magisterium in intellectual life, then we have a world that ought to be governed by a single domain of truth. It will be a world best ruled by a sole ultimate authority: whether it be the Catholic Church in 1277, or the findings of secular-minded scientists today. If instead there are multiple magisteria, then the world is a series of discontinuous patches with different methods called for in each case, whether it be the separate tasks of science and religion as for Gould and Pope Pius XII, or the multiple modes of truth called for by Latour.

An analogous approach can be found in the psychoanalysis of Jacques Lacan, who treats the psyche as split between three separate registers: the imaginary, the symbolic and the real. The imaginary is linked with the child first recognizing itself in the mirror, and pertains to matters of fantasy, personal identity and struggle with others for recognition. The symbolic concerns language and our position in the social order. The real, which is not the same thing as 'reality', is related to the shock and trauma of that which strikes us from beyond the imaginary and symbolic orders. In one instance, Lacan borrows a case study from Helene Deutsch of a boy with a phobia about hens.[41] One of the boy's symptoms suggests that he happily imagines himself to be a hen laying eggs for his mother. But in a different context, he refuses the role of a hen when his older brother treats him like one; in this case his brother assumes the role of a rooster roughing up the hen. Lacan resolves the apparent discrepancy by claiming that the boy does not mind being a hen in the *imaginary* register for his mother, but bristles at being treated as a hen by his brother in the *symbolic* order. In Lacan's case, then, we find that even within his conception of the human psyche there is not a single magisterium, but three different magisteria operating simultaneously.

Qualia and Emergence

Let's return to Dennett, that unflappable champion of science as the pre-eminent means of attaining truth. Among his many influential writings was an article entitled 'Quining Qualia', written for a 1988 collection of articles about consciousness.[42] The coined verb 'to quine' means to deny the existence of something in which there is widespread belief. It is an homage to Dennett's former Harvard professor, the American philosopher Willard van Orman Quine (1908–2000), who famously believed in continuity rather than separation between science and philosophy.[43] Philosophy cannot just sit in an armchair and perform its analyses apart from all experience, but must integrate its findings with those of empirical science. Needless to say, Dennett wholeheartedly agrees with this principle, in some respects pushing it further than Quine himself. As for 'qualia' (the singular is 'quale'), the term refers in philosophy to immediate first-person experience that supposedly cannot be explained in terms of third-person scientific description. For instance, my first-person recounting of a clear blue sky differs from a scientific account of how my eyes and nervous system make this possible. But Dennett does not accept first-person experience as a valid separate domain. As he sarcastically puts it:

> Look at a glass of milk, at sunset; *the way it looks to you* – the particular, personal, subjective visual quality of the glass of milk is the *quale* of your visual experience at the moment. The *way the milk tastes to you then* is another, gustatory *quale*, and *how it sounds to you* as you swallow is an auditory *quale* . . . Nothing, it seems, could you know more intimately than your own qualia . . .[44]

Dennett takes qualia to be a 'thoroughly confused' notion, and his article tries to demolish the idea that they even exist.[45] At least he is perfectly clear about what he wants us to reject. There are four specific features usually ascribed to qualia, none of which impresses him. Qualia are supposed to be: (1) ineffable, (2) intrinsic, (3) private and (4) directly or immediately apprehensible in consciousness.[46]

Something is *ineffable* when it cannot adequately be described in words, just as many religions have said about God. It is *intrinsic* when it exists in its own right rather than being dependent on something else: humans intrinsically have nervous systems but not identification documents, which must be provided by some sort of external bureaucratic machinery. The taste of food on my tongue seems to be *private* since it belongs only to me, while my social role as a philosophy professor is public insofar as it entails certain rights and responsibilities with respect to other humans. Finally, those who believe in qualia often think the appearance of colours and shapes in the mind is *directly or immediately apprehensible in consciousness*, whereas the knowledge that Franklin D. Roosevelt lived from 1882 to 1945 is something we have to learn from an encyclopaedia. For Dennett, of course, there is no such thing as qualia that are ineffable, intrinsic, private or immediately and directly apprehensible. For him, qualia can be scientifically reduced downward to their physical underpinnings and upward to the behaviour to which they lead, just like inanimate objects.

Some readers, feeling protective of the inviolability of personal experience, will become defensive when they hear that Dennett wants to dissolve qualia into something more basic. It might seem as if he were trying to destroy our sense of personal experience. Dennett not only anticipates this defensiveness, but celebrates it. In his own words: 'qualia seem to many people to be the last ditch defense of the inwardness and elusiveness of our minds, a bulwark against creeping mechanism'.[47] While this is no doubt true of many who oppose Dennett, the passage is more interesting for what it tells us about Dennett's own motives. Specifically, he resists the very idea that there could be any final bulwark against mechanism, and will not tolerate any theory that allows some inward, elusive sanctum of the mind to escape the grasp of science as he rather reductively conceives of it.

For Dennett there is a single ultimate magisterium: the hard sciences. He is so sure of his basic intellectual approach that, despite his admirable clarity and frequent wit, his writings often betray impatience with his opponents, as if he could hardly understand how they could be so fearful and irrational. Dennett mocks an informal sampling of his fellow philosophers whom he asked to define qualia, and enjoys reporting that 'it begins to dawn on them that they haven't known what they were talking about over the years'.[48] Consider as well his attitude towards Louis Armstrong's famous remark: 'If you got to ask what jazz is, you ain't never going to know.'[49] Far from admiring this *bon mot* by a great musician, Dennett says that he hopes to make it look 'quaint and ridiculous' by eliminating qualia, a notion he claims is 'in utter disarray'.[50] Luckily, Dennett never tries to reduce jazz to specific acoustic vibrations and behavioural manifestations, but if he turned himself to the task that would certainly be his goal. The final sentence of his article drives the point home: 'So contrary to what seems obvious at first blush, there simply are no qualia at all.'[51]

My complaint is not that Dennett is being rude; intellectual life often benefits from blunt disagreement of this sort. What is bothersome is that he seems to be one of those thinkers, so often found in the scientistic camp, who assumes the contours of reality to be so directly accessible to scientific inquiry that he necessarily pigeonholes his opponents as misty and muddle-headed thinkers. My own decades in intellectual life have led me to different conclusions. But before saying more about that, let's consider why natural science cannot be the unified magisterium of Dennett's dreams. To do this we can consider some of his own, always entertaining, examples. We begin with one that I often like to discuss: Dennett's dismissive view of the art of wine-tasting, which can take a long time to master even for those who are physically gifted for the task.[52] He asks us to imagine a wine-tasting machine that can analyse any sample and provide a 'chemical assay' (a test whose workings he does not specify) as well as the sort of quasi-poetic language we have come to expect from wine-tasting: 'a flamboyant and velvety pinot, though lacking in stamina'.

As Dennett sarcastically exclaims: '*surely* no matter how "sensitive" and "discriminating" such a system becomes, it will never have what *we* have when we taste a wine: the qualia of conscious experience'. While many of us would agree with this – we do think there's something to be said for tasting wine ourselves, rather than having a machine do it for us – Dennett wants no part of it.[53] He confidently states what he really thinks: 'such a machine might well perform better than human wine-tasters on all reasonable tests of accuracy and consistency the wine-makers could devise'. The purely chemical description of the wine, he holds, would be more accurate and reliable than any possible human taster. In this example, Dennett not only pits human against machine; he also places the imagined chemical test of the wine side by side with the imprecise metaphor 'flamboyant and velvety'. He is clearly inviting us to agree that the metaphor is the loser in this comparison. In other words, Dennett's philosophy is not only scientism, but also a form of what I call *literalism*. This is the idea that whatever qualities might exist in the world could be replaced in principle (both practically and intellectually) with an adequate prose or mathematical description of them.[54] For the literalist, the word 'flamboyant' might correspond to one sort of chemical or behavioural description and 'velvety' to another.

Dennett thinks that over the course of time, as humans grew accustomed to the tasting machine, we could get rid of wine-tasting poetics altogether and learn to read the flavour of a wine directly from its chemical formula. In other words, the supposedly discrete and self-contained human sensual experience of wine could be eliminated in favour of a continuous scientific discourse able to exhaust whatever it finds in terms of a literal master language. But why do I and many others have the sense that Dennett is missing something fundamental when he aspires to replace qualia with their physical underpinnings or behavioural outcomes? Why is he wrong to make the flavour of wine continuous (or even synonymous) with its basic chemical building blocks?

To help answer that question, we return to the concept of emergence. As we saw earlier, emergence refers to what happens when multiple things are brought together in such a way that they do not add up to a mere group, but create something that is more than the sum of their parts.[55] We know that the chemical formula for water is H_2O: two atoms of hydrogen combined with one of oxygen. We also know that hydrogen and oxygen increase the intensity of fire, while water extinguishes it: same atoms, different outcome.

Examples of this phenomenon abound in chemistry, which is why the Mexican philosopher Manuel DeLanda (b. 1952) prefers to focus his philosophy of science on chemistry rather than physics, since physics is always tempted to reduce objects to their tiniest possible elements.[56] This does not happen to the same extent in chemistry, which is generally concerned with more complex levels than individual atoms or subatomic particles.[57] Of course, a chemist might seem reductive from the standpoint of a biologist, and a biologist might look reductive from the standpoint of a zoologist. Yet this would not concern DeLanda, since what he wants to establish is the autonomy of facts at every level of analysis. Crucially, he opposes the assumption that the smallest elements of anything can explain it sufficiently.

But if there is already a gap between H_2O and water, then how much wider is the rift between scientific formulae describing William Shakespeare's brain and a piece of literary criticism exploring his plays? Consider the following pair of statements about *King Lear*, both of them disavowing any comparison between the suffering Lear and the biblical character Job. The first sentence comes from a book by eminent Shakespeare scholar Harold Bloom (1930–2019).[58] The second is a variant I invented myself.

1. The pragmatic disproportion between Job's afflictions and Lear's is rather considerable, at least until Cordelia is murdered. I suspect that a different biblical model was in Shakespeare's mind: King Solomon.

2. The pragmatic disproportion between Job's afflictions and Lear's is rather considerable, at least until Cordelia is murdered. I suspect that a different biblical model was in Shakespeare's mind: Samson.

It is true that one major difference between science and literary scholarship is that a scientific solution to a problem is more likely to be verifiable or disprovable than the words of a literary critic. While no holdout chemist will insist that water is not H_2O, some literary critics might conceivably defend the implausible Lear/ Samson parallel: a vain and deluded old monarch compared with a self-immolating muscular hero. Yes, we know that literary criticism is not one of the exact sciences. But the point is that no 'chemical assay' could possibly address this question about *King Lear*. An interpretation of Lear requires character analysis more along the lines of the 'flamboyant and velvety pinot' that Dennett dismisses out of hand. This is one indication of the reality of emergence, and of the necessity for different ways of approaching the world and the different levels and sizes of things within it. Yet Dennett is doomed in advance to miss anything other than the lowest layer of reality (tiny particles) and its highest layer (observable human behaviour). Chemical formulae are of limited use in interpreting Shakespeare's plays, and the same is true of wine-tasting.

Scientists and Coffee-Tasters

Dennett is not only interested in attacking qualia; he's looking more generally to knock down the idea that there are independent layers of reality at all. In this respect his theory of reality is continuous rather than discrete. He makes the claim that if we cannot speak clearly about something, then there is no point knowing that it exists. When Immanuel Kant claimed in the 1780s that a 'thing-in-itself' must exist beyond any of the ways our finite human selves can conceive of objects, many critics in his time made the same objection as Dennett: if something exists about which nothing can be said, then it is an inherently worthless notion. Kant's great successors, the German Idealist philosophers (including Hegel), argued that since we at least know what 'thing-in-itself' means, it is not really unknowable.[59]

A more recent example comes from the charismatic thinker Ludwig Wittgenstein, admired by Dennett and most other philosophers in the analytic tradition. In the final sentence of his *Tractatus Logico-Philosophicus*, Wittgenstein writes as follows: 'Whereof one cannot speak, thereof one must be silent.'[60] Despite its surface appearance of mysticism, this statement embodies just the sort of hard-nosed rationalism that thinkers like Dennett adore. If there were an experience about which nothing could be said, then what would be the point of claiming that one has it? Here Dennett's literalism is showing: he assumes that either we can state clearly what we mean, or we have nothing worth saying at all. Think of his earlier attack on the metaphorical 'flamboyant and velvety pinot', which he thinks can and should be replaced by a clear examination of the wine's chemical properties.

Another example, we have seen, is that no metaphor can be translated precisely into any number of literal statements, as with Max Black's example 'man is a wolf'. Elsewhere I have written about Homer's metaphor of the 'wine-dark sea', which cannot be reduced to the literal, trivial and probably false claim that the Mediterranean is the same colour as wine.[61] Instead, their peripheral similarity is used as a sort of alibi to ascribe many other qualities of wine to the sea: danger, intoxication and oblivion, for example. From the fact that we are all somewhat confused about both our immediate conscious experience and our later memories of it, it does not follow that no mental life exists. To think that something is real only if it can be described literally is to put the cart before the horse. Few defenders of qualia believe in them because they think they are easy to describe; quite the contrary. The best reason to believe in qualia is because the alternative is to think they can be exhaustively reduced both downward to their tiniest causes and upward to their behavioural results, as Dennett would have it. But neither of these operations is tenable. To reduce a thing downward is to overlook that it has properties that emerge at a higher level than the bottom-most one, as with the ability of water to extinguish fire. To reduce it upward is to miss the fact that a thing must be more than its currently visible effects, since it is also capable of different effects in the future, as seen in the Aristotelian example of the sleeping house-builder.

Another of Dennett's entertaining thought experiments concerns two fictional coffee-tasters called Chase and Sanborn, both of them stipulated to be long-time employees of Maxwell House. One day they have a conversation. Chase confesses that after six years on the job, Maxwell House still tastes the same as ever, but he no longer likes it: perhaps because he has become a more sophisticated coffee drinker in the meantime. Sanborn responds that this is ironic, since he has had the opposite experience: namely, he thinks that the taste of Maxwell House *has* changed, and to a flavour he no longer likes. The other Maxwell House tasters are consulted, and it turns out they all disagree with Sanborn: like Chase, they think the coffee tastes the same as ever. Sanborn concludes that something must be wrong with his taste buds, or some other aspect of his physiology or memory. Dennett writes that these two fictional tasters' interpretations 'exhibit, side by side, two poles between which cases of intrapersonal experiential shift can wander'.[62] That is to say, there is no way to decide the truth of the flavour simply from listening to the personal reports of Chase or Sanborn. We can say of both tasters either that they are right about their respective experiences, wrong about them or some mixture of right and wrong. What we can do that's conclusive is to *test* the coffee to see if the flavour has changed.

It's important to emphasize Dennett's reference to scientific testing as the ultimate authority on whether the taste has changed. The problem is that whatever sceptical doubts may be raised about the reliability of our memories of qualia, such doubts also apply, in principle, to the accuracy of scientific equipment and the notations found in laboratory notebooks. Equipment may be faulty, researchers may fail to enter refuting evidence into their logs, and scientists might even be taking corrupt payoffs from the Maxwell House company. Here the practices of scientific knowledge are on trial no less than human sensation and memory. Dennett does give the nice example of 'failed nostalgia', where we remember things like childhood desks as bigger than they were, probably because we ourselves have grown and the desks have not. But this is an artificially simple case of an enduring physical object whose apparent size we compare at two widely spaced intervals in time. It does not support Dennett's broader claim, which is that physical tests must always be trusted more than personal recollection.

To see this, let's play along and follow Dennett down his chosen path. He supposes that the lack of constancy of qualia can be measured by checking their correspondence with external physical fact; in this way, our personal experience of those qualia might be contradicted by the evidence as measured objectively and scientifically. But this shows neither that I am not tasting what I think I am tasting, nor that qualia are fully reducible to their physical underpinnings rather than being emergent realities. All it really proves is that discord between the physical and perceptual levels of reality can and does occur. Dennett points out that 'Chase's intuitive judgments about his qualia constancy are no better off . . . than his intuitive judgments about, say, lighting intensity constancy or room temperature constancy – or his own body temperature constancy.'[63] But this is a tautology; Dennett is making the same argument twice in different ways. All he is saying is that just as our personal estimate of temperature in a room need not match what a thermometer tells us, so too the way coffee tastes need not match the objective chemical properties of the beans. While this is meant to make qualia seem unreliable, all it shows is that qualia and physical fact exist in different registers and don't always match. It would be like someone insisting to Bruno Latour that there is only one mode, politics, and that law and religion are really just about power struggles, and that's that: followed by pounding a fist on the table. There are certainly people who think this way, but they make the world too easy for themselves.

Dennett closely resembles this imagined opponent of Latour who tries to reduce everything to politics, though in this case he reduces everything to objective physical properties. He tells us that our experience of the coffee's qualia doesn't really matter compared with the chemical properties of the beans, as if that were the only factor involved in genuine flavour. Dennett's ultimate assumption is that scientific measurement of the beans gives us something 'more real' than anything involving a human mind. But that isn't necessarily the case. One could just as easily give priority to 'subjective' sensations of temperature and lighting intensity rather than the underlying physical data. The philosopher Martin Heidegger routinely does this for such properties as spatial distance and temporal duration.[64] Sometimes a mile-long walk is easy and other times hard; sometimes an hour flies by and other times it drags. When I see a friend approaching me from one hundred metres away, she seems closer than the glasses on my face, since I notice her immediately but usually forget about my glasses unless they need cleaning. Why should we agree with Dennett that the measured quantity of all miles and hours is 'more real' than our personal experience of them?

In short, Dennett assumes the answer to a question that is still under dispute: namely, whether empirically determined facts deserve to be treated as more fundamental than the rest of experience. Tests of all kinds frequently fail through some defect in the way the apparatus is conceived or constructed. Intelligence tests frequently miss brilliant individuals while awarding high scores to run-of-the-mill teacher's pets. Crucial experiments in physics are frequently revisited by later scientists, who say in retrospect that these experiments were misinterpreted. In similar fashion the fallibility of our personal experience of qualia, and our difficulty in describing them in words, do not prove that qualia do not exist.

Qualities that appear essential to a given experience can sometimes be seen, with hindsight, as entirely accidental. For instance, a dog may have been randomly happy in all our previous encounters, but then moody and temperamental on the next few visits, indicating that we somehow missed the mark in assessing the dog's personality. My main disagreement with Dennett, who habitually gives the upper hand to empirical science, is the same as my disagreement with a very different philosophical approach from his own: phenomenology. This current of thought believes we can reach the essence of any phenomenon by using our intellect rather than our senses. But it's unclear why the intellect should be any more reliable than the senses, given that we have all made at least as many mistakes in thought as in perception.

The Problem of Untranslatability

The centuries-long dispute between popes, Horsemen, evolutionary biologists and others throws into stark relief the heated conflict between religion and science. This often explosive disagreement conceals a deeper philosophical debate over whether the world can be addressed by a single continuous method (Christianity for the medieval Bishop Tempier, natural science for the likes of Dawkins and Dennett) or whether reality is instead a many-coloured coat requiring different approaches on different occasions. Gould weighed in on the latter side with his defence of NOMA: the view that religion and science are each best equipped for one set of problems, but not for all. In this debate I stand by Gould, but only up to a point, for he makes a subtle assumption that I would encourage readers to avoid. Namely, Gould treats religion and science as dealing with taxonomically different zones of reality. In his eyes religion is meant to address matters of ultimate meaning, while science is best equipped to explore natural phenomena unaffected by the domain of human culture. In other words, he ratifies one of the central biases of modern Western culture, which we have now encountered several times: that human thought is one half of the universe (to be understood speculatively via philosophy, religion and literature), with everything else, including animals, human bodies, the sky, planets and minerals, making up the other half (to be understood with the scientific method).

An alternative model, more flexible than NOMA, is the one found in Latour's theory of modes of existence. Although Gould only speaks of two modes (science and religion), it is not hard to imagine him arguing that every zone of reality has its own magisterium (so that art would be one system working by its own rules, religion another, law yet another, and so forth).[65] By contrast,

Latour's modes are more like different radio frequencies all occupying the same airspace. Law is a mode extending far beyond the courthouse, and the same expansiveness holds for politics, religion and the dozen other modes he recognizes. The advantage of this approach by contrast with Gould's is that it breaks with the modern division of labour, in which humanists are expected to remain largely silent about nature and scientists quiet about matters of meaning. It opens the way for more biodiversity among professionals, who are thereby permitted to contribute freely to discussions instead of being hastily excluded from certain debates simply because they happen not to be lawyers or scientists or art historians. In this way, the modern reign of the specialist is called into question. Gould even showed some awareness of this, given his understanding that there are no clear boundaries between where science, religion and (by implication) other fields are permitted to tread.

The specific lesson of NOMA is less the claim that each profession is responsible for its own subject matter, and more an insight into the untranslatability or incommensurability between different faces of reality. To clarify this point, let's return briefly to Kuhn. We saw that he spoke frequently of the incommensurability between a given scientific paradigm and its successor. As the etymology of the word suggests, two things are 'incommensurable' if they have no common standard of measurement. In everyday life we rank everything from politicians to athletes to cuisines. But if we now ask: 'Which of these is the greatest? Abraham Lincoln, Pelé or Thai food?', the question falls flat because there's no shared standard among their respective domains. How could we compare the achievements of a slain American statesman with a Brazilian World Cup hero and an Asian country's culinary output?

Or maybe you're familiar with the incommensurability between the diameter and circumference of a circle, which involves the irrational number pi, or π (more on this below). You might also be aware of the geometrical fact that a globe of the earth cannot be translated into a two-dimensional map without changing the size or shape of its land masses. And then there's the incommensurability between metaphorical and literal statements, as we've seen. These are precisely the sorts of examples Kuhn has in mind when he speaks of incommensurable paradigms in the history of science: there is a rupture or discontinuity such that words often do not even mean the same thing in two different paradigms. Here, then, it is a question of discrete breaks rather than a continuously applied 'human reason'. The difference between chemical formulae for wine and poetic tasters' descriptions of it is ultimately no different from the conflict between Newton's and Einstein's differing uses of the word 'mass'. Both cases show the limits of literalism.

Although religion and qualia might seem like very different examples, they share much in common. Neither God nor qualia are strictly 'ineffable'; believers have a lot to say about both, even if not in the verifiable and literal sense that scientistic thought demands. This obviously does not prove that either God or qualia exist. What it does establish is that any claim to a single dominant magisterium faces serious difficulty. The tendency of scientism to eliminate anything that cannot be stated in its own, rather limited terms disqualifies it from speaking broadly about the universe, or even about a single day in a human life. Dennett is always very clear in his assertion that what is real can be stated in clear prose, but it is easy to be clear when you disqualify all the unclear parts of reality in advance. To criticize Leonardo da Vinci for not painting the *Mona Lisa* in full, direct sunlight would seem like a strange complaint. For the same reason, it would be crude to demand of the ancient poet Homer that he change his metaphor 'wine-dark sea' to the literal statement 'the sea, which is approximately the same colour as wine'. By the same token, it would be misguided to want to replace 'a flamboyant and velvety pinot' with a chemical formula, however informative the latter might be.

This problem is well known to professional translators, of which I was briefly one myself. Certain foreign words are easy enough to translate: dog, cat and window all mean basically the same thing in all languages. But languages also have numerous terms without equivalents in a foreign tongue. These are the cases when translators must make difficult decisions about what to retain and what to sacrifice. We rely on a translator's experience with a given language, which cannot be exactly quantified. The reason why so many important books have been repeatedly retranslated is that there seems to be no way of getting a translation perfectly right.

Incommensurability is especially robust in the arts, where it is clear that no painting or building can be replaced by prose descriptions of them, or even by the interpretations of a highly acclaimed critic. There would be something perverse about reading Bloom's commentary on *King Lear* while refusing to read the play itself. Literary critic Cleanth Brooks makes a similar point when he says that poetry cannot be summarized in prose: 'The poem is not a prose-sense decorated by sensuous imagery.'[66] We saw that Black turns his attention to metaphor, arguing that even the lame example 'man is a wolf' cannot be converted without energy loss into the notion that each of these creatures is 'fierce, hungry, engaged in constant struggle, a scavenger, and so on'.[67] As for philosophy, one need only mention Socrates and his constant search for definitions. He fully embraces the original meaning of *philosophia*, 'love of wisdom', rather than the attainment of certainty. Socrates' attempts to find definitions fail repeatedly, but not because he is incompetent. Instead, he fails because replacing love, friendship and virtue with definitions of them is inherently impossible.

The importance of this point is as follows. If the world were purely continuous, it would be easy to translate anything into the terms of another thing. A clear summary of *King Lear* could replace the actual play, and we might read descriptions of Paul Cézanne's paintings without needing any illustrations. The chemical properties of beans could tell us everything we need to know about coffee. Language would contain no ambiguity or uncertainty. There would ultimately be no difference between perfect knowledge of a tree and the tree itself, since both would consist of the same literalizable properties. Since none of this is the case, the incommensurability between the continuous and the discrete shows us that any feasible picture of reality must include both, rather than a mere reduction of one to the other. The discrete side of reality cannot be wished away, but requires multiple magisteria to do it justice.

9.

A Corner of the Great Veil: Waves and Particles

Some years ago I was at dinner in Germany with the distinguished Austrian scientist Anton Zeilinger (b. 1945), who would later win the 2022 Nobel Prize for Physics.[1] I took the opportunity to ask him his thoughts on the most promising avenue for unifying quantum theory and General relativity, two pillars of modern physics that nonetheless contradict each other. Quantum theory treats the world as being made of tiny discrete chunks, or 'quanta'; General relativity treats gravity as a continuous bending of space-time by mass. Famously, this means that quantum theory cannot explain gravity (at least not yet), while on the flipside, General relativity cannot account for quantum effects, such as the fact that particles do not have precise positions until these are measured.

In any case, when I asked Zeilinger how these two conflicting theories might be unified, he responded as follows: 'I haven't the foggiest idea.' He added that, since some of the world's greatest minds have tried and failed to unify quantum theory and General relativity for more than a century, unification was likely to require a completely unexpected idea from out of left field. The unification of physics would mean that the whole of nature would again be covered by a single fundamental theory rather than the current two. Such a unified theory would probably lead to numerous subsidiary breakthroughs in physics, just as happened in the early modern period when the difference between celestial and terrestrial physics was replaced by Newton's idea that a single law of gravitation functions both in the sky and on the earth. It is likely that a unified theory would lead to many new and powerful technologies as well, and perhaps even to dangerous new weapons. But we are not there yet. How can a field as impressive as physics continue to function despite such profound incoherence in its foundations?

The term 'physics' is familiar enough, but it might have been a while since you stopped to remind yourself what the word actually means. As mentioned in our discussion of Aristotle, physics refers to *physis*, the ancient Greek word for 'nature'. It is the branch of science that focuses on the most basic forms of matter (protons, electrons, quarks, neutrinos) and energy (kinetic, potential, chemical, thermal, electrical). This makes it distinct from other sciences such as biology, chemistry and geology, which generally deal with mid-sized or large entities rather than the smallest ones. The early twentieth century is deservedly remembered as a golden age for physics. During the period between 1900 (Max Planck's discovery of the quantum) and 1926 (Erwin Schrödinger's wave equation) the twin pillars were established on which physics still rests: General relativity (for gravity) and quantum theory (for everything else). Both theories have been experimentally proven time and again, and serve as the basis for much of contemporary technology; despite this, in a strict sense they are incompatible.

For example, General relativity is a classical theory in which the position of every particle is determined by its prior position plus the laws of motion; in quantum theory, position is a matter of probability rather than certainty. Another example is that General relativity treats space-time as curved or bent by mass; quantum theory, by contrast, assumes a basically Newtonian model of space and time as empty containers not influenced by what happens inside them. Furthermore, by 1974 the Standard Model of particle physics was completed, giving a quantum interpretation of three of the fundamental forces of nature (electromagnetism, strong nuclear force, weak nuclear force) with only gravity left out.

Between 1977 and 2012, various quarks and bosons were discovered that confirmed the Standard Model, yet gravity has never been quantized, meaning that it has not been explained in terms compatible with quantum theory's vision of a world made of tiny ultimate units. To consider a way forward we will be looking at a 1941 book by French physicist Louis de Broglie (1892–1987), *Continu et discontinu en physique moderne* (*Continuity and Discontinuity in Modern Physics*),[2] which, as far as I know, has never been translated into English.[3] Generally speaking, this book was read with greater interest by philosophers of science than by practising physicists. But before turning to de Broglie, let's summarize briefly the recent past of physics, and ask how it relates to our theme of the continuous and the discrete.

Quantum Theory

The early history of quantum theory consists of a seemingly endless series of breakthroughs. The curious reader will find many of these discoveries reviewed in the British chemist Jim Baggott's dense but readable survey of the topic.[4] As we saw when discussing Kuhn, the theory of black-body radiation was under scrutiny in the late nineteenth century. A black body is an imagined ideal object that absorbs all wavelengths of light while reflecting none. We saw earlier that when such a body reaches equilibrium temperature with its environment, it will begin to emit electromagnetic radiation. The problem was that none of the available theories about this radiation gave values in agreement with those measured by experiment; this alone meant that a new theory of black bodies was needed. Beyond that, the important Rayleigh-Jeans Law predicted that the amount of energy contained in a black body could be mathematically infinite, though anything containing infinite energy surely ought to melt down. This was known as the 'ultraviolet catastrophe', although some historians have argued that it was a less important problem than the others. At any rate, such difficulties were the main challenges confronted in the late nineteenth century by the subfield of black-body physics. As seen earlier, Planck solved the problem by supposing that the energy emitted by a black body cannot have just any arbitrary value; instead, it must be a multiple of some smallest unit of radiation.[5] In other words, energy must come in discrete chunks.

By way of analogy, consider the attendance at a World Cup match, such as the reported 88,966 spectators present at the 2022 Final in Doha, Qatar. We can imagine that this figure might be inaccurate due to some accounting or reporting failure. What is inconceivable is that the true number might not be a whole number: if the reported figure was 88,966.47 people, we would know something was wrong. There is obviously no way for forty-seven hundredths of a person to attend a football match; people come in units, not fractions. Planck discovered that the same is true of electromagnetic radiation, which appears in nature in visible light, radio waves and microwaves, among other forms. Such radiation, just like people, comes only in units. In the decades following Planck's discovery, physicists would begin to find such smallest units all around us in nature.

In one of his four legendary articles of 1905, the young Albert Einstein extended Planck's insight, determining that light itself comes in a smallest unit that Einstein christened the 'photon', a tiny particle of light.[6] Although initially rejected by such eminent scientists as Planck and Niels Bohr (1885–1962), the photon came to be widely accepted a generation or so later, and it remains an essential ingredient of physics today. In a similar spirit was another 1905 discovery by Einstein: the insight that the random 'Brownian motion' of particles in water proves the existence of atoms, whose vibrations are what ultimately cause the water to move.[7] Until then, the real existence of atoms had remained a matter of dispute. Perhaps even more startling was the Nobel Prize-winning model of the atom introduced in 1913 by Bohr.[8] A central feature of this model was that electrons (tiny particles with a negative electrical charge) cannot be located just anywhere in their orbit around the atomic nucleus. Electron location is also 'quantized', meaning that these particles can only be found in the first orbit, second orbit, third orbit, or some other exact multiple of the first. An electron cannot be found, for instance, halfway between the third and fourth orbits. This means that atomic orbits are not continuous in the manner of the number line, but discrete.

By way of analogy, consider the orbit of planets around the sun. Imagine that Mercury's orbit was considered the smallest possible distance of a planet from the sun. Now imagine further that every other planet could only orbit the sun either twice as far as Mercury, three times as far, four times as far, or some other exact multiple of Mercury's orbital distance. This is obviously not what happens with the planets, whose location has evolved over time due to various gravitational influences aside from that of the sun. But that was essentially what Bohr was saying about the electron: not only does it change orbits when it gains or loses energy, but these movements can only occur by 'quantum leaps'; the electron is never to be found anywhere between those fixed orbits that are multiples of the first. This model had its paradoxes: for example, how can electrons jump instantly across space from one orbit to another, given that Einstein had established light speed as the maximum possible speed? And why wouldn't electrons run out of energy and enter a jumpy death spiral towards the atomic nucleus? But the model also had rapid success in explaining some previously shadowy features of the periodic table of chemical elements, such as why each element displays characteristic spectral lines when heated. This alone would have sufficed to secure Bohr a permanent place in the history of science.

Once the quantum vision of nature as made up of extremely small physical units had been established, various strange results soon followed. Among them was the finding that particles, understood to be the smallest units of matter, sometimes behave like waves. This seemed puzzling because particles are discrete, like stones, while waves rise and fall in continuous fashion as seen in the ocean. The idea that electrons have wave-like properties was introduced in the 1924 doctoral work of de Broglie, to whom we will turn again shortly.[9] His thinking opened the door to the famous 'wave–particle duality', according to which even matter – and not just light – has aspects of both waves and particles. The classic manner of presenting this duality is through the famous double-slit experiment with light. The Nobel Prize-winner Richard Feynman

(1918–88), in his fascinating book *The Character of Physical Law*, tells us that this experiment can explain 'any other situation in quantum mechanics' (though he also states, unironically, that 'nobody understands quantum mechanics').[10]

Imagine that we point a light source, such as a flashlight, at an opaque screen some distance away. We now place another such screen between ourselves and the target screen, after cutting a vertical slit in this intermediate screen so that only some of the light passes through and hits the target. What happens, as expected, is that the target screen is illuminated with a bright area in the same vertical shape as the slit through which the light has passed. This is exactly what we would expect if light were made of particles. But now imagine that we cut a second vertical slit in the intermediate screen, parallel to the first slit, thereby creating a double-slit screen. We continue to point our flashlight in the direction of the screen; both slits receive an equal amount of light. If light were made only of particles, we would expect to see exactly two brightly illuminated vertical patches on the target screen, one for each of the slits. The particles passing through the left slit would cause one illuminated vertical stripe, and those passing through the right slit would cause the other. But strangely enough, that is not what happens. Instead, when our light source hits the double-slit screen, it produces a zebra-like alternating pattern of bright and dark patches on the target screen. This is called an interference pattern, and it can only happen if light is a wave rather than a particle.

Yet the truly bizarre part of the experiment is yet to come. Let's replace our clumsy flashlight with an electric gun designed to fire just a single particle of light, a photon, in the general direction of the middle screen. Imagine that we continue to fire individual photons at the double-slit screen for hours to come. Now as before, some of the photons will hit the intermediate screen and go no further; they will not show up on the target screen. Others will pass through the left slit, and others through the right slit, and these are the ones that will make some pattern or other on the target. What sort of pattern should we expect to see in this case? Obviously, we would expect a pattern showing that light is made of particles, since we are literally firing particles of light one by one. The resulting pattern should be two brightly lit vertical stripes on the target screen, surrounded by darkness on all sides. Since waves are not involved in the experiment at all, this would be a reasonable expectation.

But what actually happens seems completely unreasonable. Namely, the pattern on the target screen turns out to be the same zebra-like alternation of bright and dark stripes that we would expect if light were a wave rather than a particle – even though we are firing particles! Baffled by this result, we decide to put detectors on each of the slits to let us know whether each photon is passing through the left one or the right one. But in a final bizarre twist, as soon as detectors are added, the photons stop acting like waves and start to behave like particles again. On the target screen we now see the two bright vertical stripes we expected from the start. It seems as if light in this experiment behaves like particles when observed with our detectors, but like waves when it is not. It is almost as if the screen 'knows' that we are using the detectors. Moreover, this experiment does not work only for light; it also works for electrons. In Feynman's words: 'Electrons behave in this respect in exactly the same way as photons; they are both screwy, but in exactly the same way.'[11] Before we say more about this wave–particle duality, let's consider some other strange results of quantum theory.

Werner Heisenberg (1901–76) was a young German physicist working at Bohr's institute in Copenhagen during the 1920s. At one point Bohr was away on vacation. In his absence, Heisenberg made his most famous discovery: the uncertainty principle.[12] For certain pairs of a particle's properties called 'conjugate variables', such as its position and its momentum (mass times velocity), the more accurately we measure one of these properties, the less precisely can we measure the other. If we ascertain the exact position of a particle, for example, we will not be able to know its momentum, since the way we determine the particle's position interferes with our ability to measure the momentum. Heisenberg concluded from this that there is a built-in limitation on what we can know about the world.

Bohr eventually returned to Copenhagen and learned about his disciple's discovery. Though obviously impressed by the insight, he responded somewhat aggressively, thinking that Heisenberg had not gone far enough. In the words of historian Richard Rhodes, Bohr 'was particularly unhappy that his Bavarian protégé had not founded his uncertainty principle on the dualism between particles and waves'.[13] Perhaps more importantly, Bohr insisted that it is not only the case that two interlinked properties of a particle cannot be *measured*: rather, the particle does not even *have* specific values for those properties prior to measurement. Bohr's findings implied that the uncertainty principle was not just a limitation on human knowledge, but an indeterminacy in reality itself. Bohr called this the principle of 'complementarity' rather than uncertainty. Eventually Heisenberg went along with this more radical interpretation of his work. To this day physicists speak of the 'Copenhagen interpretation' of quantum theory, referring to the joint work of Heisenberg and Bohr. Whereas classical physics assumes that the world can be measured objectively by an observer, the Copenhagen interpretation claims that the measurement of reality is a part of reality itself, and thereby transforms it. In popular discourse Bohr's idea of complementarity has often been modified into the claim that 'the mind creates reality', though nothing like a mind needs to be involved. Any interaction creates the effect of a measurement even if no human observer is on the scene.[14]

The Austrian physicist Erwin Schrödinger (1887–1961) pushed things further with the wave equation that still bears his name; later in his career, he would pursue a controversial wave-only vision of physics without particles or quantum leaps.[15] Closely linked to Schrödinger's work is the idea that quantum physics is not about the exact location of particles in space, but is concerned instead with the 'probability' of finding a particle in one place rather than another. Once the particle interacts with something else (whether it be a measurement by human scientists or an encounter with another inanimate thing) the particle is suddenly found in a definite place, and it is no longer just a question of statistical probability. But before it was measured, the particle was not in any definite place at all. This process is often called 'the collapse of the wave-function', since after measurement we seem to be back in the world of classical physics, with the particle now in a clearly defined place.

The popular thought experiment known as 'Schrödinger's Cat' was introduced by Schrödinger himself in 1935, in an effort to show what he (like Einstein) saw as the absurdity of the Copenhagen approach.[16] But Schrödinger's Cat has entered the lore of physics as a serious example, and has made its way into our culture at large as a textbook example of the strangeness of quantum theory. It goes like this: if we place a cat in a box with radioactive material that has a 50 per cent chance of decaying within the next hour, and if such decay will trigger the release of a poisonous gas that kills the cat, then does the cat exist as simultaneously alive and dead until the state of the radioactive material is measured?

In the eyes of classical physics the entire scenario is an absurdity. The radioactive material has either decayed or it has not, and hence the cat is either dead or alive; we simply don't know which until we open the box. But in quantum physics we can only say there is a 50 per cent probability that the material has decayed, and therefore the fate of the cat is uncertain until we either measure the radioactive material or look at the cat ourselves.

So, is the cat simultaneously both alive and dead until a measurement occurs? One especially radical solution to this paradox is to claim that the universe splits into two different universes every time a measurement occurs: with the cat being alive in one universe and dead in another.[17] Although the idea that the universe splits in half whenever a measurement occurs seems bizarre, a small percentage of physicists accept this as the proper interpretation of quantum theory. A more popular option is not to have a favourite interpretation at all, and to focus instead on the power of quantum theory to perform accurate calculations.[18] The Cornell University physicist N. David Mermin humorously called this the 'Shut up and calculate!' interpretation.[19] In other words, as long as quantum theory is useful in making accurate predictions, we can avoid all philosophical speculations about its deeper meaning. But the most popular approach of all to quantum theory remains the Copenhagen interpretation, with a majority of physicists today accepting the original path traced by Heisenberg and (more radically) Bohr.[20] Hence the mainstream interpretation is still Bohr's view that a particle has no definite position or momentum until it is measured, or interacts with something else more generally. Measurement seems to *create* certain features of reality rather than just observing them objectively from the outside. This implies that the thought experiment of Schrödinger's Cat is not an absurdity that refutes the Copenhagen interpretation (as Schrödinger thought), but an actual paradox in reality itself.

General Relativity

The early 1970s saw the consolidation of what is still called the Standard Model of particle physics. As we have seen, this model successfully unifies three of the four fundamental forces of nature in a single theory. These forces are electromagnetism (itself unified in 1873 by James Clerk Maxwell), the strong nuclear force that holds atoms together (first theorized by Hideki Yukawa in 1935) and the weak nuclear force responsible for so-called beta decay, in which a neutron is converted into a proton, an electron and an antielectron neutrino (proposed by Enrico Fermi in 1933).[21] The important fourth force of nature, still missing from the quantum picture, is gravity. This is why there are so many books and television shows these days about theories of quantum gravity, even though no successful theory of this sort currently exists.[22] What this means is that despite its paradoxical disavowal of the possibility of objective observation, quantum theory continues to assume a classical vision of space and time as neutral, unchanging containers inside which everything happens. This was the view of Isaac Newton three centuries ago.

He was later refuted by Albert Einstein's theory of General relativity, which shows that space and time themselves are bent or stretched by the presence of mass.

Human beings and other animals of medium or large size have always been aware that when an object is dropped it falls to the ground, although feathers, dust and other lightweight objects might drift in the wind for a long time first. In this limited practical sense, the existence of gravity has long been known. Aristotle explained it by saying that every object tends to move towards its natural place unless otherwise obstructed. For him, that 'natural place' for most objects is the centre of the earth, and thus they tend to move downward. By contrast, stars belong naturally in the sky; so does fire, which is why under normal circumstances it moves upward rather than downward.[23] This assumption followed from the ancient opposition between terrestrial physics for the earth and celestial physics for the heavens, so that there were effectively two different kinds of physics.

This traditional picture was damaged by Galileo in the early 1600s, partly because his telescope showed unforeseen similarities between terrestrial and celestial objects.[24] It was finally destroyed by Newton in 1687. He demonstrated that the movement of falling objects on earth and orbiting objects in the sky do not obey two different laws, but one: the law of universal gravitation.[25] This changed everything. According to Newton, all masses big and small attract each other: the sun pulls on the earth, but the earth also pulls on the sun, albeit to a much lesser degree. This also meant that nothing was falling towards the centre of the earth or rising towards the heavens. Instead, everything was locked in a reciprocal tug of war with everything else. Newton's theory assumed further – though he never quite believed it – that universal gravitation happens at infinite speed. If the sun were to disappear suddenly, its gravitational pull on us should also end instantly. But this turns out not to be true, as Einstein would later show: light speed is the maximum possible speed for everything, gravity included.

In the late nineteenth century, inconsistencies were noticed between Newtonian physics and Maxwell's more recent electromagnetic theory. In this connection, let's consider a point that would become crucially important for Einstein. According to Newtonian mechanics, the apparent speed of light should be faster if we are moving towards a light source than if we are moving away from it. To understand this, consider the case of cars passing each other in opposite directions. If I am driving towards the north at fifty miles per hour and pass a car travelling southward at fifty miles per hour, the other car will appear to be rushing towards me at 100 miles per hour. Newtonian physics assumed that the same would be true for light: if I am driving at fifty miles per hour towards a light source, the light should appear to be travelling at the speed of light plus an additional fifty miles per hour. It was Einstein who elegantly demonstrated that this is not true of light.[26] Whether we move towards or away from a light source, the light seems to travel at exactly the same speed. The strange implication is that moving clocks run more slowly.

This is often explained by imagining a simple clock that employs a single photon bouncing back and forth vertically between two mirrors, keeping time by counting the number of bounces. To see how this is affected by motion, let's put one of these light-clocks on the ground and another on board a train that speeds past the first clock. From the standpoint of the first clock, the photon in the second clock has to cover more distance, since it combines two separate motions: (1) the vertical back-and-forth bouncing of the photon found in both clocks, and (2) the horizontal motion of the train on which the second clock is stationed. If we are standing outside the train, watching the light-clock on the train blow past, we will see the light travelling a longer path for the photon to bounce from one mirror to the other. As a result, the moving clock on the train appears to be ticking more slowly than the one positioned next to us on the ground. This shows one of the key results of special relativity: that time appears to run at different speeds within different frames of reference. Another key result of special relativity, formulated most clearly by Einstein's former Zurich professor Hermann Minkowski, is that space and time can be viewed as a single four-dimensional continuum called 'space-time': a familiar compound word in our own era, but completely unknown before Einstein.[27]

However, special relativity only covers cases of uniform motion in a straight line, meaning it is not applicable to accelerating trains or planets moving in orbit. Special relativity also tells us that the speed of light is the maximum possible speed, whether for physical motion or any type of information, such as gravitational force. Einstein immediately noticed that this was incompatible with Newton's theory of gravity, which was built on the assumption that gravity acts with infinite speed across any spatial distance, implying that gravity is immeasurably faster than light. This is why Einstein's next step was to rework Newton's theory of gravity into one that explained how gravity works at less-than-infinite speed. After nearly a decade of struggle, Einstein successfully formulated such a theory. He was greatly assisted when his friend Marcel Grossmann (1878–1936). called his attention to the specific non-Euclidean geometry that Bernhard Riemann (1826–66) had developed in the nineteenth century, which treats space as curved rather than flat.[28]

Einstein's theory of gravity, known as General relativity, is the theory still accepted in physics today. In a famously beautiful series of equations, it successfully describes gravity as a bending of space-time by the presence of mass. If you tried to travel in a straight line near the sun, you would actually find yourself on a curved path going around it; this explains why planets have orbits. In 1913 Einstein and Grossmann jointly published a path-breaking article on the topic, with Einstein handling the physics and Grossmann the mathematics.[29] Einstein finished the theory himself two years later.[30] General relativity was soon bolstered by its ability to explain known anomalies in the orbit of Mercury, whose closest approach to the sun was changing unexpectedly in a way that Newton's theory could not explain. Even more dramatic was Einstein's improved modelling of how gravity would affect light, which Newton had foreseen to a lesser degree, though without ever anticipating the distortion of space-time. Einstein's theory predicted an increased bending of starlight when passing near the sun, meaning that stars would appear to be in positions considerably different from their locations as known to astronomers. It is usually impossible to observe such a thing, since the sun is too bright to allow us to see the stars that are close to it. The thing to do was wait for the darkened sky of a solar eclipse, the only case in which stars near the sun are visible. Observations were finally made during the solar eclipse of 29 May 1919, when the British astronomer Sir Arthur Stanley Eddington (1882–1944) found that Einstein's predictions for starlight were accurate. That was the moment when Einstein became a household name, the herald of a new science of nature.

Incompatibility

Quantum theory and General relativity are two of the greatest achievements in intellectual history, and both remain binding for work in physics today. But what can we say about their apparent incompatibility? This is often explained as it was in a 2022 article in the *Guardian*: 'Physicists agree that the theory of quantum mechanics applies to very tiny particles, and Einstein's theory of General relativity applies to larger objects.'[31] A similar explanation is given by prominent American scientist Michio Kaku (b. 1947) in one of his helpful physics bestsellers, *The God Equation*.[32] As rules of thumb go, this one is not so bad: it is true that quantum effects (such as the indeterminacy of position and momentum) are most easily observed at the level of such tiny entities as particles. Meanwhile, the effects of General relativity are most visible at the scale of astronomical events involving gigantic masses and vast stretches of time, where the effects of gravity on space-time become most obvious.

Yet this rule isn't entirely accurate. The probabilistic effects of quantum mechanics are observable at the relatively large scale of atoms and molecules, and could even be detected at the level of mid-sized everyday objects (dogs, shoes, aeroplanes) if the wavelengths of such objects weren't too short for quantum effects to be easily measured. Conversely, General relativity applies to the bending of space-time by all masses, not just extremely large ones. Even your own body distorts nearby space with its mass, for instance, though the effect is too tiny to be noticed. Shifting the focus away from the difference between small and large objects, the tension between quantum theory and General relativity is best understood in terms of the opposition between the discrete and the continuous. Quantum theory treats the world as made up of tiny chunks, as with units of radiation; General relativity treats it as a continuous bending of space and time, as with starlight bending as it passes through distorted space-time. The central problem is actually how the continuous and the discrete *interact*; this is the reason for the deadlock faced by physics for more than a century. While quantum theory tells us we're living in a world full of sudden jumps and tiniest units, General relativity envisions the cosmos as a continuous bending. How can discrete particles move through a continuous space-time? We find ourselves once more in the realm of inquiry raised in the works of Aristotle (interface) and Henri Bergson (endosmosis), both of which try to make sense of this interaction between the continuous and the discrete.

In the words of the German physicist Sabine Hossenfelder (b. 1976): 'The origin of the contradiction is that General relativity is not a quantum theory but nevertheless must react to matter and radiation, which have quantum properties.'[33] On the same page she offers a helpful example. The tiny electron has a paltry mass of only 9.109 times 10^{-31} kilograms, compared with the massive 1.989 times 10^{30} of the sun. Even so, according to General relativity even the minuscule electron's gravitational force bends space and time. But *which* space and time does it bend, exactly? For according to quantum theory, it is impossible to know exactly where the electron is at any moment other than in terms of statistical probability. But General relativity is a classical theory, which means it is not equipped to view the location of a particle (or its effect on space-time) in merely statistical terms as quantum theory does.

For the past several decades, the leading contender for a unified 'theory of everything' in physics has been String Theory.[34] As the name suggests, the theory proposes that everything in the universe is composed of tiny strings that vibrate in a finite number of ways, thereby generating all the particles in the universe as well as the four forces of nature (electromagnetic, strong, weak, gravitational). One advantage possessed by String Theory is that it easily unifies quantum theory and General relativity; in fact, their unity emerges directly and elegantly from the theory. However, there is no known way to test String Theory experimentally, which many physicists see as a fatal flaw.

By the early 1990s there were five consistent versions of String Theory, which were then brought together by Edward Witten (b. 1951) in 1995 in the wider so-called M-theory.[35] Witten insisted that the meaning of M should remain deliberately ambiguous until the discovery of a final version of the theory. Whereas General relativity works in a four-dimensional space-time, M-theory expands this to an exotic total of eleven dimensions, with some of them rolled up and unperceivable; the equations simply require that many dimensions to work. Since the number eleven kept turning up in various pieces of research on the theory of gravity, it seemed like Witten was on the right track. If String Theory could ever be proven, it would unify quantum theory and General relativity by showing that both are byproducts of simple patterns of string vibration. M-theory has always had prominent critics, including Feynman and Stephen Hawking, followed more recently by Lee Smolin and Peter Woit.[36] Given the lack of ideas for how to test it experimentally, many have argued that String Theory belongs to mathematics rather than physics. Indeed, it is worth noting that while master string theorist Witten won a 1990 Fields Medal for mathematics, considered the most prestigious award for that discipline, a Nobel Prize in Physics has so far eluded him.

There is also a less common view that the incompatibility of quantum theory and General relativity is a 'feature' of present-day physics rather than a 'bug'. This amounts to saying that maybe there is nothing fundamentally wrong with the current situation, in which quantum theory explains three of the forces of nature while General relativity takes care of the fourth.[37] In the terms of this book, which has argued that the continuous and the discrete are equally real, it would not be entirely surprising if a discrete theory of electromagnetism and the strong and weak nuclear forces were to remain on equal footing with a continuous theory of gravity and space-time. During the editing of this book, a new theory of precisely this sort was proposed by Jonathan Oppenheim in London: one that treats the non-quantizability of space-time as a positive feature of physics rather than a nagging obstacle.[38] Despite this recent development, most physicists still view a successful theory of quantum gravity as the Holy Grail of their field.

Wave–Particle Duality

Let's return briefly to the scientific history of light. Newton's influential 1704 *Opticks* argued for a particle theory, and his views initially carried the day.[39] Wave theories of light became more popular in 1803 with the interference experiments of Thomas Young in England, were solidified in the following decade by the work of Augustin-Jean Fresnel in France, and were eventually nailed down by Maxwell's revolutionary work on electromagnetism from the 1860s onwards.[40] This trend towards waves was not reversed until Einstein was able to explain the so-called photoelectric effect by deducing that light must also exist in particle form: the photons we met earlier.[41] The wave–particle duality of light is still with us today. True enough, Feynman's popular 1979 lectures on quantum electrodynamics adopt a particle-centered standpoint:

> I want to emphasize that light comes in this form: particles – particles. It is very important to know that light behaves like particles, especially for those of you who have gone to school, where you were probably told something about light behaving like waves. I'm telling you the way it *does* behave – like particles . . .

every instrument that has been designed to be sensitive enough to detect weak light has always ended up discovering the same thing: light is made of particles.[42]

While this strategy made it easier for Feynman to explain his approach to quantum electrodynamics, wave-like aspects of light are still at work in his theory. The duality remains.

Earlier I mentioned de Broglie, whose bold doctoral thesis showed that while electrons are ostensibly particles, they also exhibit wave-like properties. After the dual character of light as both wave and particle had been discovered, de Broglie extended the same duality to matter itself. He was optimistic that quantum theory had resolved the conflict between point physics (discreteness) and field physics (continuity).[43] When the French physicist Paul Langevin sent a copy of the thesis to Einstein, he famously replied that de Broglie had 'lifted a corner of the great veil'.[44] By this he meant that the young Frenchman had made a lasting contribution to physics. A few years later de Broglie would be recognized with the 1929 Nobel Prize in Physics: 'for his discovery of the wave nature of electrons', as the prize citation read.[45]

But the main reason we're interested in de Broglie here is for his book *Continu et discontinu en physique moderne* (*Continuity and Discontinuity in Modern Physics*).[46] Consisting of a closely linked series of essays on the state of physics in his time, the work was published under the horrific conditions of the Nazi occupation of Paris. This is reflected in the cheap wartime paper on which the first edition is printed, though I've found that the binding is tough enough to withstand repeated photocopying. A spirit of proud melancholy suffuses the tone at the opening and conclusion of this 266-page book. As de Broglie writes in the preface:

Whatever role the future may have in store for France in the Europe of tomorrow, it can approach that future with head held high, proud of a glorious history, sustained by the memory of the great men it has produced over the centuries, and clearly aware of the immense contribution it has made to the spiritual and material civilization of the modern world.[47]

These words are a sort of preview for the appendix to his book, where de Broglie recounts the troubled but fruitful life of the French scientist André-Marie Ampère (1775–1836), in a movingly written tribute that ends with a salute to 'all the great figures of the glorious past of France'.[48] Written at a time when the very survival of France as a nation was endangered, de Broglie's appendix found a nuanced way to assert a spirit of nationalism.

Scientifically speaking, de Broglie leads with the central point that thanks to the wave mechanics developed by Schrödinger and others, it's been proven that both light and matter have dual aspects: they are corpuscular (or distinct) and wave-like (or connected).[49] The most direct proof of this came shortly after de Broglie's dissertation, when Clinton Davisson and Lester Germer showed that electrons yield a diffraction pattern when passed through a crystal of nickel, something only waves can do.[50] Once again, this shows the existence of wave–particle duality not just in light, but in matter itself. In de Broglie's words: 'today it is certain that, for material particles as well as for photons, it is necessary to combine the images of wave and particle, and in precisely the manner dictated by the total development of our knowledge of light'.[51] This tends to confirm Aristotle's view that both the continuous and the discrete are built into the structure of reality, so that neither is more fundamental than the other.

Importantly enough, the meaning of 'discrete' has changed from what it was in classical physics. Then it was thought that discreteness meant that a particle occupies a specific location at every moment. But as de Broglie notes, we have begun to 'suspect little by little that the essential property of a physical particle is the discontinuous and total character of its manifestation in observable phenomena, and not the possibility of localizing it in space or attributing to it a determinate trajectory over the course of time'.[52] This is a key point: we may not always know exactly where a particle is, but we do know that whenever it is measured, it will manifest as a particle. As seen, the same holds for material particles no less than for photons.[53] The wave-function means there is only a statistical probability of finding a particle in one place or another. That is to say, the wave associated with a particle 'is not the physical vibration of something', unlike most waves, 'but only a field of probability'.[54]

We recall that Aristotle divided the world taxonomically, so that certain things (substances, qualitative change) are always discrete and others (matter, space, time, number) are always continuous. The different lesson of quantum theory is that one and the same thing – a photon, an electron – can be both continuous and discrete simultaneously. Quantum theory does have its own sort of taxonomy, different from Aristotle's: one in which the unobserved is continuous and the observed is discrete. Remember the individual photons fired at the double-slit screen that made a wave pattern on the target screen until detectors were added to the slits, after which they produced a particle pattern. Recalling once more the proviso that no human is needed to do the 'observing', we can restate this to say that reality is made of continuous waves until it interacts with anything else, at which point it becomes discrete.

While some physicists are wary of philosophical reflections on quantum theory, this was never true of de Broglie himself. He was the product of a broad classical education, and always remained committed to intellectual pluralism. To invoke the authority of the hard sciences dogmatically, he believed, was the sign of a second-rate thinker. The true discoverers are always aware of what they do not yet know. As he puts it: 'The one who discovers a new theory is also most often the one most aware of its gaps and obscurities, and who best knows its limitations,' whereas 'disciples who are either imprudent or blinded by undiscerning enthusiasm transform it into rigid and definitive dogma'.[55] For all the triumphs of modern mathematical physics, de Broglie is aware of its blind spots in a way that Daniel Dennett never was: 'Without a doubt, mathematical physics necessarily leaves in shadow all qualitative aspects of things,' given its rush to quantify their properties and relations.[56] A similar point had been made by the English philosopher Bertrand Russell in his 1927 book *The Analysis of Matter*. Science explains only the relational properties of things, not their intrinsic properties: the ones they have in their own right irrespective of their relations with anything else.[57] This would not be a problem if qualitative aspects of reality, such as the flavour of wine, were merely byproducts of their successfully mathematized physical base. But if reality also has discontinuous aspects, like the qualia despised by Dennett, then every layer of reality is discrete. Each thing is an independent magisterium overlapping incompletely with its upward and downward neighbours, so that no complete reduction is possible in either direction.

The admirable openness of de Broglie also leads him to note the aesthetic dimension of scientific advance.[58] Consider the awe that physicists feel at the beauty of Einstein's equations for General relativity, or recall James D. Watson's incantations about the elegance of the DNA double helix.[59] While defending theoretical beauty, de Broglie also shows an unfashionable disregard for the widespread praise of 'economy' of thought.[60] In popular culture it is common to hear vulgarized adoration for Ockham's Razor: the famous principle of the medieval thinker William of Ockham (1287–1347) that in any theory, entities should not be multiplied beyond necessity.[61] To imagine an extreme example, any version of General relativity that added a team of ninety angels responsible for bending space-time would clearly be adding much more to the picture than the theory of gravity needs. But there is also a widespread oversimplification of this principle, one that reduces Ockham's Razor to the false credo that 'the simplest theory is usually the right one'.[62] Far from it, de Broglie would tell us. Instead, it's usually the most beautiful theory that turns out to be the right one, by which he means a complex sort of beauty like that found in 'monuments of the Gothic or Arabic style'.[63]

Quantum theory has led science 'to abandon the traditional notion of a rigorous determinism of observable physical phenomena'.[64] As we have seen, until an interaction occurs a particle is only a probability wave whose exact position is a matter of statistical probability rather than certainty. Our friend de Broglie goes so far as to call the situation a 'crisis of determinism',[65] where determinism refers to the view that everything (including human thought) works according to natural laws that function in mechanical fashion. Yet de Broglie is careful to speak only for physics when he talks about determinism, rather than applying his findings to the questions posed by philosophy or religion. This makes a refreshing contrast with the dismissive treatment of these fields sometimes found in leading scientific figures, especially in the Anglo-American world. Hawking was only the most prominent recent figure to weigh in and proclaim the superiority of physics to philosophy.[66]

Returning to the problem of determinism, de Broglie aptly defines it as the view that if we know the initial conditions of any situation, we should be able to predict with certainty what happens next.[67] However, if it is only a matter of probability whether a particle will be found in one place or another, there is a lack of determinism in the physical realm.[68] This also touches on the old philosophical problem of free will. In modern philosophy it is often the case that the deterministic character of the physical world is accepted, but it is still argued (or at least hoped) that human thought has the unique ability to break free from the chains of cause and effect and freely make decisions on its own behalf. Restated in the terms of this book, if you believe in free will then you accept a discontinuity between the conditions prior to an action and the action itself.

The most prominent thinker in this two-faced tradition is surely Immanuel Kant, who feels the weight of the evidence both for and against human freedom.[69] The question of whether human free-dom exists beyond the mechanical necessity of nature has obvious implications for ethics and the philosophy of religion. But just as often, the theme of free will recurs in the age-old question about the relative balance between individual decisions and the social environments in which they are made. For example, how severely should criminals be punished? The answer is closely linked to whether we think each person is basically responsible for their own deeds, or whether we prefer to understand their actions in the context of unlucky genetics or violent and impoverished childhood environments.

The indeterminacy yielded by quantum theory seems to throw a wrench into the mechanical vision of nature, a topic explored in numerous books, such as the physicist Henry Stapp's *Quantum Theory and Free Will*.[70] In the current philosophical climate there has been a surprising surge in support for panpsychism, the doctrine that thought does not belong only to humans and animals, but can be found even in inanimate beings, so that thermostats, dust and electrons must have some capacity for thought.[71] It is an intriguing speculative proposal, though one I reject on the grounds that it makes mind more important than necessary in the structure of the universe. But it is easy to see why the probabilistic nature of quantum mechanics would lead some observers to draw an analogy between the theorized indeterminism of human thought and the thinking capacity of matter. Working in the opposite direction, Roger Penrose and Stuart Hameroff have claimed that physical microtubules in the brain are able to capture and multiply sub-atomic quantum effects to provide the basis for human freedom,[72] although Max Tegmark has calculated that these microtubules are much too large for this to be possible.[73]

Returning to our central theme, de Broglie tells us that contemporary physics has thrown a new light on 'the traditional dilemma of "continuous or discontinuous" '.[74] As we have seen, a portion of twentieth-century physics has quantized reality, reconceiving it as made up of tiny discrete units, from Planck's theory of black-body radiation onwards. Yet with General relativity –and even in a good deal of quantum theory – we also see the importance of continuous fields that govern entire regions of space-time and vary gradually from one point in space to another. Consider electromagnetic fields, or gravitational fields. Yet it was not de Broglie's wish to reduce all particles to fields, or the reverse. He insists instead that particles and fields 'must necessarily meet, since both must find their place side by side in the overall framework of physics'.[75] If particles are analogous to the stones in our title, fields are a form of waves: continua made of gradually shifting values rather than sudden, discrete leaps.

We have seen that in recounting the history of the wave–particle duality, de Broglie lays special emphasis on the history of the theory of light, the domain where this duality first became pressing. In various periods of science either particle or wave models of light were favoured, though by 1941 de Broglie was right to say that 'the two opposed conceptions of light both contain part of the truth'.[76] Light moves in a straight line and reflects from surfaces, two of the first features of light noticed by humans. Hence it is easy to see light as made up of particles behaving according to the known laws of physical movement: bouncing off objects at predictable angles, just like billiard balls. But particle theory has a harder time explaining refraction, which means the change of light's direction when it moves from one kind of material into another. This explains why a spoon in a glass of water looks bent: when light moves from the air into the water, it creates the optical illusion that the spoon is actually changing direction.

Willebrord Snell discovered the law of refraction in 1621, although René Descartes was the first to publish it, in 1637. It was left to Snell's fellow Dutchman, the famed Christiaan Huygens (1629–95), to formulate a wave theory of light capable of explaining refraction. In doing so Huygens was premature, de Broglie argues, because at that point the particle theory was able to explain so much that an alternative theory of light as waves did not yet seem necessary. Earlier we encountered an analogous example: the premature heliocentric astronomy of Aristarchus of Samos, formulated long before Ptolemy's earth-centered theory began to face serious problems.

Although de Broglie stays in his physics lane when discussing causality and determinism, he does show subtle awareness of one of the philosophical problems that haunts any conception of individuals. Namely, if entities are fully discrete – totally enclosed in themselves and continuous with nothing else – it is unclear how they could interact at all. This is why many occasionalist philosophers claimed that God must be constantly intervening in every instant, and why

Leibniz's theory of windowless monads required God to create pre-established harmony between all things. Any physics of particles also suggests a granular and discontinuous model of reality, with no particle having direct access to another due to their total lack of continuity.[77]

Yet de Broglie insightfully conceives of fields as a site of contact between particles. This would be like saying that stones in a lake do not collide directly, but interact only through the medium of water waves. Although this is a false explanation of the collision of stones, it does provide a suggestive analogy for the interaction of physical particles.[78] Moreover, de Broglie attempts to solve the presumed incommensurability of point physics and field physics with the idea that the discrete is able to generate the continuous due to the *vagueness* of the discrete. We have seen that in quantum theory particles can no longer be said to occupy any definite position, but only have certain probabilities of being in certain locations. This means that the probable locations of multiple particles blur into each other, which de Broglie takes to be the explanation of physical continuity.[79] In a suggestive phrase, he refers to this as a reconciliation 'of the continuous with the discontinuous by the game of probabilities'.[80]

De Broglie also shows good historical awareness of the various disputes over continuity, discreteness and their implications for our understanding of nature. He rightly names the ancient Greek atomist Democritus and the modern philosopher Leibniz as partisans of the discontinuous character of reality, while listing Aristotle and Huygens as champions of continuity (though we have seen that for Aristotle this is only partly true).[81] He muses about 'how interesting the notion of interaction is, from the philosophical point of view, since it implies a certain limitation on the concept of physical individuality'.[82] Although he is merely dipping his toe into deep philosophical waters, the point he makes is profound. Interaction requires the interacting things to become one, at least temporarily, which means that interaction requires at least a temporary or partial loss of individuality.

The Cut

Generally speaking, the continental tradition in philosophy (the one guided mainly by recent French and German thinkers) has been less attentive to mathematics and science than Anglo-American analytic philosophy. A recent exception to this rule has been the increased continental interest in the mathematical work of the tortured pioneer Georg Cantor. Cantor is best known for his work on transfinite numbers, in particular his discovery that there are infinitely many infinities of infinitely many different sizes. His later years were darkened by depressive mental illness, no doubt worsened by the frequently negative reactions to his ideas. Among continental thinkers it is Alain Badiou in France who has made the most extensive use of Cantor to develop his own philosophy.[83] Badiou's protégé Quentin Meillassoux (b. 1967) has also made ingenious use of Cantor's theory of transfinite numbers, arguing on Cantorian grounds that the laws of nature can change at any moment for no reason whatsoever, and that God does not exist now but might exist in the future.[84] One of my reasons for introducing mathematics here is to show a way in which its relation to the continuous and the discrete is the opposite from that of physics. We will return to this point at the end of the section.

Here we'll be spending some time with Cantor's older friend and colleague Richard Dedekind (1831–1916), his invaluable ally in struggles with the severe mathematical critic Leopold Kronecker (1823–91). Although Dedekind spent most of his career teaching in his hometown of Braunschweig, Germany, it was during an earlier stay at the Polytechnic (now the ETH) in Zurich, Switzerland – where the young Einstein would later study – that he introduced the concept now known as the 'Dedekind cut'.[85] The cut amounts to splitting all rational numbers in half by choosing a point on the number line, and recognizing that all numbers to the left of that point are smaller and all of the numbers to its right are larger. The purpose of doing this is to shed light on how irrational numbers fit on the number line along with the rationals; we will review these terms shortly.

The famous article in which Dedekind explains this concept is titled simply 'Continuity and Irrational Numbers'.[86] He informs us with proud precision that the discovery it contains was made on 24 November 1858, and he candidly declares that Section III of his article successfully defines 'the essence of continuity'.[87] What he means by continuity in arithmetic is exactly what Aristotle meant. There is no definite number of numbers between any two numbers: between 100 and 200, for example. We cannot simply say that there are ninety-nine integers between them, because we can create an endless roster of further numbers with the use of decimal points.

Let's now recall the distinction between rational and irrational numbers, which some readers may not have thought about since secondary school. Rational numbers are those that can be written as a ratio between two integers. Think of such ratios as 4 to 1 or 173 to 28, which can also be written as 4:1 and 173:28, or as 4/1 and 173/28. It is easy to show that every integer is a rational number. Let's choose at random the integer 453. We could simply choose to rewrite it as the ratio 453/1, or as 906/2, or 1,359/3, all of these ratios being equal. The same thing can be done for any integer. Any fraction is also a rational number, since fractions are ratios by definition. If we consider any list of fractions, such as 1/3, 14/27, -6/8, 7/-9 and -13/-15, we can see that all of them are also rational numbers.

Irrational numbers were discovered, at the latest, by the Pythagorean school in ancient Greece. They are defined as numbers that cannot be expressed as a ratio of two integers. Perhaps the most famous example is the square root of 2, or √2, meaning the number which when multiplied by itself gives the answer 2. Unlike √2, many square roots are rational: the square root of 4 is 2, since 2 times 2 equals 4; the square root of 9 is 3; the square root of 16 is 4; the square root of 25 is 5, and so forth. The square root of 2.56 is 1.6, which is also a rational number since 1.6 can be rewritten as the ratio 16/10, which reduces in turn to 8/5. But the square root of 2 happens to be irrational. There is no way to put any two integers in a ratio that would equal √2. For most practical purposes it is enough to say that the square root of 2 is 1.414, but this is merely a convenient approximation. The actual square root of 2 is 1.414213562373 . . . with the decimal places continuing to infinity.

In fact, one of the easiest ways to define an irrational number is to say that it's a number where we never come to the final decimal place when writing it out. The sole exception is with repeating decimals, as for instance in the case of 0.333333333 . . . The reason is that in such cases we are actually dealing with rational numbers. After all, 0.333333333 . . . can also be written as 1/3, which is automatically a rational number since it is written as a ratio of integers. Another famous example of an irrational number is pi, or π, useful in calculating the area and circumference of a circle. We can always write out π as far as we wish: 3.14159265358 . . . though this will go on to infinity without the numbers ever repeating. There is also e, known in mathematics as 'Euler's Number' after the famed mathematician Leonhard Euler (1707–83), though he did not actually discover it. We can begin by writing this number as 2.71828182845 . . . though like all irrational numbers it never repeats or comes to an end. Although √2, π and e all have special names, most irrational numbers are obscure and ignored. What makes √2, π and e so famous is that all have important mathematical uses.

When speaking of Aristotle, we briefly discussed the so-called number line. In its modern form it depicts zero in the middle, negative numbers decreasing infinitely to the left, and positive numbers increasing infinitely to the right. Credit for this version of the number line – the first to include negative numbers – is usually given to the English mathematician and cryptographer John Wallis (1616–1703), in his 1685 treatise on algebra.[88] When it comes to the number line, several basic features stand out. For any given number there is one and only one point on the line that corresponds to it. Moreover, any such number is greater than the ones to its left on the line, and lesser than the ones to its right. And finally, for any two points on the line – no matter how close to each other – there are still infinitely many points between them. As we saw when discussing Aristotle, that is exactly what 'continuum' means: that there is no definite number of intermediate points between any two given points, since the space between them can always be divided up even further.

In the context of twentieth-century physics, quantum theory showed that many features of nature are not continua: photons, for instance. A photon is a natural unit, and cannot simply be cut into smaller segments as we please. But General relativity joins Aristotle in treating time and space as continuous, meaning infinitely divisible. The number line must also be continuous, since we can always add a new number – whether rational or not – between any two points on the line. But a number line without irrational numbers would not be continuous, since it would have holes where the irrational numbers would otherwise be. The number line therefore consists of all rational and irrational numbers, which taken together make up the so-called real numbers. The domain of real numbers only excludes imaginary and complex numbers, which involve the square root of negative one; these do not concern us here.

We have seen that the number line contains all real numbers, and that it is also continuous. Dedekind now asks as follows: 'In what then does this continuity consist? Everything must depend on the answer to this question, and only through it shall we obtain a scientific basis for the investigation of *all* continuous domains.'[89] Dedekind shows disarming humility in admitting that the answer may strike his readers as banal, though it has long since entered the canon of mathematical knowledge. We have seen that any point on the number line separates all other points into just two groups: those that lie to its left-hand side (because they are lesser) and those that lie to its right (because they are greater). For instance, the number 4.2 cuts the entire line into numbers less than or greater than 4.2. Continuity, Dedekind tells us, simply consists in the opposite or converse principle.[90]

That is to say, if there are two classes of points such that all numbers in the first class are to the left of all numbers in the second class, 'then there exists one and only one point which produces this division of all points into two classes, this severing of the straight line into two portions'.[91] The number 4.2 produces a cut of all other numbers between those that are smaller than 4.2 and those that are larger. The numbers that are smaller come increasingly close to 4.2 (4.19999999 . . .), and those that are larger come decreasingly close to 4.2 (4.20000000 . . .), but never will they reach it. The underlying point is deceptively simple. Although we can never finish writing out the square root of 2 (as 1.414213562373 . . .) we do know that it is larger than 1.414213562373 and smaller than 1.414213562374. Therefore, it can easily be placed between these two rational numbers on the number line. There is nothing new in saying so, Dedekind notes, since the ancient Greeks already knew about irrational numbers and their incommensurability with rational ones. It's important to remark that while Dedekind aspires to define the essence of all types of continuity, he admits that he is not in a position to say whether the *physical* world is continuous rather than discrete. Quantum theory did not yet exist when he made his discovery in 1858, but if it had, I suspect he would have remained agnostic on the question of whether nature is made of tiniest units. All aspirations aside, his mathematical theory of continuity is independent of the state of physics.

As mentioned, if we made a number line out of nothing but rational numbers, it would be filled with holes and would therefore not be a continuous line. This leads Dedekind to say that 'the domain of rational numbers is insufficient'. In order for the number line to be continuous, it needs many more numbers than the rational ones.[92] Cantor would push this further in his theory of transfinite numbers, showing that there are infinitely many infinities all differing in size.[93] The counting numbers 1, 2, 3, 4, 5 . . . are discrete; they are infinite but countable, since we always know how to determine the next number in the series. I mentioned crying as a small child when trying to count to 100. But how much louder would I have cried if I were trying to count the *continuum* of numbers between zero and 100? For as we have seen, this continuum is made up of all rational and irrational numbers, and these are infinite but *not* countable, since after 1 there is no definite next number. If we decide to count 1.000000001 immediately after counting 1, someone could always insert another number between these two: indeed, there is no limit to how many intermediate numbers they could insert. This means that the counting numbers are one size of infinity, and the real numbers a greater size of infinity.

Are there any sizes of infinity between these two? Cantor suspected (without knowing for sure) that the answer was no: this is known as his Continuum Hypothesis. To make a long story short, it has never been either proven or disproven. More than that, assuming we utilize the standard version of set theory (known as Zermelo–Fraenkel Set Theory with the Axiom of Choice), Kurt Gödel showed in 1940 that the Continuum Hypothesis cannot be disproven, and Paul Cohen demonstrated in 1963–4 that it cannot be proven.[94] At any rate, many more mathematical infinities have been discovered that are greater than the infinity of the real numbers.

As mentioned at the outset of this section, one of the reasons for talking about mathematics here is to shed light on an issue that's specific to physics. While the number line begins with a continuum but ends with no possibility of numbers ever touching, there is a sense in which physics works in the opposite direction. That is to say, physics historically considered its subject matter to be discrete individual things: atoms are a clear example. Although we saw that Aristotle's *Physics* seems to prioritize continua over discrete elements, this was only the case insofar as he spoke of time, space, motion and number. As soon as we speak of individual things we are speaking of substances, the central topic of the *Metaphysics*, and substances are necessarily discrete. In this respect, physics must somehow account for how individual things are able to touch, something that numbers in mathematics simply cannot do.

Here is another way of looking at it. When we consider the duality of waves and particles in physics, in an obvious sense we think of waves as continuous and particles as discrete. But in a further sense both are discrete with respect to each other, given that each has a specific individual character irreducible to the properties of the other. Yet paradoxically they must also be continuous with respect to each other, since we know that individuals move through continua on a regular basis; objects occupy different points in space and time rather than remaining frozen in one location in a single instant. Although this book has argued that both the continuous and the discrete are necessary elements of reality, that neither is reducible to the terms of the other, we also know that both belong to the same universe. This means that the two must somehow interact: there must be an interface between the continuous and the discrete. Hence the nature of touch is a genuine concern, and is something that Aristotle ought to have clarified in greater detail. One of the ancient Greek words for touch is *thixis*, and it is to this concept that we now turn.

The Problem of Thixis

One of the favourite stratagems of modern philosophers is to claim that certain perennial difficulties in their field are nothing more than 'false problems'. Consider the mind–body problem, so prominent in the philosophy of René Descartes: mind and body are such different kinds of things that we should wonder how they interact at all. There are a number of possible ways to dismiss this as a false problem. One easy option is to say that the human mind is nothing more than the material brain, so that both mind and body are governed by the same laws of nature from the start. This is what materialists think. Another way is to argue that while human thought may be different in kind from the material world, the two are entangled by their very nature, so that there is no real problem of how they make contact. For instance, the American philosopher William James ridicules the idea of a *salto mortale* (perilous leap) across what he takes to be a non-existent gap between thought and world.[95] The phenomenological philosophers Edmund Husserl and Martin Heidegger use similar strategies to dismiss the traditional problem of how thought interacts with world. Husserl does this by saying that all consciousness is always conscious *of something*, so that conscious-ness is always outside itself from the start.[96] Heidegger does it by saying that we are 'thrown' into a world rather than existing outside it and then somehow choosing to enter it.[97] But such strategies tell us little about how the two distinct elements are able to interact.

Instead of false problems, I would prefer to speak of false *non*-problems. The old mind–body dualism may have attained the status of a cliché, and thus it is no surprise that we now see a proliferation of recent philosophies that are so dismissive of it. But any problem that interests intelligent people over a long period of time probably has something to it; we may simply need to forget it for a while so that later generations can reconsider it with fresh eyes. Another false non-problem in our time, one seldom even acknowledged, is the problem of touch.[98] We have brushed against this topic on sev-eral occasions in this book, and return to it here near the end. Some readers might still see it as a non-problem, given how much ram-pant touching between things appears to occur around us at every moment. There seems to be no difficulty in shaking someone's hand; to push a car when the engine stalls seems to involve noth-ing more than a bit of muscular strain. On a regular basis we find trains coupling together in a railway station, breakfast cereal touch-ing the sides of its box, and inanimate masses slamming together in the asteroid belt. There seems to be nothing easier than touching the roof of your mouth with your tongue. Where is the supposed problem? Recall the occasionalist philosophers we discussed earlier: the ones who believed that nothing can touch anything else, so that God is needed to make all relations in the universe happen. Per-haps occasionalism strikes you as just a far-fetched and outdated theology; that's how most philosophers view it, so you certainly aren't alone. But from this example we see that at least some phi-losophers have recognized a problem with assuming that two things can touch each other directly. Before I try once more to convince you that these philosophers were in large part right, it will prove useful to introduce some new terminology.

As mentioned, there are several words for touch in the ancient

Greek language. The one I will use here is *thixis*, partly because it is used less frequently than the others, and partly because it refers to contact between surfaces, which is exactly what I mean to talk about. Having a technical term for such contact makes it easier to ask directly what any given philosophy says about touch. For example: 'What was Spinoza's concept of *thixis*? . . . How does Empedocles account for *thixis* between such different elements as air, earth, fire and water?' We can also use *thixis* to make a series of compound words, which will help us make sense of the different strategies for dealing with contact, even in those philosophies that do not acknowledge touch as puzzling at all. Let's begin with six such terms:

1. *Katholikothixis*, meaning a universal ability for anything to touch anything else. The view implied here is that touch is simply not a philosophical problem. Everything exists in the same world, and therefore anything can bump into or otherwise influence anything else. Standard materialism falls under this heading. It holds that everything is made of physical matter, and thus assumes there is no problem with one physical thing moving through space and making contact with another. There is also Bruno Latour's Actor-Network Theory, which is based on the idea that any entity of any kind can touch any other and influence it, without viewing this as any sort of problem.

2. *Athixis*, entailing that nothing can touch anything else. This is clearly a more exotic option than the previous one. One possible form of *athixis* would be the theory that entities are completely separated from each other without any way of ever making contact. Imagine that the occasionalist philosophers had not introduced God as a solution to their problem of contact; the result would have been *athixis*, or a cosmos consisting solely of non-touching entities failing to influence each other at all. Or imagine someone claiming that there are countless parallel universes, none of them able to communicate; some string theorists have been driven to this view by the fact that this theory may allow for up to 10^{500} different universes in what they call a multiverse. In this case *athixis* would apply only

to the relations between different universes; within any given universe, touch would presumably still be possible. Another possible form of *athixis* occurs if one believes only in continua, or believes instead in the pre-Socratic *apeiron*. Moreover, as we saw when discussing Dedekind, one of the features of continua is that while they can be divided infinitely downward into smaller and smaller units, those units will never touch each other as closest neighbours. In other words, the number line is a form of *athixis*, and the same holds for any continuum, even though in another sense all parts of a continuum have no distance at all from each other. We can always add a new number between any two arbitrarily close numbers; we can always add another instant of time between any two apparently neighbouring instants; we can always find another point in space between any two given points. The continuum is *actually* one, and only *potentially* many. And even once it becomes many, none of the many units can be considered as immediate neighbours, and hence no touching ever occurs.

3. *Monothixis*. According to this view, contact between entities is in fact a problem, but one exceptional entity exists that makes contact possible. This is obviously the case with the occasionalist theory of God, who is depicted as the only being permitted to touch any other. This subclass of *monothixis* can be called *theothixis* since it makes God the privileged entity. But at least one other such theory exists, which can be termed *anthropothixis*, meaning that humans take over the exceptional touching role that occasionalists give to God. Such a doctrine can be found in the philosophies of David Hume and Immanuel Kant, who both think cause-and-effect relationships cannot be proven to exist except in the human mind. It certainly seems as if fire burns our hands whenever we touch it, but we cannot ascribe to the fire any 'power to burn' with absolute certainty. It is ironic that while *theothixis* is ridiculed today, *anthropothixis* is more or less the mainstream view among Western philosophers, insofar as Hume and Kant are both still held in high esteem. In any case, both of these subclasses can be considered forms of *monothixis*.

4. *Amphithixis*. You probably know the Greek prefix *amphi-* from the word 'amphibian', referring to creatures who live both in water and on land: frogs, newts, some kinds of salamanders. If we imagine philosophies that acknowledge two different kinds of entities (such as Descartes' recognition of thinking substance and physical substance) then *amphithixis* would mean that both kinds of entities are capable of touching other entities of both kinds, which Descartes emphatically does not believe. Imagine a nursery-school teacher asking all the children in a class to form a circle while holding hands. Assuming that there are both blue-eyed and brown-eyed children in the class, we will inevitably find cases of the blue-eyed holding hands with the blue-eyed, the brown-eyed with the brown-eyed, and blue-eyed children holding hands with brown-eyed ones. The clearest example from the history of philosophy involves four terms rather than two: the air, earth, fire and water of the pre-Socratic thinker Empedocles. He notes no particular problem with any of these elements interacting, and tells us simply that they are joined by love and separated by hate.

5. *Homothixis*. This refers to cases where there are two or more kinds of entities, but each entity can only touch others of its own kind. Descartes fits well in this category. A good example from everyday life are the shopping carts found outside a supermarket. When not in use, these carts can be pushed together tightly in as long a chain as we wish, provided they are all of the same size and type. A more academically respectable example comes from the *Ethics* of the aforementioned Baruch Spinoza. Definition II of this illustrious work reads as follows: 'A thought is limited by another thought. But a body is not limited by a thought nor a thought by a body.'[99] Stated more simply, for Spinoza bodies can only touch bodies and thoughts can only make contact with other thoughts.

6. *Heterothixis*. This is the position defended by Object-Oriented Ontology, in which entities can only touch entities of the opposite kind rather than the same kind. This can easily be seen from the practical everyday example of magnets. As is widely known, the north pole of one magnet can only touch the south pole of

another. Any attempt to make two norths or two souths touch leads to repulsion; if you have ever played with magnets, you probably remember what this feels like. Another example is the famous solitaire card game called Klondike, in which red cards can only be placed on black cards, and vice versa. We will discuss the philosophical relevance of this principle shortly.

In any discussion of touch in philosophy, we should consider not only the foregoing classifcation of various types of contact, but also the question of what sorts of entities are trying to touch. Recall the case of Henri Bergson, who used the term 'endosmosis' (as well as 'simultaneity') to refer to contact between what he regarded as two entirely different realms: the qualitative continuity of *durée* inside the mind, and the quantitative discreteness of matter outside the mind. As far as we can tell, Bergson never saw any problem either with two material objects colliding in the world or with two thoughts linking together. Indeed, we recall that Bergson resembles his wife's cousin Marcel Proust in his love for weaving divergent thoughts and feelings into a single interwoven tapestry. Yet for Bergson there is also something asymmetrical about endosmosis, since it seems to be produced by the mind and not by nature: through the medium of the superficial 'second self', which acts like a compromised but effective persona hiding our deeper true nature. There is no indication that Bergson thinks the world has the same ability to translate itself into our terms; the work of diplomacy is done only by the self.

For Aristotle the relevant word was 'contact', considered as lying halfway between separation and continuity. This great ancient philosopher seems to have had no problem at all with different physical things making contact. However, he did have an early intuition of the later modern obsession with the thought–world relation. As we saw, this occurs in his remark in the *Metaphysics* that things cannot be defined, since things are concrete but definitions are made of universals. Aristotle would never say it openly, and maybe not even think it, but implicit in this conflict between the concrete and the universal is the insight that thought can never make direct contact with reality. The other problem of contact in Aristotle, one that he never quite states, is the problem of how the continuous and the discrete interact with each other. We know that they must, since otherwise discrete substances could never move through continuous time and space.

The time has now come to make a renewed case for the importance of the problem of touch. We begin with the fact that modern philosophy has been rich in so-called 'representationalist' theories of truth. While you may not know the term, you are surely familiar with what it describes: the view that when we open our eyes and look at the world we do not see things, but representations of things. The English philosopher John Locke (1632–1704) is one of the most prominent representationalist philosophers, and the more recent American philosopher Richard Rorty (1931–2007) one of the best-known anti-representationalists.[100] Despite its numerous detractors, the case for representationalism is simple: for although fire is able to burn forests and cook food, fire as an image in my mind cannot do either of these things. Thus we can think of our relation to the world as an act of translation, one in which we render the forms of the world in terms graspable by humans. But dogs also see the world in a doggish way, parrots and turtles in a manner of their own, and so forth. The result is that no living creature makes direct contact with the world, but only indirect contact of a sort that is distinct to their species, and to some degree even to each individual animal. All that we encounter directly is a representation of reality. Numerous arguments are offered against this theory: for instance, the German philosopher Markus Gabriel (b. 1980) dismisses it as a form of 'interface skepticism'.[101] But many philosophers still defend some form of a representationalist theory, myself included. Fire in the world (real) and fire in my experience (sensual) are not the same thing. Nor can they be distinguished merely by saying that the former exists in physical matter and the latter does not. Instead, the fire in my experience is a more or less adequate *translation* of the fire that burns and cooks other objects, and through translation it happens to lose the abilities to burn or to cook.

But we need to go even further down this path, to a point that strikes many contemporary philosophers as absurd. Namely, if representationalism is true, then it cannot be limited to creatures with minds. Why not? Because inanimate objects are every bit as finite as humans, even if they are not 'conscious' in the way that animals are. The term 'finitude' is introduced by Heidegger in his interpretation of Kant, as a way of describing Kant's view (as well as his own) that humans are not able to touch reality directly.[102] Let's consider for a moment why finitude is so important. The ecological philosopher Jakob Johan von Uexküll made a powerful argument that even though all living creatures in a given region share the same surroundings (*Umgebung*), they do not share the same environment (*Umwelt*).[103] Instead, the environment for each creature depends on its perceptual capacities. His most famous example is the tick, the infamous parasite that falls from trees and sucks the blood of its host. As Uexküll has it, the tick's entire environment consists of just three relevant factors: (1) it senses butyric acid found in the sweat of mammals, and this causes it to loosen its grip on the tree and fall onto the mammal; (2) it uses its sense of heat to find a place on the mammal's body where blood is close to the surface; (3) it uses its sense of touch to bite through the skin and suck the blood of its host. Such limited horizons for this creature! But while humans seemingly have a much richer existence than this, Kant contends that we are limited to perceiving three dimensions of space, a linear and irreversible time, and events that have causes rather than appearing at random.

Although many people can be persuaded to accept the existence of finitude in the cases of humans and animals, they tend to recoil in horror if asked to view inanimate things in the same way. 'What? Are you claiming that stones and dust have minds?' But that isn't the point. Instead, the point is that nothing like a 'mind' is needed to interact with the world in a finite, limited way. If the tick's existence seems sadly restricted, try to imagine the even sadder state of a flame when encountering a cotton ball. The flame is presumably not conscious, yet it still interacts only with a narrow range of the cotton ball's features: namely, those that are relevant to its flammability. In other words, the flame is still finite rather than infinite in its interactions with the world, just as living creatures are. It follows that no thing makes direct contact with another thing, but only with a restricted sample of its properties, just as occurs in the representationalist theory of truth. This limitation is not the product of 'consciousness', but of the simple fact that no relation can exhaust the full depth of the objects that come into relation. One discrete entity cannot make direct contact with another; an interface of some sort must be involved.

From this discussion of finitude we are reminded that no two discrete entities can ever touch, which means that they must somehow *become continuous* if they are ever to influence one another. Light is shed on this point in a fascinating but failed attempt by Latour to show the contrary. He was one of the rare contemporary philosophers who did sometimes see an issue with how two things touch. And despite his personal religious faith, any theocentric form of occasionalism was out of the question for him, since he generally did not like to make philosophical arguments by appealing to God. As we saw earlier, this led him to the ingenious idea that every interaction between two things requires a *local* actor as a mediator, for no interaction is possible without 'nuncios, mediators, delegates, fetishes, machines, figurines, instruments, representatives, angels, lieutenants, spokespersons, and cherubim', as he put it in one of his many colourful lists.[104] Ultimately this passion for indirect communication spurred him to develop a form of secular occasionalism in which anything at all can be a causal mediator, not just God.

The specific issue Latour raises is how politics ever makes contact with neutrons, something that would initially have seemed improbable.[105] Discovery of the neutron was announced in 1932 by James Chadwick in England, though it was found earlier the same year in Italy by Ettore Majorana (1906–1959?), who mysteriously vanished just six years later.[106] Although the world of physics was rocked by this discovery of a new particle, similar to a proton but lacking electrical charge, it hardly had clear political implications.

That would change with the discovery of the fission of uranium atoms by neutrons in 1938, by the German chemists Otto Hahn and Fritz Strassmann and the Austrian physicists Lise Meitner and Otto Robert Frisch.[107] In France, the task of establishing a link between neutrons and politics was undertaken by Frédéric Joliot, son-in-law of the Curies, who launched a failed effort to encourage a French atomic bomb project in the brief period before France was knocked out of World War II. Latour concludes from this case that every interaction requires a local mediator on the spot, instead of automatically treating God or the human mind as the universal mediator for everything.

While this did put Latour one step beyond occasionalists and their *monothixis*, it also raised a different sort of problem.[108] The motive for Latour's treatment of Joliot comes from his view that any relation between actors needs a mediator; direct contact between any two things is not possible. That is why Joliot is inserted between politics and neutrons, as the guarantor of their interaction. Yet this only pushes the problem of mediation one step further back: after all, how is Joliot himself able to touch either politics or neutrons? A new mediator will be needed between Joliot and each of these things, and so on to infinity. Latour has now fallen into a Zeno's Paradox of mediation, one in which we never actually get to the point of two things touching, since an infinite number of things will have to be placed between them. The only way to prevent an infinite regress where no two actors ever touch is to introduce a form of what we called *heterothixis*, though I was never able to persuade Latour on this point. Namely, two discrete entities must both make direct contact with the same continuum, and this is what enables them to exchange forces or even to fuse together briefly or permanently into one.

Before treating this problem further, let's recall a point from our earlier discussion of Aristotle. As concerns the continuum, he treated it as a whole with no actual parts. Its parts are present only potentially, and that is why he thought Zeno's paradoxes fail: one need not pass through an actual infinity of points to reach any doorway or pass any turtle in a footrace. We then saw that, oddly enough, he apparently said the same about substance. Aristotle seemed to think there could only be one substance per household, so to speak: in any case where multiple substances combine in a larger one, such as berries on the bush, only the larger one could remain a real substance. The others revert to a lesser status until the substance as a whole eventually breaks into pieces and liberates its parts. I challenged this claim at the time, favouring instead a model where a whole thing could be a substance and its components could be substances simultaneously.

But that is less important than another difference that Aristotle himself would have to recognize. Namely, even if it were true that both a continuum and a substance hold their parts in a state of potentiality, this potentiality would have to be very different in the two cases. As we have frequently discussed, a continuum can be cut up however we want, but the same is not true for a substance. In an organ donor situation, for instance, we cannot simply slice up a human body however we please. It needs to be 'carved at the joints', as Socrates puts it in Plato's *Phaedrus*.[109] The doctors need to know exactly where to cut so as to preserve the integrity of the bodily organs to be donated. They cannot invent imagined organs wherever they want, but must follow the underlying logic of a human body's subcomponents. Note that for Aristotle it would be impossible to carve time, space or the number line at the joints, since no such joints exist. Even if continua and substances share the feature of having their parts only potentially (or so Aristotle holds), there is still a big difference between them. This point is important for our conclusion, to which we now turn.

Waves and Stones

One of the most dangerous temptations for human thought is the seductive lure of contrarianism. Wherever there seems to be broad unspoken consensus on a given issue, the contrarian gains instant acclaim (along with numerous enemies) by negating whatever the conventional wisdom on the topic may be. Where I grew up in eastern Iowa, most baseball fans supported the Chicago Cubs, except for one sour-faced contrarian who insisted on allegiance to the St Louis Cardinals. Most of the kids in my generation loved the film *Star Wars*, but this affected rebel would frown and say that *Star Wars* was stupid and *Rocky* was better. Whatever the majority sentiment on any possible topic, he would unfailingly take a position of forced or affected originality: vanilla ice cream was the best; the other Scout group was better than our own. This meant that in every situation he was the automatic king of the hill.

We all know perfectly well that samples of this attitude are not just found among children. For instance, in a post-World War II era where capitalism plus liberal democracy has often been treated like the only acceptable political model, a number of 'Alt-right' contrarians (Aleksandr Dugin, Hans-Hermann Hoppe, Nick Land, Curtis Yarvin) have emerged to argue in favour of dictatorship or other previously unthinkable proposals. The shock such opinions provide to the liberal consensus brings these figures a good deal of immediate renown. But precisely for this reason, the authors in question are under too little pressure to account for those elements of the current situation that may have earned the right to preservation. Is there really any genuine political insight in renewed calls for the invasion of weaker neighbours, or nostalgia for casual racism and sexual harassment, or contempt for representative government, or perverse denial of scientific consensus on wildfires and polar ice? Here we find no innovative grappling with shifting world conditions, but a good deal of enjoyment from poking other people in the eye. While the simple act of denouncing the cliché-like aspects of recent consensus gives these contrarians an air of edgy iconoclasm, it also obscures their own unquestioned assumptions, which are often more hackneyed than those they denounce.

In an academic context, there are other bits of contrarian wisdom that pop up repeatedly. Plato's *Republic* famously discusses a form of government with disturbing policies one might expect of the North Korean dictatorship: forced marriages arranged by the state in the guise of randomness, and 'noble lies' told to the populace to maintain stability.[1] But no sooner do we summarize the *Republic* in this way than a nearby contrarian steps forward to assert that Plato never intended this as a political blueprint for an actual country. This may even be true; I have encountered people with sophisticated reasons for saying so.[2] Yet we should also consider the possibility that Plato meant exactly what Socrates says in the dialogue, instead of simply mocking Karl Popper's justified alarm at Platonic tyranny.[3] The Renaissance political philosopher Niccolò Machiavelli (1469–1527) is infamous for his treatise *The Prince*, which glows with admiration for the ruthless power politics of cunning strategists from ancient Greece through his own day.[4] But inevitably someone is on hand to dismiss this standard cutthroat reading. Machiavelli's true views, they will say, can be found in the more pro-republican sentiments of another of his books, the *Discourses on Livy*.[5] While this sort of table-turning has its place in intellectual life, it risks overlooking the significant dose of truth in the textbook view of Machiavelli as a coddler and champion of despots.

A further example arises in discussions of Martin Heidegger's philosophy of technology. This sombre mastermind paints an oppressively gloomy picture of the 'enframing' and 'standing reserve' brought about by modern technicity, treating it as the culmination of a forgetfulness that began in ancient Greece.[6] Yet nearly any self-respecting Heideggerian will rush to call this an oversimplification: they will inform you that Heidegger was not stupid enough to be a paranoid technophobe. They will then point to his frequent citation of the poet Friedrich Hölderlin's Janus-faced line that 'where the danger is, there too grows the saving power'.[7] The intended meaning is that technology for Heidegger is not just a poison, but also a cure: the wound is healed only by the spear that smote you.[8] Here the brute fact of Heidegger's negative attitude towards technology is effaced, and the discussion is thereby softened rather than sharpened. Not overlooking the obvious is one of the keys to a balanced intellectual life.

The philosopher Hegel, whom we met earlier, is not among my intellectual heroes. Yet he was a thinker of powerful mind and numerous merits, and one of those merits was a lifelong allergy to first-degree contrarianism of this sort. As seen in our earlier discussion of dialectic, Hegel was averse to simply denouncing our first, naive approach to any topic; instead, both the initial statement and its negation are 'sublated' in a higher standpoint that preserves both in more refined form.[9] Rather than basking in the cheap superiority of negation, we pass instead into a 'negation of the negation' in which both the naive and critical outlooks are lifted to a new state of mutual complication.

A related but trickier insight is shared by Alain Badiou when he notes that the new is not simply opposed by the old, as we all tend to think. Instead, any genuine breakthrough is opposed by what he terms a 'reactionary novelty', meaning a previously unknown reason for resisting the new.[10] I take him to mean that the gravest danger to progress is not the initial gullible view of the layperson, but the unforeseen mutation of a foe whom we thought we already knew. Once American school shootings reached a truly unbearable level of frequency, the arguments made by gun rights groups shifted from the traditional case for constitutional freedoms to an aggressive (and somewhat affected) push for the arming of teachers. By contrast with the constitutional argument, a long-standing conservative trope, the call for open weaponry in the schools counts as a reactionary novelty. In similar fashion, it is easy enough to denounce Donald Trump as a 1950s throwback or even a 1930s fascist, but we should wonder if these diagnoses really strike the mark. In 2018 it was Latour who pushed things further, reading Trumpism instead as a rare and dangerous political innovation: namely, an escapist flight from global warming and the refugee crisis it is destined to worsen.[11] Although I think Latour missed the elements of traditional American race and gender politics in Trump's rise, he was right to see something like a reactionary novelty in the Trumpian platform.[12]

Some readers may have guessed where these remarks on contrarianism are headed. Since the dawn of the twentieth century, a perpetual faction in intellectual life has argued that we must discard the traditional view of the world as made up of enduring solid objects and replace it with a fresh vision of unyielding process, flux and becoming. (Years ago, I adhered to this viewpoint myself.) Everything, we are told, is changing all the time. While it might be assumed that this perspective is politically liberating – 'change' is always the slogan of progressive movements – a good part of alteration leads to nothing but destabilization and decay. There is nothing emancipatory about a mob assaulting the United States Capitol, no matter what changes it provokes.

On a related note, it has long been customary to say that 'art as process' is superior to 'art as product'. This processual approach goes back to at least the early 1950s, in Harold Rosenberg's misinterpretation of the paintings of Jackson Pollock.[13] Given that Pollock's radical drip paintings caused initial confusion even for art world veterans, Rosenberg chose to describe them as 'action paintings': as if the emotions that drove Pollock to prance intensely around his canvases were the point, rather than the completed canvases themselves. Along the same lines, any theory of durable objects is often denounced as the unquestioned byproduct of Indo-European noun/verb grammar, with the implication that certain non-Western languages may have a direct window onto reality that is lacking in our own mother tongues. Benjamin Lee Whorf's speculations on the superiority of the Hopi language of Arizona for describing quantum mechanics are not without interest, but a Hopi scientist will be no less baffled by the double-slit experiment than any researcher from Sweden, Iran or Brazil.[14]

All of these cases are examples of first-degree contrarianism; all fail the basic Hegelian test of identifying something worth preserving in whatever one wishes to negate. This book's way of sidestepping the Contrarianism of Becoming has been to accept the permanence of both waves and stones as features of our world, with neither dissolved into the other. Throughout, I have made the dissenting point that discreteness and individual things cannot be reduced to byproducts of constant motion. We are creatures of habit, and the same holds for birds and jungle cats. New ideas are rare, and they tend to go unheard for some time even when publicized. The rarity and difficulty of genuine change must be respected if we want to do more than endlessly express opposition, painting either moustaches or swastikas on the *Mona Lisa*. When intellectual life becomes a trench war, we have as little dynamic motion as in the trenches of northern France, which gave rise to nothing more than shock troops and poison gas, at least until the tank was invented and *Blitzkrieg* replaced trench war. Badiou makes another good point when he argues that no theory of events is meaningful without a certain degree of background stasis and banality.[15] If every instant is an event, then events lose all significance.

Gathering the Problems

Before going further, we should gather the various problems considered in this book. In chronological order, we considered the problems of interface (Aristotle), dialectic (Kublai and Timur), endosmosis (Bergson), retroactivity (Kuhn), cascades (Gould), inscription (architecture), untranslatability (Dennett) and *thixis* (de Broglie and Dedekind). Let's consider how these problems fit together, as a step towards possible reconciliation. This will be easier if we put them in a different arrangement, grouping them in accordance with a small number of general themes.

Untranslatability was the problem that emerged from our discussion of Nonoverlapping Magisteria (NOMA) and Daniel Dennett's contempt for qualia. There would be no untranslatability if nothing were discrete. If individual things were just arbitrary cuts from primal flux by the human mind, then the mind would be able to exchange one thing for another without difficulty. Yet we have seen that this is untrue. A three-dimensional globe of the earth cannot be redone as a two-dimensional map without significant distortion of the size or shape of land masses; the globe and the map are incommensurable by nature, and our minds can do nothing about it. Shakespeare cannot be translated into French or Russian without painful sacrifice, but also with possible surprise improvements in one passage or another.

The problem of *cascades* arose with Gould and his theory of punctuated equilibrium, and again in Latour's story of Joliot mediating between politics and neutrons. In both cases we saw that any vagueness as to where contact occurs is impossible, since the question of touch inevitably cascades down to the level of real individuals. In Gould's case this was because species cannot be punctuated in time unless there are individual creatures who do the punctuating: there must be a first California Spotted Owl and a final ankylosaurus. In Latour's case it was because Joliot cannot continue to insert infinitely many new mediators between any two entities; somewhere there must be objects that actually touch. Our exact knowledge of the point of contact may be vague, but the point itself must exist in a specific place.[16] The problems of untranslatability and cascades are united through their shared demonstration that discrete individuals are real.

Turning now to another set of questions, the problem of *inscription* concerned the duel between the continuous and the discrete on the interior of any discrete being, as when a basically continuous building bows to the necessity of windows, doors and other specific elements placed in discrete locations. In fact, I will venture the proposition that continuous structures tend to encourage attention to contrasting moments of discreteness, and vice versa: the subordinate style is always brashly embedded within the dominant one. This creates a tension in the heart of the object itself. In our discussion of architecture I likened this tension to the one between figure and ground in Gestalt psychology; every physical interior of a building provides an excellent case study of such interplay.

Continuing further, we encountered the *dialectical* problem of Kublai and Timur. The savage Mongol conqueror wants to destroy the wealthy strongholds of civilization, yet these cities also beckon nomads by appealing to what they lack. Hence it is only a matter of time before the city lures the herder to become civilian in turn. The reverse holds for Timur, whose inheritance of a rich cosmopolitan tradition did not prevent him from violently enforcing an artificial steppe around Samarkand, through repeated annihilations of every possible rival. We might call it urbanism with a barbaric face. Just as in the case of inscription, there is something like a figure/ground relation at work here as well: we could say that for Kublai Khan, barbarism was the background and Chinese civilization the figure that captured his attention. Conversely, for Timur and his army the background was civilized and the mesmerizing figure was the nomadic project of destroying all cities. I am reminded of Friedrich Nietzsche's injunction to beware of fighting monsters lest we become monsters ourselves.[17]

But in view of the mostly positive words about the dialectic in this book, I should clarify why dialectical reasoning is insufficient to account for the coexistence of the continuous and the discrete. The Hungarian dialectician György Lukács (1885–1971) attempts such an accounting as follows: 'Objectively, however, the life of a people is a continuum. Reformism sees nothing but continuity, revolutionary modernism sees nothing but ruptures, fissures and catastrophes. History, however, is the living dialectical unity of continuity and discontinuity, of evolution and revolution.'[18] This balanced assessment is admirable. Yet simply identifying the continuous and the discrete as separate pieces of the puzzle does little to explain how they fit together. As we saw earlier, the weakness of dialectical approaches can usually be found in their overly hasty resolution of conflicts. For the skilled Hegelian, thought moves at the speed of light: it quickly dissolves whatever contradiction we encounter, advancing in newfound unity until the next bump in the road appears. But contradiction is stubborn and unyielding, perhaps taking decades or centuries to sort itself out; as often as not, there is no resolution at all. We have seen no immediate (or even delayed) settlement of the century-long impasse between quantum theory and General relativity, despite Lukács's rightful view that both the continuous and the discrete exist. The strife between democratic and authoritarian government is not settled for good by some angel of history hovering over the wheat fields of Ukraine or the cliffs of Taiwan, but may persist for centuries; its eventual outcome might be the child of luck, not of some inherent rationality in the world.

To reflect on inscription and the dialectic means to consider the somewhat unstable relation between discrete things and their non-discrete backgrounds. An architect may claim to design continuous gradients and pure non-angular flow, but they cannot escape the use of discrete forms with distinct points of articulation. And while a Kublai Khan may deem himself the legitimate heir of the steppe, he is lured and then ennervated by its opposite: the local fineries of established Chinese civilization. Zaha Hadid's curving and continuous architectural forms do not all look alike; each has a distinct identity. Peter Eisenman establishes points of disruption, but can only link them via stretches of smoothly flowing corridor. Indeed, every discrete individual can eventually reverse its properties, since whatever it ostensibly opposes remains inside its body like a virus. The Scandinavians once embraced a pillaging Viking credo featuring human sacrifice, only to flip at a later date into mild social democrats with an advanced non-patriarchal safety net. Traditionally, China was a reclusive land power that had burned all ships in its treasure fleet by the sixteenth century; today, spurred by rivalry with the maritime United States, the Chinese navy has become the world's largest.

With Thomas Kuhn we discovered something even stranger: the

object exists in tension not only with its background, but even with its own properties. The Kuhnian problem of *retroactivity* brought us face to face with the idea that discrete punctuations in history do not happen until a second punctuation occurs: both moments are needed to constitute a paradigm shift. This difference between 'discovery' and 'invention' points to a fracture within the object itself: between the fact *that* it is and the determination of *what* it is. An object is not purely continuous with itself, let alone with other things. While somewhat similar to the duel between the thing and its background, as found in the problems of inscription and the dialectic, the strife between an object and its own qualities is basically different in character.

That leaves just three remaining problems, all of them closely linked. The philosophers Aristotle and Bergson encountered the comparable problems of *interface* and *endosmosis*. In both cases it was hard to determine where the continuous and the discrete make contact so as to exchange their causal energies. For Aristotle, the site of contact is his underdeveloped notion of touch.[19] For Bergson it is endosmosis, where the continuously becoming deep self of *durée* puts on the diplomat's costume of a 'second self'. This crafty, hypocritical ambassador enters the foreign world of quantity and becomes commensurable with the non-mental things that populate homogeneous external space. Touch and endosmosis are both specific approaches to the problem of *thixis*, as described at the close of chapters 1 and 9. In Bergson's case it is a matter of how continuous time interacts with discontinuous space. For Aristotle, the question is framed instead as the problem of how time and space are involved with discontinuous individual things. Both philosophers strike at the heart of the matter: the problem of *thixis*.

Cutting and Carving

I've already proposed that a successful theory of *thixis* should take the form of *heterothixis*. As a reminder, this means that touch can only occur between two things of different kinds: that two discrete individuals can only meet in a continuum, and two continua only by intersecting in an individual that inhabits both. Here I take it for granted that readers know that the prefix *homo-* comes from the ancient Greek word for 'same', while *hetero-* comes from the word for 'other'. *Heterothixis* is a technical term in this philosophical discussion of touch, and has nothing to do with specific variants of human sexuality. The thesis here is simply that all forms of human and non-human contact entail *heterothixis*.

One of the similarities between a continuum and a discrete individual is that both are in some sense *one*. A continuum is one because its parts are only potential rather than actual, as seen in Aristotle's refutation of Zeno. An individual thing is one because it is a single thing rather than many. The latter point becomes more complicated when we recall that any individual thing is made of countless smaller things, be they neighbouring subcomponents (such as the tyres and windshield in a car) or more distant ones (the quarks and electrons in a car). It was no doubt a worry about such unchecked multiplication of substances that led Aristotle to claim, in the *Metaphysics*, that 'it is only when the whole has been dissolved' that its parts can exist as substances in their own right.[20] But we have seen that while a continuum can be cut up however we like, the same is not true of the parts of a substance. Substance must be carved at the joints. If a surgeon does not know exactly where to cut a body then disaster results, and the same for a mechanic's knowledge of the engine in your car. Although a continuum can be split up arbitrarily by an outside observer who cuts it, a discrete substance is pre-divided into parts according to which we must carve it. For this reason we can distinguish between cutting and carving as technical terms.

Aristotle's claim that bodily organs can only exist as autonomous substances once the larger whole is dissolved entails that the organs are temporarily devoured by the system (the entire body) in which they are found. As we have seen, there are problems with this view. The first is that in cases of organ failure, the failure is never of the whole, but of one or more specific failing organs; the organs cannot have lost their individuality or they would never have been able to fail. We saw in Chapter 1 that this mistake was repeated by Heidegger in the twentieth century, when he claimed that in the tool-system all tools blend into a continuous whole. But rather than the tool-system being *cut* by whatever malfunctions occur on its surface, the system must be *carved*, since individual tools pre-exist the system as a whole. Only for that reason can they go against the grain of the system by breaking.

A second problem stems from Aristotle's assumption that individuals exist by nature and are relatively easy to identify. Today we are likely to recognize a greater variety of objects, and are also more likely to be aware that they are embedded in multiple objects at once. A nation can be a member of the European Union (EU) and also part of the Schengen visa-free travel zone (or not, as with Ireland and Cyprus as of this writing). An EU state can also belong to NATO (or not, as with Ireland, Cyprus, Austria and Malta). It can adhere to one international faction in cases of trade and a completely different one as concerns carbon dioxide and methane targets. On the individual human level, it is now a commonplace insight that we assume more different identities at once than ever before: consider a human rights lawyer who becomes an obscenity-shouting hooligan at stadiums, but also frequents online chatrooms devoted to gardening, sailing and less reputable things. Aristotle himself was simultaneously a philosopher, a teacher, a son, a husband, a Macedonian and a resident of Athens, combining these roles without evident difficulty. Considered as an object – and humans are just a very special kind of object – Aristotle was a component of multiple substances at once, in the sense of belonging to a philosophical school, a family, an ethnic group and a city. If he were completely dissolved into one of these larger objects, he could not have been part of the others as well, though we know that he was. Aristotle might have lost his family through divorce, or fled from his residence in Athens (as did happen once), without losing either his Macedonian identity or his leadership of a group of students. Although bodily organs may seem to belong to a single larger object, the body itself, these other examples show what a misleading case this is. It might seem that my teeth belong solely to my body, but they were also elements in an orthodontic research project during my teenage years. By nature, substance is a multi-tasker.

All this entails that the parts of a substance are organized more loosely than Aristotle thinks: they retain their independence even prior to a collapse of the whole. His tightly linked substances are modelled too closely on organisms, with their functional integration of every piece into a single ultimate purpose. But I would rather speak of OOO's more loosely arranged substances as *disorganisms*: not because they are utterly chaotic or random, but because the parts resist excessive organization, always having one foot outside the various wholes to which they belong.[21] This sort of looseness can even lend greater force to the parts, as in Mark Granovetter's sociological concept of the strength of weak ties.[22] In cases where we are tightly involved with the wholes to which we belong – as with family, close friends and inseparable allies – relationships tend to provide support and stability rather than new horizons. Much of what can happen between ourselves and those closest to us has probably already happened; at the very least, our ongoing interactions tend to be somewhat predictable. It is our looser acquaintances, their personalities less familiar to us, who more often surprise us with news, fresh possibilities and astonishing rendezvouses of both the good and bad sort. As Manuel DeLanda puts it: 'Low-density networks, with more numerous weak links, are for this reason capable of providing their component members with novel information about fleeting opportunities.'[23]

Causation as Composition

When people talk about cause and effect, they are usually speaking of something that happens over the course of time, however slowly or quickly it might unfold. Buttons were pressed, and rockets fired, and this led to the *Apollo 11* mission landing safely on the moon. A cat pushed a drinking glass off the table, which caused it to shatter on the floor. This fixation on causality unfolding over time dictates the sorts of problems by which philosophers are most impressed. From David Hume in the eighteenth century through Quentin Meillassoux in the present, thinkers are dazzled by any suggestion that past experience is no sure guide to the future: the sun may not rise tomorrow; the next book we read might change to fire in our hands; gravity might suddenly disappear or become twice as strong.

Yet there is another prominent form of causation that would exist even if time were to freeze in its tracks. I speak of the compositional relation of parts and wholes, one that exists in every moment irrespective of change or stasis. An apple consists of parts (stem, core, flesh, skin and calyx), and these in turn are emergent things composed of molecules, with the molecules made in turn of atoms, the atoms of quarks and electrons, and presumably even further down the ladder. I am unaware of any Hume-like challenge to this relation of parts and wholes. Such a proposal would have to argue that the molecules of an apple might just as easily compose a pomegranate. Just imagine if a swarm of locusts formed a cathedral as a new emergent object, or if we tested a bag of imported rice and found it to be made of neon. So far, no philosopher has taken such a risk. The Humes of the world do not question that silver atoms join in molecules of silver rather than of gold; that silver molecules give rise to silver coins rather than gold ones; or that silver coins may be redeemed for the cash value of silver and not that of gold. Sceptics love to challenge our certainty about what will happen tomorrow morning, even while assuming the rigid necessity of composition today. For all their fascination with possible causal breakdowns over time, sceptics have little to say about potential malfunctions in the relation between a thing and its components. In philosophy, the study of parts and wholes is known as 'mereology', and I happen to think it is the right place to study causation as well.[24]

This is true not only of explicit cases of part/whole composition, but equally so for the sorts of causes that unfold over time.

The reason is that the interaction of separate objects can be re-interpreted as the temporary formation of a new object, with the original objects now functioning as the larger one's parts. Assemble enough components in the proper arrangement, and you have yourself a car. Closely related to these sorts of combinations is what we called the downward effect of a whole on its constituent elements. A strict profession such as the military special forces not only prefers to recruit those who already possess a firm sense of discipline, but tends to increase the strictness of anyone who joins. This sort of thing is often called 'retroactive' causation, but since we are already using that term for Kuhn's discovery of the two-step rhythm of paradigm shifts ('that' and 'what') we will speak here of 'downward' causation instead. Even to enter a place means to be shaped, deliberately or not, by its typical customs and behavioural styles. To be in Los Angeles means to increase driving time as well as daytime recreational activities, whereas in New York, Cairo or St Petersburg one tends to use taxis or public transit and adopt a nocturnal lifestyle. In Los Angeles we use the definite article to refer to freeways: tomorrow I'll drive 'the 710' to work. But if this were San Francisco, just up the coast, I would say that I drive '280' to work, never 'the 280', without ever stopping to think about why. Such downard causation even happens in the purely physical world. I once considered the grim scenario of a midair crash of two planes.[25] Unfortunately, during the final editing of this book, Washington, DC experienced the tragic midair collision of a commercial aircraft with a military helicopter. The way I interpret such events is not in the usual sense as a physical impact between two independent pieces of material, but as the temporary unification of the two aircraft into a single object. This substance then has grave downward impact on its parts before they separate once more, in new and severely damaged form. In fact, all causation can be interpreted in this way: as a form of composition.

Downward causation has been explored by such authors as DeLanda and philosopher of science Roy Bhaskar (1944–2014). DeLanda speaks of the 'mechanisms through which a whole provides its component parts with *constraints and resources*, placing limitations on what they can do while enabling novel performances'.[26] What he means is that we are never free to do absolutely anything; our options are constrained not only by our human nature, but also by the larger objects of which we are a part.[27] No one can function without air tanks on the moon or underwater; it is hard to prove mathematical theorems when starving, or in the midst of being publicly flogged. Bhaskar, for his part, declares that wholes are 'capable of acting back on the materials out of which they are formed'.[28]

This also provides yet another argument against undermining. Namely, we cannot undermine a thing by dissolving it into tiny pieces, since the whole also haunts the pieces of which it was made. Recall that Steve Jobs (1955–2011) was one of the three original founders who built Apple from scratch, only to see himself fired by the same corporation in 1985 and exiled for more than a decade. Or think of the way in which a bomb (the whole) destroys its components (the parts). And then there was Gould's remark about each level of evolution imposing constraints on lower levels, with cancer marking one especially terrible failure of this process.

For now, however, we can leave the question of causal necessity to those who love it most. Of greater importance to us here are two neighbouring topics. The first, we have seen, is composition. Whereas modern philosophy puzzles over the relation between appearance and a reality that may or may not lie behind it, OOO is more interested in an 'ontology in depth', one that recognizes countless descending layers of part/whole composition. The second neighbouring theme is contact, which Aristotle distinguished from the fusion between two objects. Contact is more like a form of *pre*-causation where I interact with a buffered environment: one that is perceived or taken for granted without yet leading to any causal result; the hitman must identify the target before firing the weapon. And by the same token, contact is also what we have once events have run their course: even the wildest party ends the next morning on a floor covered with motionless debris.

The Interiors of Objects

One of the two great achievements of modern European philosophy was to revive the problem of touch, which had been raised intermittently in the past. In European philosophy beginning with Descartes, this topic stood at the centre of philosophical reflection, though transmuted into the oversimplified form of a single relation between thought and world. To a large extent we remain in this position today. Philosophy's licence to talk about object–object relations has been seized and monopolized by the natural sciences, which account for these relations in mathematical terms without posing the philosophical question as to how they can ever make contact.

Since the days of Leibniz, German thinkers in particular have been fascinated by the notion of experience as occurring on the inside of something else, as if in the hollow of a tree or the interior of a building. Too often this has taken the form of an idealism where this hollow space has no clear connection with the world outside it, and where the object making contact is always a conscious human being. OOO modifies this picture by insisting on two additional points: (a) objects have ways of breaking free from any interior into new ones, and (b) human thought is not the only entity that inhabits an interior.

To give credit where it is due, the philosopher Gernot Böhme (1937–2022) notes that one thing makes contact with another only within a specific *atmosphere*, which we can identify with the inside of an object.[29] My perception of a sailboat is possible only because the sailboat and I are in a prior causal relation that unites us into a larger object, though on the inside of that object we still confront the sailboat as something alien. The specific character of this interior greatly affects our experience of the objects encountered there. A knife means something very different in such atmospheres as a supermarket deli, a cutlery store, a stage play or a beach. Alphonso Lingis describes something similar in his typically beautiful prose, in a discussion of how each thing appears within what he calls a 'level':

As we approach an outdoor café in the night, we see a volume of amber-hued glow. When we enter it, our gaze is filled with the light. We begin to make out forms discolored with an amber wash, like fish seen through troubled waters. After some moments, the luminous haze neutralizes and the faces of people emerge in the hues of their own complexions. The tone of the light has become a level about which the colors of things and faces surface according to the intensity and density of their contrast with this level.[30]

By starting this passage with the *approach* to the café, Lingis also raises the theme of movement between one level or atmosphere and another. This is precisely what was lacking in modern idealist philosophies of the interior, in which the human observer is always rooted in place rather than taking objects as cues for navigating between one hollow space and the next.

The offbeat, cigar-smoking talk-show host Peter Sloterdijk (b. 1947) has built an entire philosophy from the closely related notion of *spheres*, beginning from an initial formative space encompassing mother and child.[31] But Sloterdijk conceives of spheres intersecting with further spheres, bubbles then expanding into a global foam passing well beyond any human exceptionalism.[32] Another such German thinker is the philosophically minded sociologist Niklas Luhmann (1927–98), who conceives of various social systems as bubbles (Sloterdijk's term) barely perturbed by outside influence at all. The systems of art, law or religion replicate themselves and resist external shock: much like a stagnant academic department hiring younger clones of its current members.[33]

Let's return briefly to the representationalist theory of perception, which grabbed our attention during the earlier discussion of finitude. At this very moment I am looking at a white orchid on a table near the window. In view of the finitude of human and all other experience, we must reject the notion that I am viewing the flower directly. The flower as I perceive it may have certain features in common with the real orchid, yet it lacks most of the powers of the orchid itself. I do not interact with the full depth of the flower, but encounter it in human terms, just as the tick of Uexküll (an interior-loving Baltic-German) perceives mammals in severely limited tick terms. There would be no perception of the flower unless I were already in causal relation with it, in which the two of us have fused into a compound object, however briefly. But if my causal relation with the flower fuses us in the compositional sense, my contact with it maintains the distance between us; these are different but simultaneous mechanisms. And as mentioned, insofar as the flower and I are the two components of a larger joint object, the flower and I make contact on the interior of the larger one.

But the picture is actually even more complicated, since it is never the case that we perceive just one object at a time, even when one object commands especial attention. We encounter a great variety of things and qualities at any given moment, which suggests that we are on the interior of numerous objects simultaneously. This poses no difficulty for our theory, unlike Aristotle's, since OOO permits objects to be components in a limitless number of larger objects at once.[34] The landscape of contact has a certain immediate continuity to it; we encounter the amber glow of a night café in a single glance, and in principle we might even view it as a single lump. But we never actually perceive our environment as continuous. Conscious human awareness of multiple things is a cognitive latecomer: well before explicit thought occurs, our biological equipment has already begun to carve our environment at the joints rather than cutting it arbitrarily. Whether it be binocular vision enabling us to pick out three-dimensional objects against a more amorphous background, or an extended neurological present enabling us to track the motion of a single object through space, or metaphor removing an object from the exchange rates of literal meaning, our physiques – like our minds – are carvers rather than cutters.

In light of Uexküll's revealing remarks on ticks, snails and fighting fish, we can also rephrase the question of animal and plant cognition in terms of a creature's specific capacity to convert the continuum on an object's interior into a rough sonar display of individual substances. Related families of creatures have different clusters of skills for carving up the world, and we cannot assume that these groupings match up either with evolutionary family trees or existing human prejudice as to how living things should be ranked. This realization finally puts us in position to depart from modern philosophy's commonsensical way of distinguishing between human, animal, vegetable and mineral, with the latter three terms always treated privatively by contrast with humans. New light can be shed not only on the experience of plants – they are an increasingly popular topic of philosophical discussion – but also on such exotic entities as fungi, lichens and viruses.[35] We are also in a fresh position to speculate on possible cognitive skills still missing from human experience. Finally, we will no longer join with Descartes in supposing that animals, bricks, machines or nations have a relational life equal to zero.

The Ultimate Nature of Reality

To lose one's faith in a former hero does not amount to the 'end of innocence', but to a new life of innocent dismay; the disillusioned person is just as sincere in their hopelessness as a gourmand who enjoys a finely cooked meal. Likewise, the hardboiled cynic who claims not to be surprised by anything is just as locked into their unchanging bitter persona as the child surprised by everything is locked into her own. The modern world praises the idea of the 'critical' intellectual who does not just gullibly believe whatever they are told. We are pressured instead to become contrarians of the first degree: to negate our surroundings once, and once only, without taking the needed additional step. This goes hand in hand with one of the recurrent tropes of modern philosophy: namely, the idea that human thought 'transcends' the world rather than being naively involved with it in the manner of animals or untutored, ignorant people. The purported human gift and human mission is to convert the implicit into the explicit: to step beyond *belief* so as to *know* instead. And often enough, knowledge is equated with negation of the superstitions and fetishes of whomever we wish to denounce. Too often a spirit of negation is also taken as a sign of superior political credentials, even if it gives rise mostly to sarcastic magnificoes appalled by the naivety of others.

As mentioned, the philosophical face of this outlook is the concept of transcendence: the idea that humans have a unique and unsurpassable ability to see things 'as' they really are instead of merely being fascinated by them. Rationalist philosophers such as the phenomenologist Edmund Husserl (1859–1938) are highly confident in our ability to do so.[36] Others, like his former student Heidegger, are rendered more cautious by their awareness of everyday distractions and historically rooted assumptions, ranging from our hopeless entanglement in gossip to our tendency to live robotically in the way that 'one' is supposed to live.[37] Nonetheless, even Heidegger clings to the model of human transcendence.[38] Political reactionary though he was, he was very much a modern in his notion that humans should rise above the world and try to make it explicit, albeit with the qualification that full explicitness is neither possible nor desirable.[39]

In truth we rise above nothing, but walk the corridors of the world from one fascination to the next. The idea that humans (and human thinkers in particular) are less gullible than animals is a centuries-long cliché rather than an insight. Note that a typical animal might believe naively in a particular habitat, a handful of mortal enemies and dozens of possible food and water sources. By contrast, humans are captivated by everything from non-existent unicorns, to topological theorems in n-dimensional space, to existent but inaccessible galaxies, to counterfactual questions about whether Athenian philosophy could have existed if the Persians had prevailed at the Battle of Salamis. Moreover, each of us enters privately into more or less convincing personal fantasies and delusions, not to mention small communities where we share some eccentric enthusiam with four or five others. In the case of animals and less imaginative humans, gullible beliefs are actually much harder to come by.

In the entire history of earth life, no creature has believed in a greater number of bizarre and remote entities than the twenty-first century human. Civilization now places at our disposal a massive machinery for expanding our range of beliefs beyond all previous limits. The internet may be a time-waster, as old-school intellectuals claim, but never has there been such a tool for feeding our curiosity about every old and new thing under the sun. Recent AI chat software even extends our possible beliefs into such non-existent realms as the plot of the next fourteen sequels of *The Godfather* (my youngest brother's favourite ChatGPT query), or a *King Lear* written by Samuel Beckett rather than Shakespeare (my own favourite query). Humans want to be fascinated and entertained, neither of which happens when we aspire to rise above the world and take a cagey distance from it. While the phrase 'I don't care' still counts as today's most devastating putdown and most elevating social triumph, this faded modernist trope speaks more grimly about the humanity of the one who says it than those who are targeted by it.

Another problem with the model of human transcendence is that it puts us far too close to a supposed cognitive finish line. My point is this: the claim of rationalist philosophers is rarely just the comparative one that we humans are smarter than animals. Instead, it usually drifts towards the claim that human cognition is the highest possible kind, given that it already takes the form of explicit awareness of what lesser entities merely take for granted. After our supposed historical passage from naive belief to fully critical awareness, no further mental feat can even be imagined. Yet such claims are feeble platitudes, since even the dismissal of animals as subpar thinkers overlooks the possible exotic cognitions of dolphins or of fruit flies, not fully detected by even our best animal psychology. There are also the disturbing cognitive advances of alien species as imagined by numerous science-fiction tales, which remind us that human intellectual life lies (at best) somewhere in the middle of a landscape of trillions of possible minds.

OOO's redefinition of contact as something that occurs on the interior of entities places the model of human transcendence in jeopardy. The modern prejudice of human exceptionalism fades rapidly once we reconceive experience as burrowing through an interior space instead of rising up to the sky. Stationed in the hollow of multiple objects, qualities, figures, grounds and cafés lit with amber light, it is easy to notice our limited capacity to detect and carve the joints that run through the cosmos. This poses a seemingly open-ended task. The ultimate nature of reality, I mean to say, consists of objects and interiors of objects. Various causal events have brought our life to the point where it is right now, but our current contact with sensual objects is pre-causal rather than causal. That is to say, I am absorbed in a multitude of objects in the late-night restaurant where I finish this book, carving the joints of this setting as best I can. But I am also partly bewildered by it, no doubt missing opportunities beyond the ones I seize to eat, pay or chat with random strangers.

The space where experience unfolds is a whole, one that is partly pre-carved by our organs of sense and our bodily postures well before our supposedly rational mind comes into play. Far from standing on a mountaintop viewing things explicitly, we remain in thrall to a range of typical primate habits and motor skills. Certain objects stand out against the humming background of the world. Often enough, the contrast between an object and its rapidly shifting qualities alerts us to the fact that things cannot be literalized in terms of their properties. This means that our current atmosphere hints at paths to escape into others. The secret to navigating the world as a series of interlocking amber-hued cafés is not to 'know' it, nor to do a better job of reducing objects downward to their parts or upward to their effects. The secret, instead, is to invest ourselves in certain objects in a manner that exceeds their patent literal character. In the case of Homer's wine-dark sea we gain little knowledge of wine or sea, but enter a new interior where an unknown sea has wine-like qualities. We strain to believe in this sea precisely because our mind's eye fails us, and the new object emerges along with our feeling of effort. To make a promise is to enter a space where I resemble heroes with their reputations on the line. To love something is to cease viewing it as just a bundle of qualities; it is to live one's life in partial fusion with it, in an admirably naive manner.

The decision to invest ourselves in one object and not another isn't solely a question of willpower or individual character: some things have more inherent quality or merit than another. There are reasons why the Oedipus and Antigone myths, or the tales of Gilgamesh and Hiawatha, have endured much longer than any limerick or dirty joke. Well-developed taste will always seek quality, and this suggests that a meaningful life will be less random and more fated in its choices. We each may have several possible roads to Damascus, but never a limitless number. The same holds for the history of objects, as I have written about elsewhere.[40]

To carve one's own life at the joints is not a matter of knowledge, but of connoisseurship: the state of being passionate and insightful about some range of objects without reducing all of one's decisions to lists of explicit rules. The connoisseur is beyond any possible list of rules, capable as she is of discerning the interesting amidst the commonplace, and items of quality amidst warehouses of clone-like mediocrity. It takes tens of thousands of hours of sincere investment to become a connoisseur of Roman coins or rare books, and the same holds for supernovae or metaphysical problems. The connoisseur or curator does not stand above such objects, subjecting them to disinterested judgement like a Kantian art critic; instead, they are often puzzled and spend years chasing down specimens that frequently slip from their grasp. The OOO vision of human experience on the interior of multiple objects simultaneoulsy means that we often have a plethora of contingent options. Yes, we can often give reasons for any decision we make. But it was Enrico Fermi the experimental connoisseur, not Fermi the critical transcendent thinker, who inexplicably chose paraffin over metal for his decisive neutron experiment.[41] The ultimate nature of reality consists of objects and the interiors of objects. The hero of this world is not a machine analysing the chemical formulae for wine: the hero, instead, is the connoisseur.

Works Cited

Adamson, Peter. *Philosophy in the Islamic World: A History of Philosophy Without Any Gaps*, vol. 3. Oxford: Oxford University Press, 2016.

Adorno, Theodor, Walter Benjamin, Ernst Bloch, Bertolt Brecht and Georg Lukács. *Aesthetics and Politics*, trans. R. Livingstone. London: Verso, 1988.

Agamben, Giorgio. *What Is Real?*, trans. L. Chiesa. Stanford, CA: Stanford University Press, 2018. Kindle edition.

Ali, Ayaan Hirsi. *Infidel*. New York: Atria, 2007.

Allen, Stan. 'From Object to Field', *Architectural Design* 67.5–6 (May–June 1997), pp. 24–31.

Althusser, Louis. *For Marx*, trans. B. Brewster. London: Verso, 2006.

Anderson, R. Lanier. 'Friedrich Nietzsche', *The Stanford Encyclopedia of Philosophy* (Spring 2024 edition), Edward N. Zalta and Uri Nodelman, eds., https://plato.stanford.edu/archives/spr2024/entries/nietzsche/.

Anonymous. *The Book of the Thousand and One Nights*, 4 vols., trans. J. C. Mardrus and P. Mathers. London: Routledge, 1986.

Anonymous. *The Secret History of the Mongols*. London: Penguin, 2023.

Aquinas, St Thomas. *The Summa Theologica of St. Thomas Aquinas*, 5 vols. New York: Benzinger Brothers, 1948.

Aristophanes. 'The Clouds', in *Four Plays by Aristophanes*, trans. W. Arrowsmith, pp. 7–166. London: Penguin, 1994.

Aristotle. *De Anima: Books II and III (With Passages from Book One)*. trans. D. W. Hamlyn. Oxford: Clarendon, 1993.

—— *The Art of Rhetoric*, trans. H. Lawson-Tancred. London: Penguin, 1992.

—— *Metaphysics*, trans. C. D. C. Reeve. Indianapolis: Hackett, 2016. Citations are from the Kindle edition.

—— *Nicomachean Ethics*, trans. T. Irwin. Indianapolis: Hackett, 2019.

—— *Physics*, trans. C. D. C. Reeve. Indianapolis: Hackett, 2018. Citations are from the Kindle edition.

Arthur, Wallace. *Understanding Evo-Devo*. Cambridge: Cambridge University Press, 2021.

Austin, J. L. *How to Do Things With Words*. Oxford: Clarendon Press, 1975.

Averroës (Ibn Rushd). *Decisive Treatise and Epistle Dedicatory*, trans. C. Butterworth. Provo, UT: Brigham Young University Press, 2002.

—— *Long Commentary on the* De Anima *of Aristotle*, trans. R. C. Taylor. New Haven, CT: Yale University Press, 2011.

Bachelard, Gaston. *The Formation of the Scientific Mind: A Contribution to a Pscyhoanalysis of Objective Knowledge*, trans. M. McAllester Jones. Geneva: Clinamen, 2006.

Bacon, Francis. *The New Organon*, ed. F. H. Anderson. Indianapolis: The Library of Liberal Arts, 1960.

Badiou, Alain. *Being and Event*, trans. O. Feltham. London: Continuum, 2005.

—— *Deleuze: The Clamour of Being*, trans. L. Burchill. Minneapolis: University of Minnesota Press, 1999.

—— *Lacan: Anti-Philosophy 3*, trans. K. Reinhard and S. Spitzer. New York: Columbia University Press, 2020.

—— *Logics of Worlds: Being and Event II*, trans. A. Toscano. London: Continuum, 2009.

Baggott, Jim. *Higgs: The Invention and Discovery of the 'God Particle'*. Oxford: Oxford University Press, 2012.

—— *The Quantum Story: A History in 40 Moments*. Oxford: Oxford University Press, 2011.

Bateson, Gregory. *Steps to an Ecology of Mind: Collected Essays in Anthropology, Psychiatry, Evolution, and Epistemology*. Chicago: University of Chicago Press, 2000.

Bell, J. S. 'On the Einstein Podolsky Rosen Paradox', *Physics* 1.3 (1964), pp. 195–200.

Bennett, Jane. 'Systems and Things: A Response to Graham Harman and Timothy Morton', *New Literary History* 43 (2012), pp. 225–33.

Bergson, Henri. *Creative Evolution*, trans. A. Mitchell. Mineola, NY: Dover, 1998.

—— *Duration and Simultaneity*, trans. L. Jacobson. New York: The Library of Liberal Arts, 1965.

—— *Matter and Memory*, trans. N. M. Paul and W. S. Palmer. New York: Zone, 1990.

—— *Time and Free Will: An Essay on the Immediate Data of Consciousness*. Mineola, NY: Dover, 2001.

—— *The Two Sources of Morality and Religion*, trans. R. A. Audra. South Bend, IN: Notre Dame University Press, 1977.

Bhaskar, Roy. *A Realist Theory of Science*. London: Verso, 1987.

Black, Max. 'Metaphor', in *Models and Metaphors*, pp. 25–47. Ithaca, NY: Cornell University Press, 1962.

Blaszczyk, Piotr. 'On the Mode of Existence of the Real Numbers', in *Analecta Husserliana* 88 (2005), ed. A.-T. Tymieniecka, pp. 137–55.

Block, Ned. 'Troubles With Functionalism', in *Perception and Cognition: Issues in the Foundations of Psychology*, ed. C. W. Savage. Minneapolis: University of Minnesota Press, 1978.

Bloom, Harold. *Shakespeare: The Invention of the Human*. New York: Riverhead Books, 1998.

Blumner, Robyn E. 'Give the Four Horsemen (and Aayan) Their Due. They Changed America', *Free Inquiry* 41.1 (December 2020 / January 2021).

Böhme, Gernot. *The Aesthetics of Atmospheres*, ed. J.-P. Thibaud. London: Routledge, 2016.

Bohr, Niels. 'On the Constitution of Atoms and Molecules, Part I', *Philosophical Magazine* 26.151 (1913), pp. 1–24.

—— 'On the Constitution of Atoms and Molecules, Part II: Systems Containing Only a Single Nucleus', *Philosophical Magazine* 26.153 (1913), pp. 476–502.

—— 'On the Constitution of Atoms and Molecules, Part III: Systems Containing Several Nuclei', *Philosophical Magazine* 26.155 (1913), pp. 857–75.

Born, Max. *The Born–Einstein Letters: 1916–1955*, trans. I. Born. New York: Macmillan 2005.

Brandom, Robert B. *Making It Explicit: Reasoning, Representing, and Discursive Commitment*. 2nd edition. Cambridge, MA: Harvard University Press, 1998.

Brannen, Peter. *The Ends of the World: Volcanic Apocalypses, Lethal Oceans, and Our Quest to Understand Earth's Past Mass Extinctions*. New York: Ecco, 2017.

Brassier, Ray. 'Concepts and Objects', in *The Speculative Turn: Continental Materialism and Realism*, L. Bryant et al., eds., pp. 47–65. Melbourne: re.press, 2011.

—— *Nihil Unbound: Enlightenment and Extinction*. London: Palgrave, 2007.

Braudel, Fernand. *Civilization and Capitalism, 15–18th Century*, vol. 1: *The Structure of Everyday Life*, trans. S. Reynold. Berkeley, CA: University of California Press, 1992.

—— *Civilization and Capitalism, 15–18th Century*, vol. 2: *The Wheels of Commerce*, trans. S. Reynold. Berkeley, CA: University of California Press, 1992.

—— *Civilization and Capitalism, 15–18th Century*, vol. 3: *The Perspective of the World*, trans. S. Reynold. Berkeley, CA: University of California Press, 1992.

Broad, C. D. *The Mind and Its Place in Nature*. London: Forgotten Books, 2018.

Brooks, Cleanth. *The Well Wrought Urn*. New York: Harcourt, Brace, and World, 1947.

Bryant, Levi R. 'The Interior of Things: The Origami of Being', *Przegląd Kulturoznawczy* 29.3 (2016), pp. 290–304.

Bryant, Levi R., Nick Srnicek and Graham Harman, eds. *The Speculative Turn: Continental Materialism and Realism*. Melbourne: re.press, 2011.

Buranyi, Stephen. 'Do We Need a New Theory of Evolution?', *Guardian*, 28 June 2022, https://www.theguardian.com/science/2022/jun/28/do-we-need-a-new-theory-of-evolution. Last accessed on 25 March 2025.

Busch, Uwe. 'Claims of Priority – The Scientific Path to the Discovery of X-rays', *Zeitschrift für medizinische Physik* 33.2 (May 2023), pp. 230–42.

Butler, Judith. *Gender Trouble: Feminism and the Subversion of Identity*. New York: Routledge, 2011.

Canales, Jimena. *The Physicist and the Philosopher: Einstein, Bergson, and the Debate That Changed Our Understanding of Time*. Princeton, NJ: Princeton University Press, 2015.

Cantor, Georg. *Contributions to the Founding of the Theory of Transfinite Numbers*, trans. P. Jourdain. New York: Cosimo, 2007.

Carlip, Steven. 'Is Quantum Gravity Necessary?', *Classical and Quantum Gravity* 25.15 (2008), pp. 154010–17.

Carpo, Mario. 'Ten Years of Folding', in *Folding in Architecture*, ed. G. Lynn, pp. 14–19. London: Academy Editions, 1993.

Cartwright, Nancy. *The Dappled World: A Study in the Boundaries of Science*. Cambridge: Cambridge University Press, 1999.

Chadwick, James. 'Possible Existence of a Neutron', *Nature* 129 (1932), p. 312.

Chalmers, David. *The Conscious Mind: In Search of a Fundamental Theory*. Oxford: Oxford University Press, 1996.

Clarke, Samuel and G. W. Leibniz. *Correspondence*, ed. R. Ariew. Indianapolis: Hackett, 2000.

Cogburn, Jon and Niki Young. 'Revisiting the Notion of Vicarious Cause: Allure, Metaphor, and Realism in Object-Oriented Ontology', *Open Philosophy* 7 (2024), pp. 1–14.

Cohen, Paul. 'The Independence of the Continuum Hypothesis [part 1]', *Proceedings of the National Academy of Sciences of the United States of America* 50.6 (1963), pp. 1143–8.

—— 'The Independence of the Continuum Hypothesis [part 2]', *Proceedings of the National Academy of Sciences of the United States of America*, 51.1 (1964), pp. 105–10.

Croizat, Léon. *Panbiogeography or An Introductory Synthesis of Zoogeography, Phytogeography, Geology: With Notes on Evolution, Systematics, Ecology, Anthropology, etc.* Caracas: self-published, 1958.

Danto, Arthur. *After the End of Art: Contemporary Art and the Pale of History*. Princeton, NJ: Princeton University Press, 2014.

Darwin, Charles. *The Descent of Man*. London: Penguin, 2004.

—— *The Origin of Species*, 150th anniversary edition. New York: Signet, 2009. Kindle edition.

Davisson, C. J. and L. H. Germer, 'Reflection of Electrons by a Crystal of Nickel', *Proceedings of the National Academy of Science USA* 14 (10 March 1928), pp. 317–22.

Dawkins, Richard. *The Extended Phenotype: The Long Reach of the Gene*. Oxford: Oxford University Press, 1982.

—— *The God Delusion*. New York: Mariner Books, 2006.

—— 'Parasites, Desiderata Lists, and the Paradox of the Organism', *Parasitology* 100 (1990), pp. S63–S73.

—— *The Selfish Gene*, 40th anniversary edition. Oxford: Oxford University Press, 2016.

—— 'When Religion Steps on Science's Turf: The Alleged Separation Between the Two Is Not So Tidy', *Free Inquiry* 18.2 (Spring 1998), pp. 18–19. https://cdn.centerforinquiry.org/wp-content/uploads/sites/26/1998/04/22155918/p18.pdf. Last accessed on 25 March 2025.

Dawkins, Richard and Sam Harris, 'An Evening With Richard Dawkins – Featuring Sam Harris – Night 1', https://www.youtube.com/watch?v=7WaGETYqWCs. Last accessed on 25 March 2025.

de Broglie, Louis. *Continu et discontinu en physique moderne*. Paris: Albin Michel, 1941.

—— 'Recherches sur la théorie des quanta', *Annales de Physique* 10.3 (1925), pp. 22–128.

Dedekind, Richard. *Essays on the Theory of Numbers*, trans. W. W. Beman. Chicago: Open Court, 2001.

deGrasse Tyson, Neil. *Twitter*, 29 June 2016. https://twitter.com/neiltyson/status/748157273789300736?lang=en. Last accessed on 25 March 2025.

DeLanda, Manuel. 'Emergence, Causality and Realism', in *The Speculative Turn: Continental Materialism and Realism*, L. Bryant, N. Srnicek and G. Harman, eds., pp. 381–92. Melbourne: re.press, 2011.

—— *A New Philosophy of Society: Assemblage Theory and Social Complexity*. London: Continuum, 2006.

—— *Philosophical Chemistry: Genealogy of a Scientific Field*. London: Bloomsbury, 2015.

DeLanda, Manuel and Graham Harman. *The Rise of Realism*. Cambridge: Polity, 2017.

Deleuze, Gilles. *Bergsonism*, trans. H. Tomlinson and B. Habberjam. New York: Zone, 1990.

—— *Cinema I: The Movement-Image*, trans. H. Tomlinson and B. Habberjam. Minneapolis: University of Minnesota Press, 1986.

—— *Cinema II: The Time-Image*, trans. H. Tomlinson and R. Galeta. Minneapolis: University of Minnesota Press, 1989.

—— *Difference and Repetition*, 2nd revised edition, trans. P. Patton. London: Bloomsbury, 2014.

—— *Empiricism and Subjectivity: An Essay on Hume's Theory of Human Nature*, trans. C. Boundas. New York: Columbia University Press, 2001.

—— *Expressionism in Philosophy: Spinoza*, trans. M. Joughin. New York: Zone, 1992.

——*The Fold: Leibniz and the Baroque*, trans. T. Conley. Minneapolis: University of Minnesota Press, 1992.

—— *The Logic of Sense*, trans. M. Lester with C. Stivale. New York: Columbia University Press, 1990.

—— *Negotiations 1972–1990*, trans. M. Joughin. New York: Columbia University Press, 1997.

—— *Nietzsche and Philosophy*, trans. H. Tomlinson. New York: Columbia University Press, 1983.

—— *Proust and Signs: The Complete Text*, trans. R. Howard. Minneapolis: University of Minnesota Press, 2004.

Deleuze, Gilles and Félix Guattari. *Anti-Oedipus: Capitalism and Schizophrenia*, trans. R. Hurley, M. Seem and H. Lane. Minneapolis: University of Minnesota Press, 1983.

—— *A Thousand Plateaus: Capitalism and Schizophrenia*, trans. B. Massumi. Minneapolis: University of Minnesota Press, 1987.

de Madariaga, Isabel. *Ivan the Terrible*. New Haven, CT: Yale University Press, 2006. Kindle edition.

Dennett, Daniel. *Breaking the Spell: Religion as a Natural Phenomenon*. London: Penguin, 2006.

—— *Consciousness Explained*. London: Penguin, 1993.

—— *Darwin's Dangerous Idea: Evolution and the Meanings of Life*. New York: Simon and Schuster, 1996.

—— 'Quining Qualia', in *Consciousness in Contemporary Science*, A. J. Marcel and E. Bisiach, eds., pp. 381–414. Oxford: Oxford University Press, 1988.

—— 'The Unimagined Preposterousness of Zombies: Commentary on T. Moody, O. Flanagan and T. Polger', *Journal of Consciousness Studies* 2.4 (1995), pp. 322–6.

Derrida, Jacques. *Margins of Philosophy*, trans. A. Bass. Chicago: University of Chicago Press, 1982.

—— 'Plato's Pharmacy', in *Dissemination*, trans. B. Johnson, pp. 61–171. Chicago: University of Chicago Press, 1981.

—— 'Point de folie – Maintenant l'architecture', trans. K. Linker, in *Architecture Theory Since 1968*, ed. M. Hays, pp. 570–81. Cambridge, MA: MIT Press, 2000.

—— *On Touching – Jean-Luc Nancy*, trans. C. Irizarry. Stanford, CA: Stanford University Press, 2005.

—— *Writing and Difference*, trans. A. Bass. Chicago: University of Chicago Press, 1978.

Descartes, René. *Meditations on First Philosophy*, trans. D. Cress. Indianapolis: Hackett, 1993.

—— *The Passions of the Soul*, trans. S. Voss. Indianapolis: Hackett, 1989.

Deutsch, Helene. 'A Case of Hen Phobia', in *Neuroses and Character Types: Clinical Psychoanalytic Studies*, pp. 84–96. New York: International Universities Press, 1965.

de Vries, Hugo. *Die Mutationstheorie. Versuche und Beobachtungen über die Entstehung von Arten im Pflanzenreich*. Leipzig: Veit and Co., 1901–3.

De Witt, Bryce Seligman and Neill Graham, eds. *The Many-Worlds Interpretation of Quantum Mechanics*. Princeton, NJ: Princeton University Press, 1973.

al-Din, Rashid. *The Successors of Genghis Khan*, trans. J. A. Boyle. New York: Columbia University Press, 1971.

Donnellan, Keith. 'Reference and Definite Descriptions', *The Philosophical Review* 75.3 (July 1966), pp. 281–304.

Doolittle, W. Ford. 'Phylogenetic Classification and the Universal Tree', *Science* 284.5423 (June 1999), pp. 2124–8.

Dosse, François. *History of Structuralism*, vol. 1: *The Rising Sign, 1945–1966*, trans. D. Glassman. Minneapolis: University of Minnesota Press, 1997.

Duns Scotus. *Philosophical Writings: A Selection*, trans. A. B. Wolter. Indianapolis: Hackett, 1987.

Ehrenfest, Paul. 'Über die physikalischen Voraussetzungen der Planck'schen Theorie der irreversiblen Strahlungsvorgänge', in Paul Ehrenfest, *Collected Scientific Papers*, ed. M. J. Klein, pp. 88–101. Amsterdam: Interscience, 1959.

Einstein, Albert. *The Collected Papers of Albert Einstein*, vol. 14: *The Berlin Years: Writings and Correspondence, April 1923–May 1925*, ed. D. K. Buchwald. Princeton, NJ: Princeton University Press, 2015.

—— 'Zur Elektrodynamik bewegter Körper', *Annalen der Physik* 17.10 (1905), pp. 891–921.

—— 'Die Grundlage der allgemeinen Relativitätstheorie', *Annalen der Physik* 49.7 (1916), pp. 769–822.

—— 'Die Plancksche Theorie der Strahlung und die Theorie der spezifischen Wärme', *Annalen der Physik* 22 (1907), pp. 180–90.

—— 'Über die von der molekularkinetischen Theorie der Wärme geforderte Bewegung von in ruhenden Flüssigkeiten suspendierten Teilchen', *Annalen der Physik* 322.8 (1905), pp. 549–60.

—— 'Über einen die Erzeugung und Verwandlung des Lichtes betreffenden heuristischen Gesichtspunkt', *Annalen der Physik* 322.6 (1905), pp. 132–48.

Einstein, Albert and Marcel Grossmann, 'Entwurf einer verallgemeinerten Relativitätstheorie und einer Theorie der Gravitation', *Zeitschrift für Mathematik und Physik* 62 (1913), pp. 225–61.

Eisenman, Peter. *Eisenman Inside Out: Selected Writings, 1963–1988*. New Haven, CT: Yale University Press, 2004.

Eldredge, Niles and Stephen Jay Gould. 'Punctuated Equilibria: An Alternative to Phyletic Gradualism', in *Models in Paleobiology*, ed. T. M. Schopf, pp. 82–115. San Francisco: Freeman Cooper, 1972.

Eliot, T. S. *Selected Essays, 1917–1932*. New York: Harcourt, Brace, and Co. 1932.

Empson, William. *Seven Types of Ambiguity*. New York: New Directions, 1966.

Euclid. *The Thirteen Books of the Elements*, trans. T. Heath. New York: Dover, 1956.

Everett, Hugh. 'Relative State Formulation of Quantum Mechanics', *Reviews of Modern Physics* 29.3 (1957), pp. 454–62.

Faber, Roland. 'Touch: A Philosophic Meditation', in *The Allure of Things: Process and Object in Contemporary Philosophy*, R. Faber and A. Goffey, eds., pp. 47–67. London: Bloomsbury, 2014.

Fairbank, John King and Merle Goldman. *China: A New History*, 2nd enlarged edition. Cambridge, MA: Belknap, 2006. Kindle edition.

Fakhry, Majid. *Islamic Occasionalism: And Its Critique by Averroes and Aquinas*. New York: Routledge, 2007.

al-Farabi, Abu Nasr Muhammad. *On the Perfect State*, trans. R. Walzer. Chicago: Kazi, 1998.

Felski, Rita. 'Context Stinks!', *New Literary History* 42.4 (2011), pp. 573–91.

Fermi, Enrico and Franco Rasetti. 'Slow Neutrons', *Il Nuovo Cimento* 12 (1935), pp. 201–10.

Ferraris, Maurizio. *Hysteresis: The External World*, trans. S. De Sanctis. Edinburgh: Edinburgh University Press, 2024.

Feynman, Richard. *The Character of Physical Law*. London: Penguin, 1992.

—— *QED: The Strange Theory of Light and Matter*. Princeton, NJ: Princeton University Press, 1985.

Fichte, J. G. *The Science of Knowledge*, trans. P. Heath and J. Lachs. Cambridge: Cambridge University Press, 1982.

Flam, Jack. *Matisse and Picasso: The Story of Their Rivalry and Friendship*. New York: Icon Editions, 2003.

Foucault, Michel. *The Birth of the Clinic: An Archaeology of Medical Perception*, trans. A. Sheridan. New York: Vintage, 1994.

—— *Discipline and Punish: The Birth of the Prison*, trans. A. Sheridan. London: Penguin, 2020.

—— *History of Madness*, trans. J. Murphy. London: Routledge, 2006.

Freud, Sigmund. *Beyond the Pleasure Principle*, trans. J. Strachey. New York: Norton, 1990.

—— *The Interpretation of Dreams*, trans. J. Crick. Oxford: Oxford University Press, 2008.

Fried, Michael. *Manet's Modernism: or, The Face of Painting in the 1860s*. Chicago: University of Chicago Press, 1998.

Gabriel, Markus. *Fields of Sense: A New Realist Ontology*. Edinburgh: Edinburgh University Press, 2015.

Galilei, Galileo. *Siderius Nuncius (The Sidereal Messenger)*, trans. A. Van Helden. Chicago: University of Chicago Press, 2016.

Gerson, Lloyd P. *Aristotle and Other Platonists*. Ithaca, NY: Cornell University Press, 2005.

al-Ghazali, Abu Hamid. *Deliverance from Error: Five Key Texts Including His Spiritual Autobiography*, trans. R. J. McCarthy. Chicago: Fons Vitae, 2004.

Gibbon, Edward. *The Decline and Fall of the Roman Empire*, 3 vols., trans. D. Womersley. London: Penguin, 1996.

Giedion, Sigfried. *Space, Time, and Architecture: The Growth of a New Tradition*, 5th edition, revised and enlarged. Cambridge, MA: Harvard University Press, 1977.

Gilson, Étienne. *The Philosophy of St. Thomas Aquinas*, trans. E. Bullough. New York: Barnes and Noble, 1993.

Gödel, Kurt. *The Consistency of the Continuum-Hypothesis*. Princeton, NJ: Princeton University Press, 1940.

Goff, Philip. *Galileo's Error: Foundations for a New Science of Consciousness*. New York: Vintage, 2019.

Gould, Stephen Jay. 'Nonoverlapping Magisteria', *Natural History* 106 (March 1997), pp. 16–22. Citations refer to an online version, https://web.archive.org/web/20190403152432/http://www.stephenjaygould.org/library/gould_noma.html. Last accessed on 25 March 2025.

—— *Rocks of Ages: Science and Religion in the Fullness of Life*. New York: Ballantine Books, 1999.

—— *The Structure of Evolutionary Theory*. Cambridge, MA: Belknap, 2002.

Gould, Stephen Jay and Richard C. Lewontin. 'The Spandrels of San Marco and the Panglossian Paradigm: A Critique of the Adaptationist Program', *Proceedings of the Royal Society of London B*, 205.1161 (September 1979), pp. 581–98.

Gould, Stephen Jay and Elisabeth A. Lloyd. 'Individuality and Adaptation Across Levels of Selection: How Shall We Name and Generalize the Units of Darwinism?', *Proceedings of the National Academy of Sciences* 96.21 (1999), pp. 11904–9.

Gould, Stephen Jay and Elisabeth S. Vrba. 'Exaptation – A Missing Term in the Science of Form', *Paleobiology* 8.1 (Winter 1982), pp. 4–15.

Granovetter, Mark S. 'The Strength of Weak Ties', *American Journal of Sociology* 87.6 (1973), pp. 1360–80.

Grantham, Todd A. 'Hierarchical Approaches to Macroevolution: Recent Work on Species Selection and the "Effect Hypothesis"', *Annual Review of Ecology and Systematics* 26 (1995), pp. 301–21.

Greenberg, Clement. *Art and Culture: Critical Essays*. Boston: Beacon Press, 1989.

Greene, Brian. *The Elegant Universe: Superstrings, Hidden Dimensions, and the Quest for the Ultimate Theory*. New York: Norton, 2009.

Grousset, René. *The Empire of the Steppes: A History of Central Asia*, trans. N. Walford. New Brunswick, NJ: Rutgers University Press, 1970.

Hameroff, Stuart. 'Consciousness, Neurobiology and Quantum Mechanics', in J. Tuszynski, ed., *The Emerging Physics of Consciousness*, pp. 193–253.

Harman, Graham. *Architecture and Objects*. Minneapolis: University of Minnesota Press, 2022.

—— *Art and Objects*. Cambridge: Polity, 2020.

—— *Bells and Whistles: More Speculative Realism*. Winchester: Zero, 2013.

—— *Bruno Latour: Reassembling the Political*. London: Pluto, 2014.

—— 'Concerning Stephen Hawking's Claim that Philosophy Is Dead', *Filozofski vestnik* 33.2 (2012), pp. 11–22.

—— 'Concerning the COVID-19 Event', *Philosophy Today* 64.4 (Fall 2020), pp. 845–9.

—— 'DeLanda's Ontology: Assemblage and Realism', *Continental Philosophy Review* 41:3 (2008), pp. 367–83.

—— 'Greg Lynn on Animate Form: An Object-Oriented Response', in *Animate(d) Architecture*, Vahid Vahdat, ed., pp. 17–33. Liverpool: Liverpool University Press, 2024.

—— *Heidegger Explained: From Phenomenon to Thing*. Chicago: Open Court, 2007.

—— 'I Am Also of the Opinion That Materialism Must Be Destroyed', *Environment and Planning D: Society and Space* 28.5 (2010), pp. 772–90.

—— *Immaterialism: Objects and Social Theory*. Cambridge: Polity, 2016.

—— 'On Interface: Nancy's Weights and Masses', in *Jean-Luc Nancy and Plural Thinking: Expositions of World, Politics, Art, and Sense*, P. Gratton and M.-È. Morin, eds., pp. 95–108. Albany, NY: SUNY Press, 2012.

—— 'Latour's Interpretation of Donald Trump', in *Nonmodern Practices: Latour and Literary Studies*, E. Arnould-Bloomfield and Lyu, eds., pp. 191–215. London: Bloomsbury, 2020.

—— 'Magic Uexküll', in *Living Earth: Field Notes from the Dark Ecology Project 2014–2016*, pp. 115–30. Amsterdam: Sonic Acts Press, 2016.

—— 'A New Occasionalism?', in *Reset Modernity!*, B. Latour and P. Weibel, eds., pp. 129–38. Cambridge, MA: MIT Press, 2016.

—— *Object-Oriented Ontology: A New Theory of Everything.* London: Pelican, 2018.

—— 'Object-Oriented Ontology and Commodity Fetishism: Kant, Marx, Heidegger, and Things', *Eidos* 2 (2017), pp. 28–36. http://eidos.uw.edu. pl/files/pdf/eidos/2017-02/eidos_2_harman.pdf. Last accessed on 25 March 2025.

—— 'The Only Exit from Modern Philosophy', *Open Philosophy* 3 (2020), pp. 132–46.

—— *Prince of Networks: Bruno Latour and Metaphysics.* Melbourne: re.press, 2009.

—— 'On the Punctuation of Organisms: The Case of Helmuth Plessner', in *Life in the Posthuman Condition: Critical Responses to the Anthropocene*, S. E. Wilmer and Audronė Žukauskaitė, eds., pp. 111–29. Edinburgh: Edinburgh University Press, 2023.

—— 'Retroactivity in Science: Latour, Žižek, Kuhn', *Open Philosopy* 7 (2024), pp. 1–18.

—— 'Time, Space, Essence, and Eidos: A New Theory of Causation', *Cosmos and History* 6.1 (2010), pp. 1–17.

—— *Tool-Being: Heidegger and the Metaphysics of Objects.* Chicago: Open Court, 2002.

—— 'On the Undermining of Objects: Grant, Bruno, and Radical Philosophy', in *The Speculative Turn: Continental Materialism and Realism*, eds. L. Bryant, N. Srnicek and G. Harman, pp. 21–40. Melbourne: re.press, 2011.

—— 'Undermining, Overmining, and Duomining: A Critique', in *ADD Metaphysics*, ed. Jenna Sutela, pp. 40–51. Aalto, Finland: Aalto University Design Research Laboratory, 2013.

—— 'War, Space, and Reversal: Paul Virilio's Apocalypse', in *Philosophy After Hiroshima*, ed. E. Demenchonok, pp. 132–48. Cambridge: Cambridge Scholars Press, 2010.

—— 'The Well-Wrought Broken Hammer: Object-Oriented Literary Criticism', *New Literary History* 43 (2012), pp. 183–203.

—— 'Zero-Person and the Psyche', in *Mind That Abides: Panpsychism in the New Millennium*, ed. D. Skrbina, pp. 253–82. Amsterdam: Benjamins, 2009.

Harman, Graham and Christopher Witmore. *Objects Untimely: Object-Oriented Philosophy and Architecture.* Cambridge: Polity, 2023.

Harris, Sam. *The End of Faith: Religion, Terror, and the Future of Reason.* New York: W. W. Norton, 2005.

—— *Letter to a Christian Nation.* New York: Vintage, 2006.

Hartmann, Nicolai. 'The Megarian and the Aristotelian Concept of Possibility: A Contribution to the History of the Ontological Problem of Modality', trans. F. Tremblay and K. Peterson, *Axiomathes* 27 (2017), pp. 209–23.

Hegel, G.W. F. *Hegel: The Letters*, trans. C. Butler and C. Seiler. Bloomington, IN: Indiana University Press, 1985.

—— *Phenomenology of Spirit*, trans. A.V . Miller. Oxford: Oxford University Press, 1976.

—— *Science of Logic*, trans. A. V. Miller. New York: Humanities Press, 1969.

Heidegger, Martin. *The Basic Problems of Phenomenology*, trans. A. Hofstadter. Bloomington, IN: Indiana University Press, 2021.

—— *Being and Time*, trans. J. Macquarrie and E. Robinson. New York: Harper, 1962.

—— *The Essence of Reasons*, trans. T. Malick. Evanston, IL: Northwestern University Press, 1969.

—— *Fundamental Concepts of Metaphysics: World–Finitude–Solitude*, trans. W. McNeill and A. Walker. Bloomington, IN: Indiana University Press, 2001.

—— *Kant and the Problem of Metaphysics*, 5th edition, enlarged, trans. R. Taft. Bloomington, IN: Indiana University Press, 1997.

—— *The Question Concerning Technology: And Other Essays*, trans. W. Levitt. New York: Harper, 1977.

—— *What Is Called Thinking?*, trans. J. G. Gray. New York: Harper and Row, 1968.

Heisenberg, Werner. 'Über den anschaulichen Inhalt der quantentheoretischen Kinematik und Mechanik', 43.3–4 (1927), pp. 172–98.

Herodotus. *The Histories*, trans. T. Holland. New York: Penguin, 2013.

Hesiod. *Theogony and Works and Days*, trans. M. L. West. Oxford: Oxford University Press, 2008.

Hitchens, Christopher. *God Is Not Great: How Religion Poisons Everything.* New York: Twelve, 2007.

Hobbes, Thomas. *Leviathan.* Oxford: Oxford University Press, 2009.

Hofstadter, Douglas. *Gödel, Escher, Bach: An Eternal Golden Braid*. New York: Vintage, 1989.

Hölderlin, Friedrich. 'Patmos', in *Seleted Poems and Fragments*, trans. M. Hamburger. London: Penguin, 1998.

Hossenfelder, Sabine. *Lost in Math: How Beauty Leads Physics Astray*. New York: Basic Books, 2018. Kindle edition.

Hull, David L. 'Individuality and Selection', *Annual Review of Ecological Systematics* 11, pp. 311–32.

Hume, David. *An Enquiry Concerning Human Understanding*. Indianapolis: Hackett, 1993.

Husserl, Edmund. *Cartesian Meditations: An Introduction to Phenomenology*, trans. D. Cairns. The Hague: Martinus Nijhoff, 1977.

—— *Logical Investigations*, 2 vols., trans. J. N. Findlay. London: Routledge and Kegan Paul, 1970.

Huxley, Julian. *Evolution: The Modern Synthesis*. London: Allen and Unwin, 1942.

Ibn Khaldun. *The Muqaddimah: An Introduction to History*, trans. F. Rosenthal. Princeton, NJ: Princeton University Press, 2005. Abridged Kindle edition.

Inglehart, Ronald F. 'Giving Up on God: The Global Decline of Religion', *Foreign Affairs*, September/October 2020, pp. 110–18.

Jablonski, David. 'Heritability at the Species Level: Analysis of Geographic Ranges of Cretaceous Mollusks', *Science* 238 (1987), pp, pp. 360–63.

Jacobsen, Annie. *Nuclear War: A Scenario*. London: Dutton, 2024. Kindle edition.

Jacquette, Dale. *Alexius Meinong, the Shepherd of Non-Being*. Cham, Switzerland: Springer Synthese, 2015.

James, William. *Essays in Radical Empiricism*. New York: Longman, Greens and Co., 1958.

Jencks, Charles A. *The Language of Post-Modern Architecture*. New York: Random House, 1977.

John Paul II, Pope. 'Message to the Pontifical Academy of Sciences on Evolution', *The Quarterly Review of Biology* 72.4 (December 1997), pp. 381–3. The citation is to an online version, https://www.jstor.org/stable/3037603. Last accessed on 25 March 2025.

Johnson, Philip and Wark Wigley. *Deconstructivist Architecture*. New York: The Museum of Modern Art, 1988.

Jones, Steve. 'A Wonderful Life by Leaps and Bounds', *Nature* 456 (2008), pp. 873–4.

Jung, Carl Gustav. *The Portable Jung*, trans. R. F. C. Hull; ed. J. Campbell, London: Penguin, 1976.

Kaku, Michio. *The God Equation: The Quest for a Theory of Everything*. New York: Doubleday, 2021. Kindle edition.

Kant, Immanuel. *Critique of Pure Reason: Unified Edition*, trans. W. Pluhar. Indianapolis: Hackett, 1996.

—— *Grounding for the Metaphysics of Morals*, trans. J. Ellington. Indianapolis: Hackett, 1993.

Kazantzakis, Nikos. *The Last Temptation of Christ*, trans. P. A. Bien. New York: Simon and Schuster, 2012.

—— *The Odyssey: A Modern Sequel*, trans. K. Friar. New York: Simon and Schuster, 1958. Kindle edition.

—— *Zorba the Greek*, trans. C. Wildman. New York: Simon and Schuster, 1953.

Keele, Rondo. *Ockham Explained: From Razor to Rebellion*. Chicago: Open Court, 2010.

Kierkegaard, Søren. *Either/Or, Part I*, trans. H. Hong and E. Hong. Princeton, NJ: Princeton University Press, 1987.

Kimura, Motoo. 'Evolutionary Rate at the Molecular Level', *Nature* 217 (1968), pp. 624–6.

Kirk, Robert and Roger Squires. 'Zombies v. Materialists', *Proceedings of the Aristotelian Society, Supplementary Volumes*, vol. 48 (1974), pp. 135–63.

Klein, Melanie. *The Psychoanalysis of Children*, trans. A. Strachey. New York: Vintage, 1997.

Kojève, Alexandre. *Introduction to the Reading of Hegel: Lectures on the Phenomenology of Spirit*, trans. J. Nichols Jr. Ithaca, NY: Cornell University Press, 1969.

Koolhaas, Rem and Bruce Mau. *S, M, L, XL*. New York: The Monacelli Press, 1997.

Koyré, Alexandre. *From the Closed World to the Infinite Universe*, trans. R. E. W. Maddison. Baltimore: Johns Hopkins University Press, 1957.

—— *Galileo Studies*, trans. J. Mepham. Atlantic Highlands, NJ: Humanities Press, 1978.

Kripke, Saul. *Naming and Necessity*. Cambridge, MA: Harvard University Press, 1980.

Kuhn, Thomas. *Black-Body Theory and the Quantum Discontinuity, 1894–1912*. Chicago: University of Chicago Press, 1978.

—— *The Copernican Revolution: Planetary Astronomy in the Development of Western Thought*. Cambridge, MA: Harvard University Press, 1957.

—— *The Road Since Structure: Philosophical Essays, 1970–1993, with an Autobiographical Interview*. Chicago: University of Chicago Press, 2000.

—— *The Structure of Scientific Revolutions*, 50th anniversary edition. Chicago: University of Chicago Press, 2012.

Lacan, Jacques. *From an Other to the other: The Seminar of Jacques Lacan, Book XVI*, trans. B. Fink. Cambridge: Polity, 2024.

Ladyman, James and Don Ross, with David Spurrett and John Collier. *Every Thing Must Go: Metaphysics Naturalized*. Oxford: Oxford University Press, 2007.

Lakatos, Imre and Paul Feyerabend. *For and Against Method: Including Lakatos's Lectures on Scientific Method and the Lakatos–Feyerabend Correspondence*. Chicago: University of Chicago Press, 2000.

Landau, L. D. and E. M. Lifshitz, *The Classical Theory of Fields*, vol. 2., 4th edition. Oxford: Butterworth-Heinemann, 1980.

Lando, Giorgio. *Mereology: A Philosophical Introduction*. London: Bloomsbury, 2018.

Lao Tzu. *Tao Te Ching*, trans. D. C. Lau. London: Penguin, 1964.

Latour, Bruno. *Aramis, or The Love of Technology*, trans. C. Porter. Cambridge, MA: Harvard University Press, 1996.

—— *Down to Earth: Politics in the New Climactic Regime*, trans. C. Porter. Cambridge: Polity, 2018.

—— *An Enquiry into Modes of Existence: An Anthropology of the Moderns*, trans. C. Porter. Cambridge, MA: Harvard University Press, 2013.

—— 'Irreductions', in *The Pasteurization of France*, trans. A. Sheridan and J. Law, pp. 153–237. Cambridge, MA: Harvard University Press, 1988.

—— *The Making of the Law: An Ethnography of the Conseil d'État*, trans. M. Brilman and A. Pottage, revised by the author. Cambridge: Polity, 2010.

—— *Pandora's Hope: Essays on the Reality of Science Studies*. Cambridge, MA: Harvard University Press, 1999.

—— *The Pasteurization of France*, trans. A. Sheridan and J. Law. Cambridge, MA: Harvard University Press, 1988.

—— *Reassembling the Social: An Introduction to Actor-Network Theory*. Oxford: Oxford University Press, 2005.

—— *Rejoicing: Or the Torments of Religious Speech*, trans. J. Rose. Cambridge: Polity, 2013.

—— *We Have Never Been Modern*, trans. C. Porter. Cambridge, MA: Harvard University Press, 1993.

Latour, Bruno and Émilie Hermant. *Paris Invisible City*, trans. L. Carey-Libbrecht, corrected V. Pihet. http://bruno-latour.fr/virtual/index.html. Last accessed on 25 March 2025.

Lavoisier, Antoine. *Elements of Chemistry*. Mineola, NY: Dover, 1984.

Lawrence, Bruce. 'Introduction to the 2005 Edition', in Ibn Khaldun. *The Muqaddimah: An Introduction to History*, trans. F. Rosenthal, pp. vii–xxv. Princeton, NJ: Princeton University Press, 2005. Abridged Kindle edition.

Le Corbusier. *Towards a New Architecture*, trans. F. Etchell. New York: Dover, 1986.

Leibniz, G. W. *New Essays on Human Understanding*, P. Remnant and J. Bennett, eds. Cambridge: Cambridge University Press, 1996.

—— 'The Principles of Philosophy, or, the Monadology', in *Philosophical Essays*, trans. R. Ariew and D. Garber, pp. 213–25. Indianapolis: Hackett, 1989.

—— *Theodicy: Essays on the Goodness of God, the Freedom of Man, and the Origin of Evil*, trans. E. M. Huggard. Chicago: Open Court, 1996.

Léroi-Gourhan, André. *Gesture and Speech*, trans. A. B. Berger. Cambridge, MA: MIT Press, 2018.

Lesser, Wendy. '"You Say to Brick": Louis Kahn Begins to Articulate the Ideas that Define His Architecture', *Humanities: The Magazine of the National Endowment for the Humanities* 38.2 (Spring 2017). https://www.neh.gov/humanities/2017/spring/feature/louis-kahn-visionary-0. Last accessed on 25 March 2025.

Levinas, Emmanuel. *Ethics and Infinity: Conversations with Philippe Nemo*, trans. R. Cohen. Pittsburgh: Duquesne University Press, 1985.

—— *Existence and Existents*, trans. A. Lingis. The Hague: Martinus Nijhoff, 1988.

Lévi-Strauss, Claude. *The Elementary Structures of Kinship*, trans. J. H. Bell and J. R. von Sturmer. Boston: Beacon Press, 1969.

—— *Structural Anthropology*, trans. C. Jacobson and B. Grundfest Schoepf. New York: Basic Books, 1963.

Lewin, Roger. 'Evolution's New Heretics: A Growing Number of Evolutionary Biologists Think That the Interests of Groups Sometimes Supersede Those of Individuals', *Natural History* 105 (1996), pp. 12–17.

Lind, William S., Keith Nightengale, John F. Schmitt, Joseph W., Sutton and Gary I. Wilson. 'The Changing Face of War: Into the Fourth Generation', *Marine Corps Gazette*, October 1989, pp. 22–6. http://www.d-n-i.net/fcs/4th_gen_war_gazette.htm. Last accessed on 25 March 2025.

Lingis, Alphonso. *The Imperative*. Bloomington, IN: Indiana University Press, 1998.

Lloyd, Elisabeth A. and Stephen Jay Gould. 'Species Selection on Variability', *Proceedings of the National Academy of Sciences* 90.2 (1993), pp. 595–9.

Locke, John. *An Essay Concerning Human Understanding*, 2 vols. Mineola, NY: Dover, 1959.

London, Jack. *Novels and Stories*. New York: Library of America, 1982.

Lovelock, James. *Gaia: A New Look at Life on Earth*. Oxford: Oxford University Press, 2000.

Lovelock, James and Lynn Margulis. *Writing Gaia: The Scientific Correspondence of James Lovelock and Lynn Margulis*, B. Clarke and S. Dutreuil, eds. Cambridge: Cambridge University Press, 2022. Kindle edition.

Lukács, Georg. 'Realism in the Balance', trans. R. Livingstone, in Theodor Adorno et al., *Aesthetics and Politics*, pp. 28–59. London: Verso, 1988.

Luhmann, Niklas. *Social Systems*, trans. J. Bednarz Jr. with D. Baecker. Stanford, CA: Stanford University Press, 1996.

Lynn, Greg. *Animate Form*. New York: Princeton Architectural Press, 1999.

—— 'Architectural Curvilinearity: The Folded, the Pliant, and the Supple', in *Folding in Architecture*, ed. G. Lynn, pp. 24–31. London: Academy Editions, 1993.

—— 'Blobs, or Why Tectonics Is Square and Topology Is Groovy', *ANY: Architecture New York* 14 (1996), pp. 58–61.

Macey, David. *The Lives of Michel Foucault*. London: Verso, 2019.

Machiavelli, Niccolò. *Discourses on Livy*, trans. H. Mansfield and N. Tarcou. Chicago: University of Chicago Press, 1996.

—— *The Prince*, trans. P. Bondanella. Oxford: Oxford University Press, 2008.

Malebranche, Nicolas. *The Search After Truth: With Elucidations of the Search After Truth*, ed. T. Lennon and P. Olscamp. Cambridge: Cambridge University Press, 2010.

Mandelbrot, Benoit B. *Fractals: Form, Chance and Dimension*, trans. A. E. Pitcher. Brattleboro, VT: Echo Point Books and Media, 2020.

Marder, Michael. *Plant-Thinking: A Philosophy of Vegetal Life*. New York: Columbia, 2013.

Margulis, Lynn. *Origin of Eukaryotic Cells*. New Haven, CT: Yale University Press, 1970.

—— *Symbiotic Planet: A New Look at Evolution*. New York: Basic Books, 1999.

Marías, Julián. *Generations: A Historical Method*, trans. H. Raley. Tuscaloosa, AL: University of Alabama Press, 1970.

—— *History of Philosophy*, trans. S. Appelbaum and C. Strowbridge. Mineola, NY: Dover, 1967.

Marx, Karl. *Capital: Volume One*, trans. B. Fowkes. London: Penguin, 2004. Kindle edition.

—— *A Contribution to the Critique of Political Economy*, trans. N. I. Stone. Chicago: Charles H. Kerr and Co., 1904.

—— *The First Writings of Karl Marx*, ed. P. Schafer. New York: Ig Publishing, 2006.

Mattingly, James. 'Is Quantum Gravity Necessary?', in *The Universe of General Relativity*, A. J. Knox and J. Eisenstaedt, eds., pp. 327–38. Boston: Birkhäuser, 2005.

Maxwell, James Clerk. *A Treatise on Electricity and Magnetism*, vol. 1, 3rd edition. Mineola, NY: Dover, 1954.

Mayr, Ernst. *Populations, Species, and Evolution: An Abridgment of Animal Species and Evolution*. Cambridge, MA: Harvard University Press, 1970.

Meillassoux, Quentin. *After Finitude: An Essay on the Necessity of Contingency*, trans. R. Brassier. London: Continuum, 2008.

—— 'Appendix: Excerpts from *L'Inexistence divine*', in Graham Harman,

Quentin Meillassoux: Philosophy in the Making, 2nd edition, pp. 224–87. Edinburgh: Edinburgh University Press, 2015.

Meinong, Alexius. *On Assumptions*, trans. J. Heanue. Berkeley, CA: University of California Press, 1983.

Meitner, Lise and Otto R. Frisch. 'Disintegration of Uranium by Neutrons: A New Type of Nuclear Fission', *Nature* 143 (1939), pp. 239–40.

Mendel, Gregor. 'Experiments in Plant Hybridization', trans. C. T. Druery and W. Bateson. *Journal of the Royal Horticultural Society.* 26 (1901), pp. 1–32.

Merleau-Ponty, Maurice. *The Visible and the Invisible*, trans. A. Lingis. Evanston, IL: Northwestern University Press, 1998.

Mermin, N. David. 'Could Feynman Have Said This?', *Physics Today* 57.5 (2004), pp. 10–11.

Metzinger, Thomas and Wolf Singer. 'The Unity of Consciousness: A Conversation with Wolf Singer', in Thomas Metzinger, *The Ego Tunnel: The Science of the Mind and the Myth of the Self*, pp. 66–74. New York: Basic Books, 2009.

Mickey, Sam. 'Touching without Touching: Objects of Post-Deconstructive Realism and Object-Oriented Ontology', *Open Philosophy* 1 (2018), pp. 290–98.

Milgrom, Mordehai. 'A Modification of the Newtonian Dynamics as an Alternative to the Hidden Mass Hypothesis', *Astrophysical Journal* 270 (1983), pp. 365–70.

Miller, Mark P., Thomas D. Mullins, Eric D. Forsman and Susan M. Haig. 'Genetic Differentiation and Inferred Dynamics of a Hybrid Zone Between Northern Spotted Owls (*Strix occidentalis caurina*) and California Spotted Owls (*S. o. occidentalis*) in Northern California', *Ecology and Evolution* 7.17 (September 2017), pp. 6871–83. https://doi.org/10.1002/ece3.3260. Last accessed on 25 March 2025

Minkowski, Herrmann. 'Raum und Zeit', *Jahresbericht der deutschen Mathematiker-Vereinigung* 18 (1909), pp. 75–88.

Mishima, Yukio. *Runaway Horses*, trans. M. Gallagher. New York: Vintage, 2000.

Nail, Thomas. *Being in Motion*. Oxford: Oxford University Press, 2018.

—— *The Figure of the Migrant*. Stanford, CA: Stanford University Press, 2015.

—— 'Is Nature Continuous or Discrete? How the Atomist Error Was Born', *Aeon*, 18 May 2018. https://aeon.co/ideas/is-nature-continuous-or-discrete-how-the-atomist-error-was-born. Last accessed on 25 March 2025.

Nancy, Jean-Luc. *Corpus*, trans. R. Rand. New York: Fordham University Press, 2008.

Neef, Gerrit. 'Notes on the Subgenus *Pelicaria*', *New Zealand Journal of Geology and Geophysics* 13.2 (1970), pp. 436–76.

Newton, Isaac. *Opticks: Or a Treatise of the Reflections, Refractions, Inflections and Colours of Light – Based on the Fourth Edition, London, 1730*. Mineola, NY: Dover 2012.

—— *The Principia: Mathematical Principles of Natural Philosophy*, trans. I. B. Cohen, A. Whitman and J. Budenz. Berkeley, CA: University of California Press, 2016.

Nietzsche, Friedrich. *Beyond Good and Evil*, trans. J. Norman. Cambridge: Cambridge University Press, 2001.

—— *Ecce Homo: How One Becomes What One Is*, trans. D. Large. Oxford: Oxford University Press, 2009.

Ockham, William of. *Philosophical Writings: A Selection*, trans. P. Boehner. Indianapolis: Hackett, 1990.

Oerter, Robert. *The Theory of Almost Everything: The Standard Model, the Unsung Triumph of Modern Physics*. New York: Plume, 2006.

Oppenheim, Jonathan. 'A Postquantum Theory of Classical Gravity?', *Physical Review X* 13 (2023), 041040, pp. 1–37.

Ortega y Gasset, José. *Man and Crisis*, trans. M. Adams. New York: Norton, 1958.

—— *Mission of the University*, trans. H. L. Nostrand. London: Routledge, 2017.

Parkinson, Gavin. *The Duchamp Book*. London: Tate, 2008.

Parmenides of Elea. *Fragments*, trans. D. Gallop. Toronto: University of Toronto Press, 1991.

Pascal, Blaise. *Pensées*, trans. A. J. Krailsheimer. London: Penguin, 1995.

Peirce, C. S. *Philosophical Writings of Peirce*, ed. J. Buchler. Mineola, NY: Dover, 2011.

Penrose, Roger. *Shadows of the Mind: A Search for the Missing Science of Consciousness*. Oxford: Oxford University Press, 1994.

Pirsig, Robert M. *Zen and the Art of Motorcycle Maintenance*. New York: Mariner Book Classics, 2005. Kindle edition.

Pius XII, Pope. 'Encylical: *Humani Generis*', Vatican Website, 1950. https://www.vatican.va/content/pius-xii/en/encyclicals/documents/hf_p-xii_enc_12081950_humani-generis.html. Last accessed on 25 March 2025.

Planck, Max. 'Über das Gesetz der Energieverteilung im Normalspektrum', *Annalen der Physik* 309.3 (1901), pp. 553–63.

Plato. 'Crito', in *Five Dialogues*, trans. G. M. A. Grube, pp. 45–57. Indianapolis: Hackett, 2002.

—— *The Laws*, trans. T. Saunders. London: Penguin, 2005.

—— *Meno*, trans. E. Brann, P. Kalkavage and E. Salem. Indianapolis: Focus, 2021.

—— *Parmenides*, trans. A. K. Whitaker. Newburyport, MA: Focus, 1996.

—— 'Phaedrus', trans. B. Jowett, in *Plato: The Complete Works*, pp. 868–929. London: TitanRead Classics, 2015. Kindle edition.

—— *Republic*, trans. G. M. A. Grube, revised C. D. C. Reeve. Indianapolis: Hackett, 1992.

Plessner, Helmuth. *Die Einheit der Sinne: Grundlinien einer Aesthesiologie des Geistes*. Bonn: Friedrich Cohen, 1923.

—— *Levels of Organic Life and the Human: An Introduction to Philosophical Anthropology*, trans. M. Hyatt. New York: Fordham University Press, 2019.

Polanyi, Michael. *Personal Knowledge: Towards a Post-Critical Philosophy*. Chicago: University of Chicago Press, 2015.

Polo, Marco. *The Travels*, trans. N. Cliff. London: Penguin, 2015.

Polybius. *The Histories*, trans. R. Waterfield. Oxford: Oxford University Press, 2010.

Popper, Karl. *The Logic of Scientific Discovery*. London: Routledge, 1992.

—— *The Open Society and Its Enemies*. Princeton, NJ: Princeton University Press, 2020.

Proust, Marcel. *In Search of Lost Time*, 6 vols., trans. D. J. Enright and C. K. S. Moncrieff. New York: Modern Library, 2003.

Ptolemy. *Ptolemy's Almagest*, trans. G. J. Toomer. Princeton, NJ: Princeton University Press, 1998.

Putnam, Hilary. 'The Meaning of Meaning', in *Mind, Language, and*

Reality: Philosophical Papers, vol. 2, pp. 215–71. Cambridge: Cambridge University Press, 1975.

Quine, Willard van Orman. *From a Logical Point of View: Nine Logico-Philosophical Essays*, 2nd revised edition. Cambridge, MA: Harvard University Press, 1980.

Rand, Ayn. *The Fountainhead*. New York: Signet, 1996.

Raud, Rein. *Being in Flux: A Post-Anthropocentric Ontology of the Self.* Cambridge, MA: Polity, 2021.

Rhodes, Richard. *The Making of the Atomic Bomb*, 25th anniversary edition. New York: Simon and Schuster, 2012.

Riemann, Bernhard. 'Über die Hypothesen, welche der Geometrie zu Grunde legt', *Aus dem dreizehnten Bande der Abhandlungen der königlichen Gesellschaft der Wissenschaften in Göttingen*. Göttingen, Germany: 1854. https://archive.org/details/UeberDieHypothesenWelcheDerGeometrieZuGrundeLiegen/page/n1/mode/1up. Last accessed on 25 March 2025.

Rorty, Richard. *Philosophy and the Mirror of Nature*. Princeton, NJ: Princeton Unviersity Press, 1981.

Rosen, Stanley. *Plato's Republic: A Study*. New Haven, CT: Yale University Press, 2008.

Rosenberg, Harold. 'The American Action Painters', in *The Tradition of the New*. Cambridge, MA: Da Capo Press, 1994.

Ross, W. D. *Aristotle*. London: Methuen, 1923.

Rossi, Aldo. *The Architecture of the City*, trans. L. Venuti. Cambridge, MA: MIT Press, 1982.

Rozema, Lee A., Ardavan Darabi, Dylan H. Mahler, Alex Hayat, Yasaman Soudagar and Aephraim M. Steinberg. 'Violation of Heisenberg's Measurement–Disturbance Relationship by Weak Measurements', *Physical Review Letters* 189.100404 (6 September 2012).

Russell, Bertrand. *The Analysis of Matter*. London: Routledge, 2023.

—— *History of Western Philosophy*. London: Routledge, 2009.

—— *Mysticism and Logic*. Garden City, NY: Doubleday and Co., 1957.

—— 'The Philosophy of Bergson', *Monist* 22 (1912), pp. 321–47.

Ruy, David. 'Returning to (Strange) Objects', *tarp: Architecture Manual* 10 (Spring 2012), pp. 38–42.

Salvi, Francesco, L. R. Galante and Andrea Ricciardi. *The Impressionists: The Origins of Modern Painting*. New York: Peter Bedrick Books, 2001.

Saussure, Ferdinand de. *Course in General Linguistics*, trans. R. Harris. Chicago: Open Court, 1998.

Schmitt, Carl. *The Concept of the Political: Expanded Edition*, trans. G. Schwab. Chicago: University of Chicago Press, 2007.

Schrödinger, Erwin. 'Are There Quantum Jumps? Part I', *The British Journal for the Philosophy of Science* 3.10 (1952), pp. 109–23.

—— 'Are There Quantum Jumps? Part II', *The British Journal for the Philosophy of Science* 3.11 (1952), pp. 233–42.

—— 'Die gegenwärtige Situation in der Quantenmechanik', *Naturwissenschaften* 23 (1935), pp. 807–12.

—— 'Quantisierung als Eigenwertproblem', *Annalen der Physik* 384.4 (1926), pp. 273–376.

Schulze, Franz. *Philip Johnson: Life and Work*. New York: Knopf, 1994.

Schumacher, Patrik. *The Autopoiesis of Architecture*, 2 vols. Chichester: John Wiley and Sons, 2010/2011.

Scully, Vincent. 'Introduction', in Robert Venturi, *Complexity and Contradiction in Architecture*, 2nd edition, pp. 9–11. New York: Museum of Modern Art, 1977.

Serafin, Lauren. 'Surviving Peritoneal Mesothelioma: Stephen Jay Gould's Cancer Journey', mesothelioma.com, 19 May 2017. https://www.mesothelioma.com/blog/surviving-peritoneal-mesothelioma-stephen-jay-goulds-cancer-journey/. Last accessed on 25 March 2025.

Shklovsky, Viktor. *Theory of Prose*, trans. B. Sher. Normal, IL: Dalkey Archive Press, 1991.

Simondon, Gilbert. *Individuation in Light of Notions of Form and Information*, 2 vols., trans. T. Adkins. Minneapolis: University of Minnesota Press, 2020.

Sivasundaram, Sujeevam and Kristian Hvidtfelt Nielsen, 'Surveying the Attitudes of Physicists Concerning Foundational Issues of Quantum Mechanics', *arXiv* preprint, 2 December 2016. https://arxiv.org/abs/1612.00676. Last accessed on 25 March 2025.

Skrbina, David. *Panpsychism in the West*, revised edition. Cambridge, MA: MIT Press, 2017.

Skrbina, David, ed., *Mind That Abides: Panpsychism in the New Millennium*, Amsterdam: Benjamins, 2009.

Sloterdijk, Peter. *Spheres*, vol. 1: *Bubbles*, trans. W. Hoban. New York: Semiotext(e), 2011.

—— *Spheres*, vol. 2: *Globes*, trans. W. Hoban. New York: Semiotext(e), 2014.

—— *Spheres*, vol. 3: *Foam*, trans. W. Hoban. New York: Semiotext(e), 2016.

Smith, Barry, et al. 'Open Letter Against Derrida Receiving an Honorary Degree from Cambridge University', *The Times* (London), 9 May 1992.

Smolin, Lee. *Three Roads to Quantum Gravity*. New York: Basic Books, 2008.

—— *The Trouble with Physics: The Rise of String Theory, the Fall of a Science, and What Comes Next*. New York: Houghton Mifflin, 2006.

Snow, C. P. *The Two Cultures*. Cambridge: Cambridge University Press, 2001.

Sober, Elliott and David Sloan Wilson. *Unto Others: The Evolution and Psychology of Unselfish Behavior*. Cambridge, MA: Harvard University Press, 1998.

Spinoza, Baruch. *Ethics: Proved in Geometrical Order*, trans. M. Silverthorne. Cambridge: Cambridge University Press, 2018.

Stanislavski, Konstantin. *An Actor's Work*, trans. J. Benedetti. London: Routledge, 2010.

Stanley, Steven M. 'A Theory of Evolution Above the Species Level', *Proceedings of the National Academy of Sciences* 72.2 (February 1975), pp. 646–50.

Stapp, Henry. *Quantum Theory and Free Will: How Mental Intentions Translate into Bodily Action*. Cham, Switzerland: Springer, 2017.

Strawson, Galen. 'Realistic Monism: Why Physicalism Entails Panpsychism', in ed. D. Skrbina, *Mind That Abides*, pp. 33–65.

Sullivan, Louis. 'The Tall Office Building Artistically Considered', in *Kindergarten Chats and Other Writings*, pp. 202–13. Eastford, CT: Martino Fine Books, 2014.

Tafuri, Manfredo. *Theories and History of Architecture*, trans. G. Verrecchia. New York: Harper and Row, 1980.

Tegmark, Max. 'The Importance of Quantum Decoherence in Brain Processes', *Physical Review E* 61.4 (2000), pp. 4194–206.

Thijssen, Hans, 'Condemnation of 1277', *The Stanford Encyclopedia of Philosophy* (Winter 2018 edition), ed. Edward N. Zalta, https://plato.stanford.edu/archives/win2018/entries/condemnation/.

Toulmin, Stephen. 'Does the Distinction Between Normal and Revolutionary Hold Water?', in *Criticism and the Growth of Knowledge*, I. Lakatos and A. Musgrave, eds., pp. 39–47. Cambridge: Cambridge University Press, 1970.

Toynbee, A. J. *A Study of History*, vol. 3: *The Growths of Civilizations*. London: Oxford University Press, 1935.

Tschumi, Bernard. 'Derrida: An Ally et un Ami', *Log* 4 (Winter 2005), pp. 117–19.

Turner, John. 'Why We Need Evolution by Jerks', *New Scientist* 101.1396 (9 February 1984), pp 34–5.

Uexküll, Jakob von. *A Foray into the Worlds of Animals and Humans With a Theory of Meaning*, trans. J. D. O'Neil. Minneapolis: University of Minnesota Press, 2010.

van Inwagen, Peter, Meghan Sullivan and Sara Bernstein. 'Metaphysics', *The Stanford Encyclopedia of Philosophy* (Summer 2023 edition), Edward N. Zalta and Uri Nodelman, eds. https://plato.stanford.edu/archives/sum2023/entries/metaphysics/.

Venturi, Robert. *Complexity and Contradiction in Architecture*, 2nd edition. New York: Museum of Modern Art, 1977.

Venturi, Robert, Denise Scott Brown and Steven Izenour. *Learning From Las Vegas: The Forgotten Symbolism of Architectural Form*, revised edition. Cambridge, MA: MIT Press, 1977.

Vitruvius. *The Ten Books on Architecture*. Mineola, NY: Dover, 1960.

Vrba, Elisabeth S. and Stephen Jay Gould. 'The Hierarchical Expansion of Sorting and Selection: Sorting and Selection Cannot Be Equated', *Paleobiology* 12.2 (Spring 1986), pp. 217–28.

Wallis, John. *A Treatise of Algebra, Both Theoretical and Practical*. London: Richard Davis, 1685.

Warman, Matt. 'Stephen Hawking Tells Google "Philosophy is Dead"', *Telegraph*, 17 May 2011. https://www.telegraph.co.uk/technology/google/8520033/Stephen-Hawking-tells-Google-philosophy-is-dead.html?. Last accessed on 25 March 2025.

Watkin, William. *Badiou and Communicable Worlds: A Critical Introduction to Logics of Worlds*. London: Bloomsbury, 2021.

Watson, James D. *The Double Helix: A Personal Account of the Discovery of the Structure of DNA*. New York: Scribner, 2011.

Whitehead, Alfred North. *Process and Reality*. New York: Free Press, 1978.

Whorf, Benjamin Lee. *Language, Thought, and Reality: Selected Writings of Benjamin Lee Whorf*, J. B. Carroll, ed. Cambridge, MA: MIT Press, 1956.

Wigley, Mark. *The Architecture of Deconstruction: Derrida's Haunt*. Cambridge, MA: MIT Press, 1993.

Williams, George C. *Natural Selection: Domains, Levels, and Challenges*. Oxford: Oxford University Press, 1992.

—— *Sex and Evolution*. Princeton, NJ: Princeton University Press, 1975.

Williams, S. T., N. Knowlton, L. A. Weigt and J. A. Jara. 'Evidence for Three Major Clades Within the Snapping Shrimp Genus *Alpheus* Inferred from Nuclear and Mitochondrial Gene Sequence Data', *Molecular Phylogenetics and Evolution* 20.3 (September 2001), pp. 375–89.

Williamson, Timothy. *Vagueness*. London: Routledge, 1996.

Wilson, David Sloan and Elliott Sober. 'Reviving the Superorganism', *Journal of Theoretical Biology* 136 (1989), pp. 337–56.

Witten, Edward. 'String Theory Dynamics in Various Dimensions', *Nuclear Physics B* 443.1 (1995), pp. 381–402.

Wittgenstein, Ludwig. *Philosophical Investigations*, trans. G. E. M. Anscombe, P. M. S. Hacker, and J. Schulte. Oxford: Blackwell Scientific, 1980.

—— *Tractatus Logico-Philosophicus*, trans. C. K. Ogden. New York: Harcourt, Brace and Company, 1922.

Woese, Carl R. 'A New Biology for a New Century', *Microbiology and Molecular Biology Reviews* 68.2 (2004), pp. 173–86.

Woit, Peter. *Not Even Wrong: The Failure of String Theory and the Search for Unity in Physical Law*. New York: Basic Books 2007.

Wright, Sewall. 'Evolution in Mendelian Populations', *Genetics* 16 (1931), pp. 97–159.

Xenophon. *Conversations of Socrates*, trans. R. Waterfield. London: Penguin, 1990.

Young, Niki. 'Object, Reduction, and Emergence: An Object-Oriented View', *Open Philosophy* 4 (2021), pp. 83–93.

Zeilinger, Anton. *Dance of the Photons: From Einstein to Quantum Teleportation*. New York: Farar, Straus and Giroux, 2010.

Works Cited

Zeller, Eduard. *Outlines of the History of Greek Philosophy*, trans. S. F. Alleyne and E. Abbott. Mineola, NY: Dover, 1980.

Žižek, Slavoj. 'Philosophy, the "Unknown Knowns", and the Public Use of Reason', *Topoi* 25 (2006), pp. 137–42.

—— *Tarrying with the Negative: Kant, Hegel, and the Critique of Ideology*. Durham, NC: Duke University Press, 1993.

Notes

Prologue: The Continuous and the Discrete

1 Nikos Kazantzakis, *The Last Temptation of Christ*; Nikos Kazantzakis, *Zorba the Greek*.
2 Nikos Kazantzakis, *The Odyssey: A Modern Sequel*, p. 933.
3 Yukio Mishima, *Runaway Horses*, p. 6.
4 See Judith Butler, *Gender Trouble*.
5 Maurizio Ferraris, *Hysteresis*, p. 296. The reference is to the close of G. W. Leibniz, *New Essays on Human Understanding*.

Chapter 1: Skipping Stones: The Riddle of Thixis

1 See Graham Harman, *Object-Oriented Ontology*.
2 See Graham Harman, *Tool-Being*. For a simplified treatment of the problem see Graham Harman, *Heidegger Explained*.
3 Martin Heidegger, *Being and Time*.
4 Timothy Morton, personal communication, 14 November 2011.
5 Alexius Meinong, *On Assumptions*; Dale Jacquette, *Alexius Meinong, the Shepherd of Non-Being*. I am indebted to Tristan Garcia for this way of describing the difference between Meinong and OOO.
6 See Graham Harman, 'On the Undermining of Objects'.
7 C. S. Peirce, *Philosophical Writings of Peirce*, pp. 23–40.
8 One classic work on emergence is C. D. Broad, *The Mind and Its Place in Nature*.
9 ChatGPT query, 7 January 2025.
10 Douglas Hofstadter, *Gödel, Escher, Bach*.
11 Alfred North Whitehead, *Process and Reality*.
12 See Graham Harman, 'Undermining, Overmining, and Duomining'.

13 Bruno Latour, *Aramis*, p. 52.

14 Carl R. Woese, 'A New Biology for a New Century', p. 176.

15 Thomas Nail, *Being and Motion*, p. 6, Kindle edition; Thomas Nail, *The Figure of the Migrant*.

16 Nail, *Being and Motion*, p. 2, Kindle edition.

17 Rein Raud, *Being in Flux*.

18 I am indebted to Ian Bogost of Washington University in St Louis for the firehose image.

Chapter 2: Physics and Metaphysics: Aristotle's Hidden Duality

1 Julián Marías, *History of Philosophy*, p. 372.

2 Bertrand Russell, *History of Western Philosophy*, p. 147; Aristotle, *Nicomachean Ethics*.

3 Robert M. Pirsig, *Zen and the Art of Motorcycle Maintenance*.

4 Aristotle, *De Anima*, p. 76.

5 Aristotle, *The Art of Rhetoric*, 1362a.

6 W. D. Ross, *Aristotle*, p. 7.

7 Aristotle, *Metaphysics*, p. 120.

8 See Eduard Zeller, *Outlines of the History of Greek Philosophy*.

9 R. Lanier Anderson, 'Friedrich Nietzsche'.

10 Hesiod, *Theogony and Works and Days*.

11 Karl Marx, *The First Writings of Karl Marx*.

12 Parmenides of Elea, *Fragments*.

13 Plato, *Parmenides*.

14 René Descartes, *Meditations on First Philosophy*, p. 53.

15 Plato, 'Crito'.

16 Xenophon, *Conversations of Socrates*; Aristophanes, 'The Clouds'.

17 The one dialogue in which Socrates does not appear is Plato, *The Laws*.

18 Aristotle, *Physics*.

19 See for instance Averroës (Ibn Rushd), *Long Commentary on the De Anima of Aristotle*.

20 Peter van Inwagen, Meghan Sullivan and Sara Bernstein, 'Metaphysics'.

21 Étienne Gilson, *The Philosophy of St. Thomas Aquinas*, pp. 16–17.

22 See for instance Lloyd P. Gerson, *Aristotle and Other Platonists*.

23 Aristotle, *Physics*, p. 4.

24 Aristotle, *Physics*, p. 45.

25 Aristotle, *Physics*, p. 49.

26 On the modern replacement of this theory see Alexandre Koyré, *From the Closed World to the Infinite Universe*.

27 Georg Cantor, *Contributions to the Founding of the Theory of Transfinite Numbers*.

28 Aristotle, *Physics*, pp. 52–3.

29 Aristotle, *Metaphysics*, p. 189.

30 Aristotle, *Metaphysics*, p. 199.

31 Aristotle, *Physics*, p. 164.

32 Aristotle, *Physics*, p. 164; emphasis added.

33 Aristotle, *Physics*, p. 120.

34 Euclid, *The Thirteen Books of the Elements*, p. 153.

35 Aristotle, *Physics*, p. 102.

36 Aristotle, *Physics*, p. 76.

37 Aristotle, *Metaphysics*, p. 145.

38 Aristotle, *Metaphysics*, p. 145.

39 Bruno Latour, *Reassembling the Social*. For an admiring but critical commentary see Graham Harman, *Prince of Networks*.

40 Bruno Latour, *Pandora's Hope*, p. 122.

41 Fernand Braudel, *Civilization and Capitalism*, 3 vols.

42 For another interesting defence of the Megarians see Nicolai Hartmann, 'The Megarian and the Aristotelian Concept of Potentiality'.

43 Edward Gibbon, *The Decline and Fall of the Roman Empire*, 3 vols.

44 See Graham Harman, 'Undermining, Overmining, and Duomining'.

45 Aristotle, *Metaphysics*, p. 79.

46 Aristotle, *Metaphysics*, p. 141.

47 Aristotle, *Metaphysics*, p. 132.

48 See Manuel DeLanda, 'Emergence, Causality, and Realism'; Niki Young, 'Object, Reduction, and Emergence'.

49 Aristotle, *Metaphysics*, p. 105.

50 Aristotle, *Metaphysics*, p. 82.

51 Thomas Nail, 'Is Nature Continuous or Discrete?'.

52 Aristotle, *Metaphysics*, p. 131.

53 G. W. Leibniz, *Theodicy*.

54 See René Descartes, *The Passions of the Soul*.

Chapter 3: Dynasty: From Ibn Khaldun to Mongolia

1 Gaston Bachelard, *The Formation of the Scientific Mind*.

2 Louis Althusser, *For Marx*; Michel Foucault, *History of Madness*.

3 On these topics see, respectively, Foucault, *History of Madness*; Michel Focuault, *Discipline and Punish*; Michel Foucault, *The Birth of the Clinic*.

4 Bruno Latour, *We Have Never Been Modern*.

5 See Graham Harman, 'Concerning the COVID-19 Event'.

6 See Annie Jacobsen, *Nuclear War: A Scenario*.

7 Ibn Khaldun, *The Muqaddimah*.

8 A. J. Toynbee, *A Study of History*, vol. 3: *The Growths of Civilizations*, p. 322.

9 G. W. F. Hegel, *Phenomenology of Spirit*.

10 Letter of Hegel to Niethammer, 13 October 1806, in *Hegel: The Letters*.

11 René Grousset, *The Empire of the Steppes*, p. 477.

12 See Bruce B. Lawrence, 'Introduction to the 2005 Edition' of Ibn Khaldun, *The Muqaddimah*, pp. vii–xxv.

13 Ibn Khaldun, *The Muqaddimah*, p. 105.

14 Ibn Khaldun, *The Muqaddimah*, p. 105.

15 Ibn Khaldun, *The Muqaddimah*, p. 105.

16 Ibn Khaldun, *The Muqaddimah*, p. 105.

17 Ibn Khaldun, *The Muqaddimah*, p. 106.

18 Cited in Grousset, *The Empire of the Steppes*, p. 249.

19 Ibn Khaldun, *The Muqaddimah*, p. 25.

20 Ibn Khaldun, *The Muqaddimah*, p. 175.

21 On the Mamelukes see Ibn Khaldun, *The Muqaddimah*, p. 135.

22 Ibn Khaldun, *The Muqaddimah*, p. 106.

23 Ibn Khaldun, *The Muqaddimah*, p. 106. Formatting altered.

24 For more on this theme see José Ortega y Gasset, *Mission of the University*.

25 Ibn Khaldun, *The Muqaddimah*, p. 110.

26 José Ortega y Gasset, *Man and Crisis* ; Julián Marías, *Generations*.

27 Ibn Khaldun, *The Muqaddimah*, p. 161.

28 Ibn Khaldun, *The Muqaddimah*, p. 162.

29 Ibn Khaldun, *The Muqaddimah*, p. 191.

30 Anonymous, *The Book of the Thousand and One Nights*, 4 vols.

31 Ibn Khaldun, *The Muqaddimah*, p. 115.

32 Ibn Khaldun, *The Muqaddimah*, p. 135.

33 Ibn Khaldun, *The Muqaddimah*, p. 176.

34 Ibn Khaldun, *The Muqaddimah*, p. 250.

35 Ibn Khaldun, *The Muqaddimah*, pp. 255, 256.

36 Ibn Khaldun, *The Muqaddimah*, p. 244.

37 Ibn Khaldun, *The Muqaddimah*, p. 249; Niccolò Machiavelli, *The Prince*, §12.

38 Ibn Khaldun, *The Muqaddimah*, pp. 232–3.

39 Ibn Khaldun, *The Muqaddimah*, pp. 238–40.

40 Ibn Khaldun, *The Muqaddimah*, p. 205.

41 Ibn Khaldun, *The Muqaddimah*, p. 43.

42 Ibn Khaldun, *The Muqaddimah*, p. 245.

43 Ibn Khaldun, *The Muqaddimah*, p. 129.

44 Ibn Khaldun, *The Muqaddimah*, p. 252.

45 Ibn Khaldun, *The Muqaddimah*, p. 252.

46 Grousset, *The Empires of the Steppe*, pp. xi, 474. See also Isabel de Madariaga, *Ivan the Terrible*.

47 Ibn Khaldun, *The Muqaddimah*, pp. 92–3.

48 Ibn Khaldun, *The Muqaddimah*, p. 94.

49 Plato, *Republic*; Abu Nasr Muhammad al-Farabi, *On the Perfect State*.

50 Machiavelli, *The Prince*; Thomas Hobbes, *Leviathan*.

51 Ibn Khaldun, *The Muqaddimah*, p. 94.

52 Ibn Khaldun, *The Muqaddimah*, p. 117.

53 Gibbon, *The Decline and Fall of the Roman Empire*, Chapter 1.

54 Ibn Khaldun, *The Muqaddimah*, p. 224.

55 An early anticipation of this reversal into pre-modern forms of warfare can be found in a 1989 article by William S. Lind et al., 'The Changing

Face of War'. For a philosophical assessment see Graham Harman, 'War, Space, and Reversal'.

56 Ibn Khaldun, *The Muqaddimah*, p. 224.

57 Ibn Khaldun, *The Muqaddimah*, p. 118.

58 Ibn Khaldun, *The Muqaddimah*, p. 118.

59 Ibn Khaldun, *The Muqaddimah*, p. 93.

60 René Grousset, *The Empire of the Steppes*.

61 In this section, I will generally follow Grousset's spellings of places and proper names rather than contemporary scholarly usage.

62 Grousset, *The Empire of the Steppes*, p. xxv.

63 Herodotus, *The Histories*; Polybius, *The Histories*.

64 Grousset, *The Empire of the Steppes*, p. 9.

65 Grousset, *The Empire of the Steppes*, p. 32.

66 John King Fairbank and Merle Goldman, *China: A New History*, p. 56, Kindle edition.

67 Grousset, *The Empire of the Steppes*, p. 195.

68 Grousset, *The Empire of the Steppes*, p. 195.

69 Grousset, *The Empire of the Steppes*, p. 196.

70 Grousset, *The Empire of the Steppes*, p. 199.

71 Grousset, *The Empire of the Steppes*, p. 199.

72 Rashid al-Din, *The Successors of Genghis Khan*; Anonymous, *The Secret History of the Mongols*.

73 Grousset, *The Empire of the Steppes*, p. 200.

74 Grousset, *The Empire of the Steppes*, pp. 207, 208.

75 Grousset, *The Empire of the Steppes*, p. 208.

76 Grousset, *The Empire of the Steppes*, p. 209.

77 Grousset, *The Empire of the Steppes*, p. 211.

78 Grousset, *The Empire of the Steppes*, p. 213.

79 Grousset, *The Empire of the Steppes*, p. 216.

80 Grousset, *The Empire of the Steppes*, p. 249.

81 Grousset, *The Empire of the Steppes*, p. 250.

82 Grousset, *The Empire of the Steppes*, pp. 360–61.

83 Grousset, *The Empire of the Steppes*, pp. 243–4.

84 Grousset, *The Empire of the Steppes*, pp. 226–33.

85 Grousset, *The Empire of the Steppes*, pp. 233–6.

86 Grousset, *The Empire of the Steppes*, pp. 233–44.

87 Grousset, *The Empire of the Steppes*, p. 241.

88 Grousset, *The Empire of the Steppes*, pp. 264–8.

89 Grousset, *The Empire of the Steppes*, p. 356.

90 Grousset, *The Empire of the Steppes*, pp. 364–5.

91 Grousset, *The Empire of the Steppes*, p. 288.

92 Marco Polo, *The Travels*.

93 Grousset, *The Empire of the Steppes*, pp. 293, 304.

94 Grousset, *The Empire of the Steppes*, p. 323.

95 Ibn Khaldun, *The Muqaddimah*, pp. 118, 93.

96 Jack London, *Novels and Stories*.

97 Grousset, *The Empire of the Steppes*, p. 414.

98 Grousset, *The Empire of the Steppes*, pp. 443–6.

99 Grousset, *The Empire of the Steppes*, p. 423.

100 Grousset, *The Empire of the Steppes*, pp. 414–15.

101 Grousset, *The Empire of the Steppes*, p. 416.

102 Grousset, *The Empire of the Steppes*, pp. 421, 441.

103 Grousset, *The Empire of the Steppes*, p. 420.

104 Grousset, *The Empire of the Steppes*, pp. 453–6.

105 Grousset, *The Empire of the Steppes*, p. 421.

106 Grousset, *The Empire of the Steppes*, p. 428.

107 Grousset, *The Empire of the Steppes*, p. 428.

108 Grousset, *The Empire of the Steppes*, p. 428.

109 Grousset, *The Empire of the Steppes*, p. 429.

110 Grousset, *The Empire of the Steppes*, p. 430.

111 Grousset, *The Empire of the Steppes*, p. 431.

112 Grousset, *The Empire of the Steppes*, p. 431.

113 Grousset, *The Empire of the Steppes*, p. 432.

114 Grousset, *The Empire of the Steppes*, p. 442.

115 Grousset, *The Empire of the Steppes*, p. 445.

116 Grousset, *The Empire of the Steppes*, p. 446.

117 Grousset, *The Empire of the Steppes*, p. 434.

118 Grousset, *The Empire of the Steppes*, p. 452.

119 Grousset, *The Empire of the Steppes*, p. 451.

120 Lao Tzu, *Tao Te Ching*.

121 Immanuel Kant, *Critique of Pure Reason*.

122 G. W. F. Hegel, *Science of Logic*.

123 For an influential interpretation see Alexandre Kojève, *Introduction to the Reading of Hegel*.

124 Karl Marx, *Capital: Volume One*.

125 Slavoj Žižek, 'Philosophy, the "Unknown Knowns", and the Public Use of Reason'.

Chapter 4: Flows and Instants: Bergson and the Occasionalists

1 Emmanuel Levinas, *Existence and Existents*.

2 Majid Fakhry, *Islamic Occasionalism*.

3 Abu Hamid al-Ghazali, *Deliverance from Error*. For a clear and brief summary of Ash'arīte doctrine, see Peter Adamson, *Philosophy in the Islamic World*, pp. 106–12. There is some scholarly controversy over whether al-Ghazali should be called an occasionalist, but for me the case is clear.

4 Whitehead, *Process and Reality*; Bruno Latour, 'Irreductions'. See also Graham Harman, 'A New Occasionalism?'.

5 Nicolas Malebranche, *The Search After Truth*, Book III.

6 See Latour, *Pandora's Hope*, pp. 80–112.

7 Bertrand Russell, *Mysticism and Logic*, p. 402. For similar thoughts see also Bertrand Russell, 'The Philosophy of Bergson', p. 339.

8 Heidegger, *Being and Time*; Emmanuel Levinas, *Ethics and Infinity*, pp. 37–8.

9 Levinas, *Ethics and Infinity*, pp. 37–8; Plato, 'Phaedrus'; Kant, *Critique of Pure Reason*; Hegel, *Phenomenology of Spirit*; Henri Bergson, *Time and Free Will*.

10 Plato, *Meno*.

11 Henri Bergson, *Matter and Memory*; Henri Bergson, *Creative Evolution*; Henri Bergson, *The Two Sources of Morality and Religion*.

12 From the Nobel Prize website, https://www.nobelprize.org/prizes/literature/1927/summary/, last accessed on 25 February 2024.

13 Jimena Canales, *The Physicist and the Philosopher*.

14 Gilles Deleuze, *Bergsonism*.

15 Bergson, *Time and Free Will*, p. 108.

16 Bergson, *Time and Free Will*, p. 131.

17 Bergson, *Time and Free Will*, p. 164.

18 Bergson, *Time and Free Will*, pp. 164–5.

19 Marcel Proust, *In Search of Lost Time*, 6 vols.

20 Bergson, *Time and Free Will*, p. 1.

21 Bergson, *Time and Free Will*, p. 25.

22 Cited in Bergson, *Time and Free Will*, p. 27.

23 Cited in Bergson, *Time and Free Will*, p. 28.

24 Bergson, *Time and Free Will*, p. 40.

25 Bergson, *Time and Free Will*, p. xix.

26 Bergson, *Time and Free Will*, p. 104.

27 Bergson, *Time and Free Will*, p. 166.

28 Bergson, *Time and Free Will*, p. 135.

29 Bergson, *Time and Free Will*, p. 132.

30 Bergson, *Time and Free Will*, p. 166.

31 Heidegger, *Being and Time*, pp. 163–8.

32 Bergson, *Time and Free Will*, p. 97.

33 Bergson, *Time and Free Will*, p. 96.

34 Bergson, *Time and Free Will*, p. 96.

35 Bergson, *Time and Free Will*, p. 143.

36 For a critique of this assumption see Graham Harman, 'The Only Exit from Modern Philosophy'. My favourite dismantling of the underlying framework of modern philosophy can be found in Bruno Latour, *We Have Never Been Modern*.

37 Bergson, *Time and Free Will*, pp. 167–8.

38 Bergson, *Time and Free Will*, p. 169.

39 Bergson, *Time and Free Will*, p. 169.

40 Richard Rhodes, *The Making of the Atomic Bomb*, p. 218.

41 Enrico Fermi and Franco Rasetti. 'Slow Neutrons'.

42 Bergson, *Time and Free Will*, p. 170.

43 Bergson, *Time and Free Will*, p. 176.

44 Bergson, *Time and Free Will*, p. 75.

45 Bergson, *Time and Free Will*, p. 84.

46 Bergson, *Time and Free Will*, p. 82.

47 Bergson, *Time and Free Will*, p. 82.

48 Bergson, *Time and Free Will*, p. 86.

49 Hermann Minkowski, 'Raum und Zeit'.

50 Bergson, *Time and Free Will*, p. 110.

51 Bergson, *Time and Free Will*, p. 110.

52 Bergson, *Time and Free Will*, p. 110. See also Henri Bergson, *Duration and Simultaneity*.

53 Bergson, *Time and Free Will*, p. 109.

54 Bergson, *Time and Free Will*, p. 228.

55 Concerning an-Nazzam see Adamson, *Philosophy in the Islamic World*, pp. 16–17.

56 Bergson, *Time and Free Will*, pp. 120–21.

57 C. P. Snow, *The Two Cultures*.

58 Bergson, *Time and Free Will*, p. 115.

59 Martin Heidegger, *What Is Called Thinking?*, p. 8.

60 Aristotle, *De Anima (On the Soul)*.

61 Thomas Aquinas, *Summa Theologica*, Q. 18, Art. 2, Part 1.

62 Descartes, *Meditations on First Philosophy*.

63 Martin Heidegger, *Fundamental Concepts of Metaphysics*.

64 Helmuth Plessner, *Levels of Organic Life and the Human*. For a critical commentary see Graham Harman, 'On the Punctuation of Organisms'.

65 Quentin Meillassoux, 'Appendix: Excerpts from *L'Inexistence divine*', p. 238.

66 Bergson, *Time and Free Will*, p. 138.

67 Bergson, *Time and Free Will*, p. 116.

68 Jakob von Uexküll, *A Foray Into the World of Animals and Humans*, p. 70. See also Graham Harman, 'Magic Uexküll'.

69 It is difficult to imagine Heidegger offering similar praise for abstract spatialization in *Being and Time*, where he consistently emphasizes the primacy of the kind of space that guides lived experience: for instance, the glasses on my face are in some sense further away from me than the friend glimpsed two blocks away.

70 Bergson, *Time and Free Will*, p. 87.

Chapter 5: Paradigm Shift: On Scientific Revolutions

1 Thomas Kuhn, *The Structure of Scientific Revolutions*, p. 2.
2 Kuhn, *The Structure of Scientific Revolutions*, p. 2.
3 Thomas Kuhn, *The Road Since Structure*, pp. 15–20.
4 See Alexandre Koyré, *Galileo Studies*.
5 Kuhn, *The Structure of Scientific Revolutions*, p. 90.
6 Kuhn, *The Structure of Scientific Revolutions*, p. 92.
7 Kuhn, *The Structure of Scientific Revolutions*, pp. 85, 122.
8 Kuhn, *The Structure of Scientific Revolutions*, p. 114. See also Ludwig Wittgenstein, *Philosophical Investigations*, pp. 194–206.
9 Kuhn, *The Structure of Scientific Revolutions*, p. 10.
10 Kuhn, *The Structure of Scientific Revolutions*, p. 11.
11 Kuhn, *The Structure of Scientific Revolutions*, p. 44.
12 Kuhn, *The Structure of Scientific Revolutions*, p. 51.
13 Kuhn, *The Structure of Scientific Revolutions*, p. 11.
14 Kuhn, *The Structure of Scientific Revolutions*, p. 41.
15 Kuhn, *The Structure of Scientific Revolutions*, p. 25.
16 Kuhn, *The Structure of Scientific Revolutions*, p. 15.
17 See James D. Watson, *The Double Helix*.
18 Kuhn, *The Structure of Scientific Revolutions*, p. 15.
19 See the second half of Francis Bacon, *The New Organon*.
20 Bruno Latour, *The Pasteurization of France*.
21 Kuhn, *The Structure of Scientific Revolutions*, p. 13.
22 Kuhn, *The Structure of Scientific Revolutions*, p. 15.
23 Kuhn, *The Structure of Scientific Revolutions*, p. 18.
24 Max Born, *The Born-Einstein Letters*.
25 Aristotle, *Physics*; Ptolemy, *Ptolemy's Almagest*; Isaac Newton, *The Principia*; Antoine Lavoisier, *Elements of Chemistry*; James Clerk Maxwell, *A Treatise on Electricity and Magentism*, vol. 1.
26 Albert Einstein, 'Zur Elektrodynamik bewegter Körper'; Albert Einstein, 'Die Grundlage der allgemeinen Relativitätstheorie'.
27 Kuhn, *The Structure of Scientific Revolutions*, p. 36.
28 Kuhn, *The Structure of Scientific Revolutions*, p. 37.

29 Kuhn, *The Structure of Scientific Revolutions*, p. 37.

30 Kuhn, *The Structure of Scientific Revolutions*, p. 38.

31 Kuhn, *The Structure of Scientific Revolutions*, p. 53.

32 Kuhn, *The Structure of Scientific Revolutions*, p. 65.

33 Kuhn, *The Structure of Scientific Revolutions*, p. 81. For more detail see Jim Baggott, *Higgs*.

34 Kuhn, *The Structure of Scientific Revolutions*, p. 80.

35 Kuhn, *The Structure of Scientific Revolutions*, p. 68.

36 See Thomas Kuhn, *The Copernican Revolution*.

37 Kuhn, *The Structure of Scientific Revolutions*, p. 69.

38 Kuhn, *The Structure of Scientific Revolutions*, p. 75.

39 Kuhn, *The Structure of Scientific Revolutions*, p. 71.

40 Kuhn, *The Structure of Scientific Revolutions*, p. 72.

41 Mordehai Milgrom, 'A Modification of the Newtonian Dynamics as an Alternative to the Hidden Mass Hypothesis'.

42 Kuhn, *The Structure of Scientific Revolutions*, p. 77.

43 Kuhn, *The Structure of Scientific Revolutions*, p. 78.

44 Kuhn, *The Structure of Scientific Revolutions*, p. 78.

45 Samuel Clarke and G. W. Leibniz, *Correspondence*.

46 Kuhn, *The Structure of Scientific Revolutions*, p. 55.

47 Kuhn, *The Structure of Scientific Revolutions*, p. 56.

48 The case of Lenard is covered in Uwe Busch, 'Claims of Priority'.

49 Kuhn, *The Structure of Scientific Revolutions*, p. 58.

50 Kuhn, *The Structure of Scientific Revolutions*, p. 59.

51 Kuhn, *The Structure of Scientific Revolutions*, p. 149.

52 Thomas Kuhn, *Black-Body Theory and the Quantum Discontinuity, 1894–1912*.

53 See Max Planck, 'Über das Gesetz der Energieverteilung im Normalspektrum'.

54 Paul Ehrenfest, 'Über die physikalischen Voraussetzungen der Planck'schen Theorie der irreversiblen Strahlungsvorgänge'; Albert Einstein, 'Die Plancksche Theorie der Strahlung und die Theorie der spezifischen Wärme'.

55 Kuhn, *The Road Since Structure*, p. 204.

56 Kuhn, *The Road Since Structure*, p. 204.

57 Kuhn, *The Structure of Scientific Revolutions*, p. 102.

58 Kuhn, *The Structure of Scientific Revolutions*, p. 149.

59 Kuhn, *The Road Since Structure*, p. 205.

60 Saul Kripke, *Naming and Necessity*; Hilary Putnam, 'The Meaning of Meaning'; Keith Donnellan, 'Reference and Definite Descriptions'.

61 Max Black, 'Metaphor.'

62 Kuhn, *The Road Since Structure*, p. 102.

63 Harman, *Object-Oriented Ontology*, pp. 61–94.

64 Kuhn, *The Road Since Structure*, p. 197.

65 Kuhn, *The Road Since Structure*, p. 203.

66 Kuhn, *The Structure of Scientific Revolutions*, p. 151.

67 Kuhn, *The Road Since Structure*, p. 207.

68 Kuhn, *The Road Since Structure*, p. 50.

69 David Hume, *An Enquiry Concerning Human Understanding*.

70 Michael Polanyi, *Personal Knowledge*.

71 Kuhn, *The Structure of Scientific Revolutions*, p. 158.

72 J. L. Austin, *How to Do Things With Words*.

73 While this is not the place to engage with his views on the matter, see also Jacques Derrida, *Margins of Philosophy*, pp. 309–30.

74 See Konstantin Stanislavski, *An Actor's Work*.

75 Karl Popper, *The Logic of Scientific Discovery*, p. 418.

76 Alain Badiou, *Lacan: Anti-Philosophy 3*; Blaise Pascal, *Pensées*; Søren Kierkegaard, *Either/Or, Part I*; Friedrich Nietzsche, *Ecce Homo*.

77 Alain Badiou, *Being and Event*.

78 See also Graham Harman, 'Retroactivity in Science'.

79 Kuhn, *The Road Since Structure*, p. 202.

Chapter 6: Creeps and Jerks: Evolutionary Theory

1 John Turner, 'Why We Need Evolution by Jerks', p. 34; Steve Jones, 'A Wonderful Life by Leaps and Bounds', p. 873.

2 Stephen Jay Gould, *The Structure of Evolutionary Theory*, p. 674.

3 Gould, *The Structure of Evolutionary Theory*, pp. 675–6.

4 Richard Dawkins, *The Selfish Gene*.

5 Carl Gustav Jung, *The Portable Jung*, p. 166.

6 Niles Eldredge and Stephen Jay Gould, 'Punctuated Equilibria'.

7 Charles Darwin, *The Origin of Species*, p. 314, Kindle edition.

8 This is discussed in Stephen Toulmin, 'Does the Distinction Between Normal and Revolutionary Hold Water?', pp. 41–3.

9 Hugo de Vries, *Die Mutationstheorie*.

10 Julian Huxley, *Evolution*; Gregor Mendel, 'Experiments in Plant Hybridization'.

11 Eldredge and Gould, 'Punctuated Equilibria', p. 90.

12 Eldredge and Gould, 'Punctuated Equilibria', p. 89.

13 G. Neef, 'Notes on the Subgenus *Pelicaria*'. Cited in Eldredge and Gould, 'Punctuated Equilibria', p. 90.

14 Kuhn, *The Structure of Scientific Revolutions*, p. 5. Cited in Eldredge and Gould, 'Punctuated Equilibria', p. 91.

15 Eldredge and Gould, 'Punctuated Equilibria', p. 93.

16 Eldredge and Gould, 'Punctuated Equilibria', pp. 93–4.

17 Léon Croizat, *Panbiogeography*.

18 Ernst Mayr, *Populations, Species, and Evolution*.

19 Eldredge and Gould, 'Punctuated Equilibria', p. 94.

20 See for example S. T. Williams et al., 'Evidence for Three Major Clades within the Snapping Shrimp Genus *Alpheus* Inferred from Nuclear and Mitochondrial Gene Sequence Data'.

21 Mark P. Miller et al., 'Genetic Differentiation and Inferred Dynamics of a Hybrid Zone Between Northern Spotted Owls (*Strix occidentalis caurina*) and California Spotted Owls (*S. o. occidentalis*) in Northern California', p. 8.

22 Miller er al., 'Genetic Differentiation and Inferred Dynamics of a Hybrid Zone Between Northern Spotted Owls (*Strix occidentalis caurina*) and California Spotted Owls (*S. o. occidentalis*) in Northern California', p. 9.

23 Eldredge and Gould, 'Punctuated Equilibria', p. 95.

24 Eldredge and Gould, 'Punctuated Equilibria', p. 95.

25 Eldredge and Gould, 'Punctuated Equilibria', p. 97.

26 Eldredge and Gould, 'Punctuated Equilibria', pp. 99–104.

27 See Peter Brannen, *The Ends of the World*.

28 Eldredge and Gould, 'Punctuated Equilibria', p. 108.

29 Eldredge and Gould, 'Punctuated Equilibria', p. 112.

30 Eldredge and Gould, 'Punctuated Equilibria', p. 114.

31 Gould, *The Structure of Evolutionary Theory*, p. 691.

32 Gould, *The Structure of Evolutionary Theory*, p. 684.

33 Gould, *The Structure of Evolutionary Theory*, p. 686.

34 Sewall Wright, 'Evolution in Mendelian Populations'.

35 Motoo Kimura, 'Evolutionary Rate at the Molecular Level'; Gould, *The Structure of Evolutionary Theory*, p. 686.

36 Gould, *The Structure of Evolutionary Theory*, p. 689.

37 Gould, *The Structure of Evolutionary Theory*, p. 695.

38 Gould, *The Structure of Evolutionary Theory*, p. 695.

39 Lauren Serafin, 'Surviving Peritoneal Mesothelioma'.

40 Gould, *The Structure of Evolutionary Theory*, p. 695.

41 Gould, *The Structure of Evolutionary Theory*, p. 699.

42 Daniel Dennett, *Darwin's Dangerous Idea*; Gould, *The Structure of Evolutionary Theory*, p. 710.

43 Gould, *The Structure of Evolutionary Theory*, p. 701.

44 Gould, *The Structure of Evolutionary Theory*, p. 703.

45 Gould, *The Structure of Evolutionary Theory*, p. 682.

46 Gould, *The Structure of Evolutionary Theory*, pp. 648, 678.

47 Gould, *The Structure of Evolutionary Theory*, p. 131; Charles Darwin, *The Descent of Man*.

48 Gould, *The Structure of Evolutionary Theory*, p. 701.

49 Gould, *The Structure of Evolutionary Theory*, p. 703. On the decoupling of macroevolution from microevolution, and hence the discontinuity between the two, Gould often cites Steven M. Stanley, 'A Theory of Evolution Above the Species Level'.

50 Gould, *The Structure of Evolutionary Theory*, p. 705–9; see also David Jablonski, 'Heritability at the Species Level'.

51 Gould, *The Structure of Evolutionary Theory*, p. 710.

52 Gould, *The Structure of Evolutionary Theory*, p. 712.

53 Gould, *The Structure of Evolutionary Theory*, p. 713.

54 George C. Williams, *Sex and Evolution*.

55 David L. Hull, 'Individuality and Selection'.

56 Gould, *The Structure of Evolutionary Theory*, p. 619.

57 See Wallace Arthur, *Understanding Evo-Devo*.

58 Dawkins, *The Selfish Gene*, p. 39.

59 Gould, *The Structure of Evolutionary Theory*, p. 615.

60 Gould, *The Structure of Evolutionary Theory*, p. 672.

61 Gould, *The Structure of Evolutionary Theory*, p. 616.

62 Gould, *The Structure of Evolutionary Theory*, p. 621.

63 Gould, *The Structure of Evolutionary Theory*, p. 623. He cites Elliott Sober and David Sloan Wilson, *Unto Others*; Roger Lewin, 'Evolution's New Heretics'; Stephen Jay Gould and Elisabeth A. Lloyd, 'Individuality and Adaptation Across Levels of Selection'.

64 Todd A. Grantham, 'Hierarchical Approaches to Macroevolution'.

65 Stephen Jay Gould and Elisabeth S. Vrba, 'Exaptation'; Elisabeth S. Vrba and Stephen Jay Gould. 'The Hierarchical Expansion of Sorting and Selection'; Elisabeth A. Lloyd and Stephen Jay Gould. 'Species Selection on Variability'; Gould and Lloyd. 'Individuality and Adaptation Across Levels of Selection'.

66 Gould, *The Structure of Evolutionary Theory*, p. 657.

67 This example comes from Gould, *The Structure of Evolutionary Theory*, p. 708.

68 Gould, *The Structure of Evolutionary Theory*, p. 658.

69 Gould, *The Structure of Evolutionary Theory*, p. 658.

70 Gould, *The Structure of Evolutionary Theory*, pp. 655–6.

71 Gould and Vrba, 'Exaptation'.

72 Gould, *The Structure of Evolutionary Theory*, p. 1277.

73 Gould, *The Structure of Evolutionary Theory*, p. 1278.

74 Stephen Jay Gould and Richard Lewontin, 'The Spandrels of San Marco and the Panglossian Paradigm'.

75 Richard Dawkins, 'Parasites, Desiderata Lists, and the Paradox of the Organism'.

76 Gould, *The Structure of Evolutionary Theory*, p. 700. See also Benoit B. Mandelbrot, *Fractals*.

77 Gould, *The Structure of Evolutionary Theory*, p. 680.

78 W. Ford Doolittle, 'Phylogenetic Classification and the Universal Tree', p. 2124.

79 Gould, *The Structure of Evolutionary Theory*, p. 683.

80 Lynn Margulis, *Symbiotic Planet*.

81 Lynn Margulis, *Origin of Eukaryotic Cells*.

82 David Sloan Wilson and Elliott Sober, 'Reviving the Superorganism'.

83 Gould, *The Structure of Evolutionary Theory*, p. 683.

84 James Lovelock, *Gaia*. For the role of Lynn Margulis as a partner in developing Gaia theory, see the collected correspondence in James Lovelock and Lynn Margulis, *Writing Gaia*.

85 For the first case see Graham Harman, *Bruno Latour: Reassembling the Political*; for the second see Graham Harman, *Immaterialism*.

86 For the classic anthropological case see André Léroi-Gourhan, *Gesture and Speech*.

87 Richard Dawkins, *The Extended Phenotype*, p. 1.

88 Gould, *The Structure of Evolutionary Theory*, p. 667.

89 For a similar argument see Timothy Williamson, *Vagueness*.

Chapter 7: Fractures and Folds: Architectural Theory, 1988–93

1 See David Ruy, 'Returning to (Strange) Objects'. For further discussion see Graham Harman, *Architecture and Objects*, p. 2.

2 Greg Lynn, 'Architectural Curvilinearity', p. 24.

3 Lynn, 'Architectural Curvilinearity', p. 24.

4 Lynn, 'Architectural Curvilinearity', p. 24.

5 Lynn, 'Architectural Curvilinearity', p. 24.

6 Robert Venturi, *Complexity and Contradiction in Architecture*.

7 Gilles Deleuze, *The Fold*.

8 See Graham Harman, *Art and Objects*.

9 Clement Greenberg, *Art and Culture*, p. 154.

10 Michael Fried, *Manet's Modernism*.

11 Francesco Salvi, L. R. Galante and Andrea Ricciardi, *The Impressionists*; Arthur Danto, *After The End of Art*, Chapter 1.

12 Jack Flam, *Matisse and Picasso*.

13 See Gavin Parkinson, *The Duchamp Book*.

14 Ayn Rand, *The Fountainhead*.

15 Louis Sullivan, 'The Tall Office Building Aristically Considered'.

16 Bruno Latour and Émilie Hermant, *Paris Invisible City*. See also Graham Harman and Christopher Witmore, *Objects Untimely*, p. 13.

17 Le Corbusier, *Towards a New Architecture*. See also Sigfried Giedion, *Space, Time, and Architecture*.

18 Wendy Lesser, ' "You Say to Brick" '.

19 Charles Jencks, *The Language of Post-Modern Architecture*, p. 9.

20 Aldo Rossi, *The Architecture of the City*; Manfredo Tafuri, *Theories and History of Architecture*.

21 Robert Venturi, Denise Scott Brown and Steven Izenour, *Learning From Las Vegas*.

22 Venturi, *Complexity and Contradiction in Architecture*, p. 16.

23 Vitruvius, *The Ten Books on Architecture*.

24 For the reference to Gropius see Venturi, Scott Brown and Izenour, *Learning From Las Vegas*, p. 134. For Venturi's own view on the matter see Venturi, *Complexity and Contradiction in Architecture*, p. 16.

25 Venturi, *Complexity and Contradiction in Architecture*, p. 16.

26 Venturi, Scott Brown and Izenour, *Learning From Las Vegas*, p. 7.

27 Venturi, *Complexity and Contradiction in Architecture*, p. 23.

28 Cleanth Brooks, *The Well Wrought Urn*; William Empson, *Seven Types of Ambiguity*.

29 Venturi, *Complexity and Contradiction in Architecture*, pp. 82, 34.

30 Venturi, *Complexity and Contradiction in Architecture*, p. 70.

31 Venturi, Scott Brown and Izenour, *Learning From Las Vegas*, p. 129.

32 Venturi, Scott Brown and Izenour, *Learning From Las Vegas*, p. 8.

33 Venturi, Scott Brown and Izenour, *Learning From Las Vegas*, p. xvii.

34 Venturi, Scott Brown and Izenour, *Learning From Las Vegas*, p. 137.

35 Venturi, Scott Brown and Izenour, *Learning From Las Vegas*, p. 103.

36 Venturi, Scott Brown and Izenour, *Learning From Las Vegas*, p. 135.

37 Venturi, Scott Brown and Izenour, *Learning From Las Vegas*, p. 161.

38 Vincent Scully, 'Introduction', p. 11.

39 Venturi, *Complexity and Contradiction in Architecture*, p. 16.

40 Venturi, *Complexity and Contradiction in Architecture*, p. 104.

41 T. S. Eliot, *Selected Essays*, pp. 3–4. Cited by Venturi in *Complexity and Contradiction in Architecture*, p. 13.

42 Eliot, *Selected* Essays, p. 243; emphasis added.

43 Venturi, Scott Brown and Izenour, *Learning From Las Vegas*, p. 100.

44 Venturi, Scott Brown and Izenour, *Learning From Las Vegas*, p. 13 (signs in Las Vegas), and the photos on pp. 95, 98–9 (Guild House).

45 Venturi, Scott Brown and Izenour, *Learning From Las Vegas*, p. 130.

46 Venturi, *Complexity and Contradiction in Architecture*, p. 78.

47 Venturi, *Complexity and Contradiction in Architecture*, p. 96.

48 See Franz Schulze, *Philip Johnson*.

49 Philip Johnson and Mark Wigley, *Deconstructivist Architecture*.

50 Viktor Shklovsky, *Theory of Prose*.

51 Johnson and Wigley, *Deconstructivist Architecture*, p. 11.

52 Johnson and Wigley, *Deconstructivist Architecture*, p. 12.

53 Johnson and Wigley, *Deconstructivist Architecture*, p. 12.

54 Johnson and Wigley, *Deconstructivist Architecture*, p. 11.

55 Johnson and Wigley, *Deconstructivist Architecture*, p. 16.

56 Johnson and Wigley, *Deconstructivist Architecture*, p. 16.

57 Johnson and Wigley, *Deconstructivist Architecture*, p. 10.

58 Johnson and Wigley, *Deconstructivist Architecture*, p. 16.

59 Johnson and Wigley, *Deconstructivist Architecture*, p. 17.

60 Rita Felski, 'Context Stinks!' I am aware that Marshall McLuhan is often credited with this phrase, but am unable to find such a passage in his written works.

61 Johnson and Wigley, *Deconstructivist Architecture*, p. 17.

62 Johnson and Wigley, *Deconstructivist Architecture*, p. 18.

63 See François Dosse, *History of Structuralism*, vol. 1.

64 Ferdinand de Saussure, *Course in General Linguistics*.

65 Claude Lévi-Strauss, *Structural Anthropology*.

66 Jacques Derrida, *Writing and Difference*, pp. 278–94.

67 Mark Wigley, *The Architecture of Deconstruction*, p. xi; Jacques Derrida, 'Point de folie – Maintenant l'architecture'.

68 Barry Smith et al., 'Open Letter Against Derrida Receiving an Honorary Doctorate from Cambridge University'.

69 Derrida, 'Point de folie – Maintenant l'architecture', p. 570.

70 Bernard Tschumi, 'Derrida', p. 117.

71 Tschumi, 'Derrida', p. 118.

72 Peter Eisenman, *Eisenman Inside Out*, pp. 3–9 (pro-Euclidean), p. 219 (anti-Euclidean).

73 Eisenman, *Eisenman Inside Out*, p. 170.

74 Eisenman, *Eisenman Inside Out*, p. 187.

75 Eisenman, *Eisenman Inside Out*, p. 210.

76 For a discussion see Harman, *Architecture and Objects*, pp. 134–5.

77 Gilles Deleuze, *Empiricism and Subjectivity*.

78 Deleuze, *Bergsonism*; Gilles Deleuze, *Expressionism in Philosophy*; Gilles Deleuze, *Nietzsche and Philosophy*; Gilles Deleuze, *Proust and Signs*.

79 Gilles Deleuze, *Negotiations*, p. 6.

80 Gilles Deleuze, *Difference and Repetition*; Gilles Deleuze, *The Logic of Sense*.

81 Gilles Deleuze and Félix Guattari, *Anti-Oedipus*; Gilles Deleuze and Félix Guattari, *A Thousand Plateaus*.

82 Gilles Deleuze, *The Fold*; Gilles Deleuze, *Cinema I*; Gilles Deleuze, *Cinema II*.

83 Duns Scotus, *Philosophical Writings*, pp. 1–14.

84 See Alain Badiou, *Deleuze: The Clamour of Being* for a focused critique of Deleuze as a philosopher of the One.

85 Levi R. Bryant, 'The Interior of Things'.

86 Deleuze, *The Fold*.

87 G. W. Leibniz, 'The Principles of Philosophy, or the Monadology'.

88 Mario Carpo, 'Ten Years of Folding', p. 16.

89 Carpo, 'Ten Years of Folding', p. 16.

90 Stan Allen, 'From Object to Field', p. 24.

91 Allen, 'From Object to Field', p. 24.

92 Allen, 'From Object to Field', pp. 24–5.

93 Allen, 'From Object to Field', p. 28.

94 Allen, 'From Object to Field', p. 29.

95 Jane Bennett, 'Systems and Things', p. 227.

96 Lynn, 'Architectural Curvilinearity', p. 25.

97 Lynn, 'Architectural Curvilinearity', p. 27.

98 Lynn, 'Architectural Curvilinearity,' p. 24.

99 David Ruy, 'Returning to (Strange) Objects'.

100 Lynn, 'Architectural Curvilinearity', p. 26.

101 Lynn, 'Architectural Curvilinearity', p. 26.

102 Lynn, 'Architectural Curvilinearity', p. 26. See also Lao-Tzu, *Tao Te Ching*.

103 Greg Lynn, *Animate Form*. For a critical assessment see Graham Harman, 'Greg Lynn on Animate Form'.

104 For a critique of the cinematic conception of time see Henri Bergson, *Creative Evolution*.

105 Lynn, *Animate Form*, p. 11.

106 Lynn, *Animate Form*, p. 14.

107 Lynn, *Animate Form*, p. 35.

108 Lynn, *Animate Form*, p. 17.

109 Greg Lynn, 'Blobs'.

110 Gilbert Simondon, *Individuation in Light of the Notions of Form and Information*, 2 vols.

111 On the link between information and difference see Gregory Bateson, *Steps to an Ecology of Mind*.

112 Lynn, *Animate Form*, p. 31.

113 Lynn, *Animate Form*, p. 30.

114 Rem Koolhaas and Bruce Mau, *S, M, L, XL*.

115 Karl Marx, *A Contribution to the Critique of Political Economy*.

116 See the opening pages of Marx, *Capital: Volume One*. For a critical treatment see Graham Harman, 'Object-Oriented Ontology and Commodity Fetishism'.

117 Sigmund Freud, *The Interpretation of Dreams*.

118 Sigmund Freud, *Beyond the Pleasure Principle*.

119 Patrik Schumacher, *The Autopoiesis of Architecture*, vol. 1, p. 297.

120 For further discussion see Harman, 'Greg Lynn on Animate Form'.

Chapter 8: The Pope and the Horseman: How Many Magisteria?

1 Ronald F. Inglehart, 'Giving Up on God', p. 110.

2 Robyn E. Blumner, 'Give the Four Horsemen (and Ayaan) Their Due'.

3 Ayaan Hirsi Ali, *Infidel*.

4 Sam Harris, *The End of Faith*; Sam Harris, *Letter to a Christian Nation*.

5 Daniel Dennett, *Breaking the Spell*.

6 Christopher Hitchens, *God Is Not Great*.

7 Richard Dawkins, *The God Delusion*.

8 Richard Dawkins and Sam Harris, 'An Evening With Richard Dawkins – Featuring Sam Harris – Night 1', YouTube video, from 10:24 to 15:27.

9 Stephen Jay Gould, 'Nonoverlapping Magisteria'; Stephen Jay Gould, *Rocks of Ages*.

10 Gould, 'Nonoverlapping Magisteria,' p. 2.

11 Gould, 'Nonoverlapping Magisteria', p. 1.

12 Gould, 'Nonoverlapping Magisteria', p. 4.

13 Gould, 'Nonoverlapping Magisteria', p. 3; Pope Pius XII, 'Encyclical: *Humani Generis*'.

14 Cited in Hans Thijssen, 'Condemnation of 1277'.

15 Averroës, *Decisive Treatise and Epistle Dedicatory*.

16 St Thomas Aquinas, *Summa Theologica*, 5 vols.

17 Dawkins, 'When Religion Steps on Science's Turf', p. 1.

18 Dawkins, 'When Religion Steps on Science's Turf', p. 1.

19 Dawkins, 'When Religion Steps on Science's Turf', p. 1.

20 Dawkins, 'When Religion Steps on Science's Turf', p. 1.

21 Dawkins, 'When Religion Steps on Science's Turf', p. 2; Pope John Paul II, 'Message to the Pontifical Academy of Sciences on Evolution', p. 4. The term 'ontological difference' was no doubt borrowed by the philosophically literate John Paul II from Martin Heidegger, *The Basic Problems of Phenomenology*.

22 Dawkins, 'When Religion Steps on Science's Turf', p. 2.

23 Dawkins, 'When Religion Steps on Science's Turf', p. 2.

24 Descartes, *Meditations on First Philosophy*. See also Kant, *Critique of Pure Reason*. More recently we have the surprisingly anthropocentric version of nihilism in Ray Brassier, *Nihil Unbound*, where humans are treated simultaneously as both transient nothings and as bearers of a uniquely powerful rationality.

25 Dawkins and Harris, 'An Evening With Richard Dawkins – Featuring Sam Harris – Night 1'.

26 Daniel Dennett, *Consciousness Explained*.

27 David Chalmers, *The Conscious Mind*.

28 Robert Kirk and Roger Squires, 'Zombies v. Materialists'.

29 Daniel Dennett, 'The Unimagined Preposterousness of Zombies'.

30 See Graham Harman, 'Zero-Person and the Psyche'.

31 James Ladyman and Don Ross, *Every Thing Must Go*. For a critique of their position see Graham Harman, 'I Am Also of the Opinion That Materialism Must Be Destroyed'.

32 Thomas Metzinger and Wolf Singer, 'The Unity of Consciousness'.

33 Ray Brassier, 'Concepts and Objects', p. 64.

34 For a good introduction to Latour's philosophy see Latour, *We Have Never Been Modern*. For an overview of his thinking see Harman, *Prince of Networks*; and Harman, *Bruno Latour: Reassembling the Political*. He discusses his religious views in Bruno Latour, *Rejoicing*.

35 Bruno Latour, *An Enquiry Into Modes of Existence*.

36 By my count there are really only fourteen modes in Latour's book, but that is a topic for another occasion.

37 See also Bruno Latour, *The Making of the Law*. An earlier version of this argument about law was made in Polanyi, *Personal Knowledge*, pp. 276–8.

38 Polanyi, *Personal Knowledge*, p. 278.

39 Neil deGrasse Tyson, Twitter, 29 June 2016.

40 Carl Schmitt, *The Concept of the Political*.

41 Jacques Lacan, *From an Other to the other*, pp. 264–6. He draws this example from Helene Deutsch, 'A Case of Hen Phobia'.

42 Daniel Dennett, 'Quining Qualia'.

43 Willard van Orman Quine, *From a Logical Point of View*.

44 Dennett, 'Quining Qualia', p. 381.

45 Dennett, 'Quining Qualia', p. 382.

46 Dennett, 'Quining Qualia', p. 385.

47 Dennett, 'Quining Qualia', p. 386.

48 Dennett, 'Quining Qualia', p. 403.

49 Cited from Ned Block, 'Troubles With Functionalism', p. 281.

50 Dennett, 'Quining Qualia', pp. 382, 403 respectively.

51 Dennett, 'Quining Qualia', p. 409.

52 Harman, *Object-Oriented Ontology*, pp. 37–8.

53 Dennett, 'Quining Qualia', p. 384.

54 Harman, *Object-Oriented Ontology*, pp. 38–41.

55 Harman, *Object-Oriented Ontology*, pp. 30–32.

56 Manuel DeLanda, *Philosophical Chemistry*; see also Manuel DeLanda and Graham Harman, *The Rise of Realism*, pp. 82–3.

57 DeLanda, 'Emergence, Causality and Realism'.

58 Harold Bloom, *Shakespeare*, p. 477.

59 Hegel, *Phenomenology of Spirit*; J. G. Fichte, *The Science of Knowledge*.

60 Ludwig Wittgenstein, *Tractatus Logico-Philosophicus*, p. 90.

61 Harman, *Object-Oriented Ontology*, p. 149.

62 Dennett, 'Quining Qualia', p. 390.

63 Dennett, 'Quining Qualia', p. 396.

64 Heidegger, *Being and Time*.

65 Such a view is actually defended in Niklas Luhmann, *Social Systems*.

66 Cleanth Brooks, *The Well Wrought Urn*, p. 204. For a fuller discussion see Graham Harman, 'The Well-Wrought Broken Hammer'.

67 Black, 'Metaphor', p. 41.

Chapter 9: A Corner of the Great Veil: Waves and Particles

1 For a readable sample of his work see Anton Zeilinger, *Dance of the Photons*.

2 Louis de Broglie, *Continu et discontinu en physique moderne*.

3 In what follows, all translations from de Broglie's book from French into English are my own.

4 Jim Baggott, *The Quantum Story*.

5 Planck, 'Über das Gesetz der Energieverteilung im Normalspektrum'.

6 Albert Einstein, 'Über einen die Erzeugung und Verwandlung des Lichtes betreffenden heuristischen Gesichtspunkt'.

7 Einstein, 'Über die von der molekularkinetischen Theorie der Wärme geforderte Bewegung von in ruhenden Flüssigkeiten suspendierten Teilchen'.

8 Niels Bohr, 'On the Constitution of Atoms and Molecules', published in three parts.

9 Louis de Broglie, 'Recherches sur la théorie des quanta'.

10 Richard Feynman, *The Character of Physical Law*, Kindle locations 1721, 1707.

11 Feynman, *The Character of Physical Law*, Kindle location 1705.

12 Werner Heisenberg, 'Über den anschaulichen Inhalt der quantentheoretischen Kinematik und Mechanik'.

13 Rhodes, *The Making of the Atomic Bomb*, p. 131.

14 See Lee A. Rozema et al., 'Violation of Heisenberg's Measurement–Disturbance Relationship by Weak Measurements'.

15 Erwin Schrödinger 'Quantisierung als Eigenwertproblem'; Erwin Schrödinger, 'Are there Quantum Jumps?', published in two parts.

16 Erwin Schrödinger, 'Die gegenwärtige Situation in der Quantenmechanik'.

17 Hugh Everett, 'Relative State Formulation of Quantum Mechanics'; Bryce Seligman DeWitt and Neill Graham, eds., *The Many-Worlds Interpretation of Quantum Mechanics*.

18 Feynman, *The Character of Physical Law*, Kindle location 2289. Concerning Feynman's reputation as a philosophy-hater see Paul Feyerabend's remark in Imre Lakatos and Paul Feyerabend, *For and Against Method*, p. 385.

19 N. David Mermin, 'Could Feynman Have Said This?'.

20 Sujeevam Sivasundaram and Kristian Hvidtfelt Nielsen, 'Surveying the Attitudes of Physicists Concerning Foundational Issues of Quantum Mechanics', p. 11.

21 Robert Oerter, *The Theory of Almost Everything*.

22 See for instance Lee Smolin, *Three Roads to Quantum Gravity*.

23 Aristotle, *Physics*.

24 Galileo Galilei, *Siderius Nuncius*.

25 Newton, *The Principia*.

26 Einstein, 'Zur Elektrodynamik bewegter Körper'.

27 Minkowski, 'Raum und Zeit'.

28 Bernhard Riemann, 'Über die Hypothesen, welche der Geometrie zu Grunde legt'.

29 Albert Einstein and Marcel Grossmann, 'Entwurf einer verallgemeinerten Relativitätstheorie und einer Theorie der Gravitation'.

30 Einstein, 'Die Grundlage der allgemeinen Relativitätstheorie'.

31 Stephen Buranyi, 'Do We Need a New Theory of Evolution?'.

32 Michio Kaku, *The God Equation*, p. 141, Kindle edition.

33 Sabine Hossenfelder, *Lost in Math*, p. 178, Kindle edition.

34 For a clear introduction see Brian Greene, *The Elegant Universe*.

35 Edward Witten, 'String Theory Dynamics in Various Dimensions'.

36 Lee Smolin, *The Trouble with Physics*; Peter Woit, *Not Even Wrong*.

37 Steven Carlip, 'Is Quantum Gravity Necessary?'; James Mattingly, 'Is Quantum Gravity Necessary?'

38 Jonathan Oppenheim, 'A Postquantum Theory of Classical Gravity?'.

39 Isaac Newton, *Opticks*.

40 Broglie, *Continu et discontinu en physique moderne*, pp. 20–21.

41 Broglie, *Continu et discontinu en physique moderne*, pp. 25–9.

42 Richard Feynman, *QED*, p. 15.

43 Broglie, *Continu et discontinu en physique moderne*, p. 131.

44 Albert Einstein, *The Collected Papers of Albert Einstein*, vol. 14, letter 398.

45 From The Nobel Prize website, https://www.nobelprize.org/prizes/physics/1929/broglie/biographical/, last accessed on 1 March 2024.

46 Broglie, *Continu et discontinu en physique moderne*.

47 Broglie, *Continu et discontinu en physique moderne*, p. 12. All translation from this book are my own.

48 Broglie, *Continu et discontinu en physique moderne*, p. 266.

49 Broglie, *Continu et discontinu en physique moderne*, p. 7.

50 C. J. Davisson and L. H. Germer, 'Reflection of Electrons by a Crystal of Nickel'.

51 Broglie, *Continu et discontinu en physique moderne*, p. 41.

52 Broglie, *Continu et discontinu en physique moderne*, p. 48; see also p. 75.

53 Broglie, *Continu et discontinu en physique moderne*, p. 50.

54 Broglie, *Continu et discontinu en physique moderne*, p. 56.

55 Broglie, *Continu et discontinu en physique moderne*, p. 87.

56 Broglie, *Continu et discontinu en physique moderne*, p. 68.

57 Bertrand Russell, *The Analysis of Matter*.

58 Broglie, *Continu et discontinu en physique moderne*, p. 87.

59 L. D. Landau and E. M. Lifshitz, *The Classical Theory of Fields*, vol. 2, p. 228; Watson, *The Double Helix*.

60 Broglie, *Continu et discontinu en physique moderne*, p. 88.
61 William of Ockham, *Philosophical Writings*.
62 For a debunking of this misunderstanding see Rondo Keele, *Ockham Explained*.
63 Broglie, *Continu et discontinu en physique moderne*, p. 88. For a counter-argument to this link between beauty and truth see Hossenfelder, *Lost in Math*.
64 Broglie, *Continu et discontinu en physique moderne*, p. 37.
65 Broglie, *Continu et discontinu en physique moderne*, p. 72.
66 Matt Warman, 'Stephen Hawking Tells Google "Philosophy is Dead"'. For a response see Graham Harman, 'Concerning Stephen Hawking's Claim That Philosophy Is Dead'.
67 Broglie, *Continu et discontinu en physique moderne*, pp. 59–60.
68 Broglie, *Continu et discontinu en physique moderne*, p. 62.
69 Immanuel Kant, *Grounding for the Metaphysics of Morals*.
70 Henry Stapp, *Quantum Theory and Free Will*.
71 For some examples see Chalmers, *The Conscious Mind*; Galen Strawson, 'Realistic Monism'; Philip Goff, *Galileo's Error*. For a response to Chalmers and Strawson see Harman, 'Zero-Person and the Psyche'. For an elightening history of panpyschism in Western philosophy see David Skrbina, *Panpsychism in the West*.
72 For a positive treatment of this idea see Roger Penrose, *Shadows of the Mind*; and Stuart Hameroff, 'Consciousness, Neurobiology and Quantum Mechanics'.
73 Max Tegmark, 'The Importance of Quantum Decoherence in Brain Processes'.
74 Broglie, *Continu et discontinu en physique moderne*, pp. 7–8.
75 Broglie, *Continu et discontinu en physique moderne*, p. 8.
76 Broglie, *Continu et discontinu en physique moderne*, p. 16.
77 Broglie, *Continu et discontinu en physique moderne*, pp. 139, 132.
78 Broglie, *Continu et discontinu en physique moderne*, pp. 135, 170.
79 Broglie, *Continu et discontinu en physique moderne*, p. 140.
80 Broglie, *Continu et discontinu en physique moderne*, p. 145.
81 Broglie, *Continu et discontinu en physique moderne*, p. 130.
82 Broglie, *Continu et discontinu en physique moderne*, p. 115.

83 Badiou, *Being and Event*.

84 Quentin Meillassoux, *After Finitude*; Quentin Meillassoux, 'Appendix: Excerpts from *L'Inexistence divine'*.

85 For an interesting philosophical meditation on Dedekind from a phenomenological standpoint see Piotr Blaszczyk, 'On the Mode of Existence of the Real Numbers'.

86 The article can be found in Richard Dedekind, *Essays on the Theory of Numbers*, pp. 1–13.

87 Dedekind, *Essays on the Theory of Numbers*, pp. 1–2.

88 John Wallis, *A Treatise of Algebra, Both Historical and Practical*.

89 Dedekind, *Essays on the Theory of Numbers*, p. 5.

90 Dedekind, *Essays on the Theory of Numbers*, p. 5.

91 Dedekind, *Essays on the Theory of Numbers*, p. 5.

92 Dedekind, *Essays on the Theory of Numbers*, p. 4.

93 Georg Cantor, *Contributions to the Founding of the Theory of Transfinite Numbers*.

94 Kurt Gödel, *The Consistency of the Continuum-Hypothesis*; Paul Cohen, 'The Independence of the Continuum Hypothesis', 2 parts.

95 William James, *Essays in Radical Empiricism*, p. 67. For a critical discussion see Graham Harman, *Bells and Whistles*, pp. 48–59.

96 Edmund Husserl, *Logical Investigations*, 2 vols.

97 Heidegger, *Being and Time*.

98 It should be noted that the French philosophers Jean-Luc Nancy and Jacques Derrida have conducted their own public dialogue on the theme of touch, but their concerns and philosophical orientation are fundamentally different from my own. See Jean-Luc Nancy, *Corpus*; and Jacques Derrida, *On Touching*. For a provisional critique of Nancy's approach see Graham Harman, 'On Interface'. Other interesting articles on this topic include Roland Faber, 'Touch'; and Sam Mickey, 'Touching without Touching'.

99 Baruch Spinoza, *Ethics*, p. 3.

100 John Locke, *An Essay Concerning Human Understanding*, 2 vols.; Richard Rorty, *Philosophy and the Mirror of Nature*.

101 Markus Gabriel, *Fields of Sense*, p. 14.

102 Martin Heidegger, *Kant and the Problem of Metaphysics*.

103 Uexküll, *A Foray into the Worlds of Animals and Humans*. For an assessment see Harman, 'Magic Uexküll'.

104 Latour, *We Have Never Been Modern*, p. 129.

105 Latour, *Pandora's Hope*, pp. 80–112.

106 James Chadwick, 'Possible Existence of a Neutron'. On Majorana see Giorgio Agamben, *What Is Real?*

107 Lise Meitner and Otto Frisch, 'Disintegration of Uranium by Neutrons'.

108 See Harman, *Prince of Networks*, pp. 73–5.

109 Plato, 'Phaedrus', 265e–266a.

Chapter 10: Waves and Stones

1 Plato, *Republic*.

2 See for example Stanley Rosen, *Plato's Republic*.

3 Karl Popper, *The Open Society and Its Enemies*.

4 Machiavelli, *The Prince*.

5 Niccolò Machiavelli, *Discourses on Livy*.

6 Martin Heidegger, *The Question Concerning Technology*.

7 Friedrich Hölderlin, 'Patmos'.

8 Credit for these phrases is due, respectively, to Jacques Derrida, 'Plato's Phamacy'; and Slavoj Žižek, *Tarrying with the Negative*, Chapter 5.

9 Hegel, *Science of Logic*, pp. 106–8.

10 See the discussion of the 'obscure subject' in Alain Badiou, *Logics of Worlds*, pp. 58–62.

11 Bruno Latour, *Down to Earth*.

12 For a critical assessment see Graham Harman, 'Latour's Interpretation of Donald Trump'.

13 See Harold Rosenberg, 'The American Action Painters'.

14 Benjamin Lee Whorf, *Language, Thought, and Reality*, pp. 102–24.

15 Badiou, *Logics of Worlds*. The point is made clearly by William Watkin, *Badiou and Communicable Worlds*, p. 10.

16 See Williamson, *Vagueness*.

17 Friedrich Nietzsche, *Beyond Good and Evil*, p. 69.

18 Georg Lukács, 'Realism in the Balance', p. 55.

19 Yet Aristotle was well aware of the special causal role of touch, as seen from his humorous observation that touch is the only sense that can kill us if we do it too hard. See Aristotle, *De Anima*.

20 Aristotle, *Metaphysics*, p. 82.

21 My searches so far have turned up no prior use of this term, except as the stage name for a London-born musician named Thébru Čelet (b. 1963).

22 Mark S. Granovetter, 'The Strength of Weak Ties'.

23 Manuel DeLanda, *A New Philosophy of Society*, p. 35.

24 See Giorgio Lando, *Mereology*.

25 Graham Harman, 'Time, Space, Essence, and Eidos'.

26 DeLanda, *A New Philosophy of Society*, pp. 34–5. See also Graham Harman, 'DeLanda's Ontology'.

27 For an interesting discussion of such larger objects under the heading 'superordinate objects', see Jon Cogburn and Niki Young, 'Revisiting the Notion of Vicarious Cause'.

28 Roy Bhaskar, *A Realist Theory of Science*, p. 114. Cited in DeLanda, *A New Philosophy of Society*, p. 34.

29 Gernot Böhme, *The Aesthetics of Atmospheres*.

30 Alphonso Lingis, *The Imperative*, p. 5.

31 Peter Sloterdijk, *Spheres*, 3 vols. His theorization of the mother–child relationship makes it as crucial for philosophy as Melanie Klein (1882–1960) did for psychoanalysis. See Melanie Klein, *The Psychoanalysis of Children*.

32 See also the related notion of levels in Maurice Merleau-Ponty, *The Visible and the Invisible*, p. 114.

33 Niklas Luhmann, *Social Systems*.

34 For an intriguing theory of how we perceive multiple entities in a unified experience see Helmuth Plessner, *Die Einheit der Sinne*.

35 See Michael Marder, *Plant-Thinking*.

36 Edmund Husserl, *Cartesian Meditations*.

37 Heidegger, *Being and Time*.

38 Martin Heidegger, *The Essence of Reasons*.

39 For an analytic philosophy version of the argument see Robert B. Bran-
    dom, *Making It Explicit*.
40 Harman, *Immaterialism*.
41 Rhodes, *The Making of the Atomic Bomb*, p. 218.

Acknowledgements

Above all, thanks are due to Ananda Pellerin, my editor at Penguin. She performed Herculean labours with this manuscript over an extended period of time, and also made a number of fine suggestions that have been incorporated into the final book.

Thomas Penn at Penguin showed the patience of Job in facing various delays in delivery of the manuscript due to the COVID-19 pandemic and other factors. He also made a marvellous list of final suggestions that helped tighten up the book considerably.

My friend Micah Tewers continued his recent pattern of reading the manuscript thoroughly and providing important feedback. Micah is wise beyond his years, and I'm always learning something new from him.

Index